The Hierarchies in Biological Systematics

The Hierarchies in Biological Systematics: A Conceptual Perspective examines the diverse hierarchies shaping the field of biology. From the overarching cognitive frameworks to the specific classifications that define biological taxonomy, this book is a comprehensive analysis of hierarchical structures and their inherent complexities.

Through a conceptual lens, the author explores the fundamental parameters of hierarchies, their fuzziness, and the principles guiding their substantiation, elaboration, and implementation. The book also delves into the ongoing debate between ranked and rankless taxonomic hierarchies, highlighting the author's view of the unique rationality of ranked systems while critically assessing their limitations.

Whether you are a researcher, student, or professional in biological systematics, this book offers valuable insights into the conceptual underpinnings of hierarchical classifications and their role in advancing scientific understanding.

Species and Systematics

Phylogenetic Systematics: Haeckel To Hennig,
by Olivier Rieppel

Evolution By Natural Selection: Confidence, Evidence and the Gap,
by Michaelis Michael

The Evolution of Phylogenetic Systematics,
edited by Andrew Hamilton

Molecular Panbiogeography on the Tropics,
by Michael Heads

Beyond Cladistics,
edited by David M. Williams and Sandra Knapp

Comparative Biogeography: Discovering and Classifying Bio-Geographical Patterns of a Dynamic Earth,
by Lynee R. Parenti and Malte C. Ebach

Species: A History of the Idea,
by John S. Wilkins

What Species Mean: Understanding the Units of Biodiversity,
by Julia Sigwart

What, if anything, are species?
by Brent D. Mishler

Biological Systematics,
by Igor Ya. Pavlinov

Species Problems and Beyond: Contemporary Issues in Philosophy and Practice,
by John S. Wilkins, Frank E. Zachos, Igor Ya. Pavlinov

From Observations to Optimal Phylogenetic Trees: Phylogenetic Analysis of Morphological Data: Volume 1,
by Pablo Goloboff

Refining Phylogenetic Analyses: Phylogenetic Analysis of Morphological Data: Volume 2,
by Pablo Goloboff

The Hierarchies in Biological Systematics: A Conceptual Perspective,
by Igor Ya. Pavlinov

For more information about this series, please visit:
https://www.crcpress.com/Species-and-Systematics/book-series/CRCSPEANDSYS

The Hierarchies in Biological Systematics

A Conceptual Perspective

Igor Ya. Pavlinov

CRC Press
Taylor & Francis Group
Boca Raton London New York

CRC Press is an imprint of the
Taylor & Francis Group, an **informa** business

Designed cover image: Ernst Haeckel

First edition published 2026
by CRC Press
2385 NW Executive Center Drive, Suite 320, Boca Raton FL 33431

and by CRC Press
4 Park Square, Milton Park, Abingdon, Oxon, OX14 4RN

CRC Press is an imprint of Taylor & Francis Group, LLC

ISBN: 978-1-032-98212-0 (hbk)
ISBN: 978-1-032-98112-3 (pbk)
ISBN: 978-1-003-59709-4 (ebk)

DOI: 10.1201/9781003597094

Typeset in Times
by SPi Technologies India Pvt Ltd (Straive)

Contents

About the Author

Igor Ya. Pavlinov was, until his retirement in 2018, the leading researcher and the chief of the Mammal Division at the Zoological Museum, Lomonosov Moscow State University. He is still affiliated with the Zoological Museum, where he is a curator of mammals. His doctoral dissertation was *Cladistic approach to phylogenetics and systematics: Theoretical foundations of evolutionary cladistics* (1997). His principal research interests are in theoretical systematics and phylogenetics, systematics of mammals (mainly rodents), and morphometrics. He is the author of several dozen books on natural history, including *Cladistic Analysis: Methodological Issues* (1990), *Evolution of Life* (2001), *Systematics of Mammals of the World* (2003), *Foundations of the Contemporary Phylogenetics* (2005), *Nomenclature in Systematics* (2015) and its successor *Taxonomic Nomenclature—What's in a Name: Theory and History* (2022), *Foundations of Biological Systematics* (2018) and its successor *Biological Systematics: History and Theory* (2021), *Mammals of Russia: A Guide* (2019), *The Species Problem. A Conceptual History* (2023), and several books on particular orders and families of mammals.

Preface

This volume finalizes the series of books on the philosophically considered history and theory of biological systematics, fortunately published by me with CRC. This one specifically addresses the hierarchies, both cognizable and cognitive, that this natural science branch deals with.

The character of taxonomic hierarchies is heatedly debated in contemporary systematics, though in a rather specific and simplified manner. The focal point of the ongoing discussions includes one quite particular question, namely, whether it should be ranked or not. With this, many more general and problematic issues that have been raised and discussed in-depth by systems and cognitive sciences over the last decades appeared to be relegated to a distant cognitive periphery of systematics.

Consideration of hierarchies in this book, like other subjects published by me earlier, is based on a conceptualism standpoint, as indicated by its title. This means that it concerns not only the hierarchies as such but also, to a large extent, the theoretical and even philosophical backgrounds of the hierarchy issues. When planning and writing it, I had two principal objectives in mind. One of them was to provide an overview of the contemporary ideas about hierarchies viewed from a broad historical-philosophical perspective. The other was to demonstrate how these general ideas are relevant to those particular ones that are debated in biological systematics. Accordingly, when considering the hierarchy issues, I made a special emphasis on the contribution of their broad philosophical consideration to their particular understanding in systematics. Among the most intriguing were two points: the fuzzy nature of the hierarchies studied by this discipline and the controversy over their objective *vs.* subjective status.

Taking all this into account, I hope that this enterprise of mine will encourage fellow systematicians to realize the truly broad content of the contemporary hierarchy problematics, which is far more complicated than it usually appears to those who prefer simple solutions.

Igor Ya. Pavlinov
Research Zoological Museum at Lomonosov
Moscow State University
Russia

Introduction

<table><tr><td>Но бойтесь единственно только того,
Кто скажет: 'Я знаю, КАК НАДО!'
Кто скажет: 'Идите, люди за мной!
Я вас научу, КАК НАДО!'
Александр Галич</td><td>There's only one thing that you need to fear:
The man who says 'I know THE ANSWERS!'
Who says, 'Come on, people, and follow me,
And I'll teach you all THE ANSWERS!'
Alexander Galich[1]</td></tr></table>

The hierarchies are everywhere—they are in the world around us, in ourselves, including our knowing capacities, and in the knowledge about that world and ourselves. They are assumed to be among the most impressive and enigmatic manifestations of the inherent orderliness of any complex systems of any kind and scale. Being their specific attributes, hierarchies are endowed with certain features in common, and with this they are very diverse in their own properties. Due to the ubiquity of hierarchically ordered systems of various kinds, which are encountered by humans in both their professional activity and everyday life, the very word 'hierarchy' is now universally recognized and widely used in quite different contexts. This makes the notion in question very polysemous and *nomadic* (in the sense of Minelli 2022), referring to quite distinct natural and artificial phenomena. To eliminate such a wide-ranging undesirable homonymy, this general notion is specified by adding particular adjectives and/or prefixes to it, which yields an extensive vocabulary with a lot of terms, and it is not an easy task to distinguish precisely among those that are used with specific meanings in various situations (Lane 2006; Corominas-Murtra et al. 2013). All these ambiguities constitute a noticeable portion of the so-called *hierarchy problem*, which was initially raised in quantum physics (e.g., Dvali 2013; Koren 2020; Arkani-Hamed et al. 2024) and now promises to expand over other sciences dealing with the complex systems.

The vast majority of authors, when discussing various issues concerning hierarchical structure of both the material and ideal worlds, apply the word 'hierarchy' in the singular. But actually they deal with specific partial manifestations of complexly organized hierarchical structures, resulted from the multi-aspectual character of both the hierarchies themselves and their consideration aspects. An excellent illustration of such a biased apprehension, which is most close to the main topic of this book, is the collection *Interpreting the Hierarchy of Nature* (Grande and Rieppel 1994), which in fact, contrary to its very broad title, discusses mostly phylogenetic pattern and its relation to the evolutionary processes. That is why this key word is applied in the plural throughout this book, beginning from its title.

Special attention to the hierarchies in their broad sense began to rise in the mid-20th century due to the emergence of systems and cognitive sciences and the understanding of the systemic character of all phenomena that the human cognitive activity addresses (Bertalanffy 1972; Laszlo 1972a, 1972b; Pattee 1973a; Salthe 1985, 2012; Checkland 1988; Anderson and Johnson 1997; Ruse 2009; Weinberg 2011; Hester and Adams 2014). Their development entailed a specific comprehension that the systemically organized universe, as well as its particular manifestations and fragments, is not as simple as positivists were convinced themselves and tried to assure other scientists of this. Such an understanding implies that the cognition of hierarchical structures is not such a simple task and that the 'complex problems demand approaches that can account for their inherent complexity, rather than ignore it and hope it goes away' (Hester and Adams 2014, p. 19). Both the ubiquitous character of hierarchies and their diversity entail great attention paid to them by natural philosophers, philosophers of science, natural scientists, sociologists, mathematicians, linguists, and so on. Many hundreds of articles and dozens of books have been published during the last decades, specifically addressing various hierarchy issues: what is 'the nature of hierarchies, how they originate, how they evolve, how their levels interact, and how failure occurs' (Pattee 1973b, p. 131). All such cognitive aspirations shape a specific *hierarchical thinking*, due to which the persons mastering it become 'hard-line hierarchists', thus tending to perceive any complex entity as a hierarchically ordered system (Ahl and Allen 1996).

A specific 'envelope' for all these considerations is provided by a philosophical background elaborated by contemporary conceptual philosophy of science. One of the latter's most noticeable hallmarks is the acknowledgment that the complexly structured surrounding world cannot be comprehended based on its simplified reductionist models. Of the general ideas actively developed by conceptualism, which inevitably involve all hierarchy issues, the most significant are ontic relativity, the irremovable presence of the cognizing subject in any cognitive situation, a fuzzy character of both the cognized world and knowledge about it, and so on. In particular, it makes clear that any judgments about the 'objective' or 'subjective' status of hierarchies are not 'absolute' but rather depend on the particular onto-epistemic frameworks (e.g., Wilson 1969; Allen and Starr 1982; Ahl and Allen 1996; Wu 2013).

* * *

Biological systematics is perhaps unique among the natural science branches in that it emerged in the 16th century precisely with the mastering of the inclusive hierarchy borrowed from scholasticism and continued with its refining in a specific manner (Mayr 1982; Lyubarsky 2018; Pavlinov 2018, 2021a).[2] The latter involved the ranking of hierarchical classifications, including the elaboration of a system of fixed taxonomic ranks, their specific designations, and the rules of naming taxa allocated to them. Such an approach to the description of the diversity of organisms was 'officially' fixed in the 19th century by the special rulebooks (nomenclatural codes), which became a characteristic feature of systematics, differing it noticeably from most other natural sciences. This traditional ranked taxonomic hierarchy became subject to severe criticisms from several schools of systematic thought in the 20th century. Closer to the latter's end, this caused heated debates among systematicians, who appeared divided into two camps, supporters and opponents of either ranked or rankless taxonomic hierarchy, and such a specific 'schism' was not smoothed out to this day.

However, such a controversial character of the most recent development of the taxonomic hierarchy in systematics seemed to play a bad joke with it. Indeed, when the rest of the science branches were discussing the principal issues concerning the fundamentals of hierarchies, systematics appeared focused on a very particular and technical question of ranked *vs.* rankless hierarchy of taxonomic classifications. The serious problem here is that the vast majority of argumentations *pro* or *con* each of these two standpoints have a heavy philosophical background, which was not given due attention. With this, when discussants apply philosophically burdened categories, such as 'objective' and 'subjective', they hardly take into account their ambiguous, context-dependent character but rather pretend to possess the ultimate truth and 'teach [others] all the answers' to intricate questions and summon all to follow them.

In this regard, it seems quite pertinent and timely to consider the hierarchies that biological systematics deals with in their broadest understanding and scope. Viewed from such a perspective, the discipline in question appears as permeated with the hierarchies of various levels of generality and content. The most significant of them are characteristic of the entire cognitive situation of systematics, which represents a global hierarchically ordered three-partitioned conceptual construct. More particular hierarchies are implied by the general and particular taxonomic theories, the onto-epistemic models underlying them, particular conceptions (such as species and homology), particular classifications, etc. All these hierarchies require a careful analysis from a broad scientific-philosophical perspective in order to uncover both general properties common to them and, within this broad conceptual framework, certain features specific to the partial hierarchies of various kinds. It is such a perspective that defines the main theme of this book as a scientific-philosophical consideration of hierarchies in biological systematics; accordingly, it would be futile to look for practical recommendations concerning hierarchical arrangements of classifications in it.

* * *

Such a broad scientific-philosophical perspective, from which the hierarchies in biological systematics are here discussed, is framed by the aforementioned contemporary conceptualism, including ideas of cognitive science (see Carruthers et al. 2002; Stepin 2005; Swoyer 2006; Velichkovsky 2006; Thagard 2023 on them). The same approach was applied in previous author's books devoted to the history and theory of biological systematics (Pavlinov 2018, 2021a). This implies that the basic ideas considered here reiterate to an extent those already outlined in those volumes, which may give a reason to accuse the author of self-repetition. But in fact, these ideas do not simply replicate the author's earlier thoughts and formulations but rather refine them in a way that makes them most relevant to the main topic of this book.

The broad context adopted in this book implies a kind of deductive reasoning in consideration of all issues of the hierarchies concerned. The first go some general ideas about hierarchies developed by other disciplines, which happened to be more successful in this regard (ecology, geography, sociology, etc.). This provides a sound background for the analyses of the hierarchies in systematics as the particular cases of such general considerations. Correspondingly, this book is full of conceptions and terms developed in other disciplines, and one of the principal authors' tasks was to show how these 'alien' ideas might help to understand and clarify those commonly applied in systematics. It has been found that some conceptions developed outside the subject area of systematics appeared to be of great relevance for a more thorough comprehension of the general hierarchy theme in our discipline, such as *monotone hierarchy* in computer science (e.g., Chattopadhyay et al. 2022) and *panarchy* in economics (Gunderson and Holling 2001; Gunderson et al. 2022). However, this was not a pursuit for a 'near-exhaustion' of the generalizations pertinent to this theme, and therefore, some interesting conceptions developed in mathematics and physics were omitted, as they did not seem to have immediate application in systematics, such as *ramified hierarchy* in Russel's type theory, *integrable hierarchy*, and *ergodic hierarchy* (e.g., Aratyn and Sorin 2001; Simmons 2011; Frigg et al. 2020; Coquand 2022).

* * *

Chapter 1 outlines the historical development of the ideas about hierarchy issues addressed in biological systematics. It briefly overviews the conceptual history of this discipline, which is followed by a critical analysis of the ideas of ethnobiology regarding hierarchies in folk systematics. The elaboration of rational foundations for cognitive activity by ANTIQUE natural philosophers and MEDIEVAL scholastics is discussed, focusing on the emergence of the hierarchical genus-species scheme. Two sections describe the deployment of debates over ranked *vs.* rankless inclusive taxonomic hierarchies. The history of ideas about meronomic hierarchies is cursorily sketched.

Chapter 2 provides an overview of the relevant general ideas that have been developed by contemporary conceptualism applied to biological systematics. The following issues are addressed: what constitutes a cognitive situation in systematics and how it is structured; how theoretical knowledge can be developed in systematics; what the systemic worldview is and how it concerns the hierarchical vision of both the biological diversity and the knowledge about it; and what the principal outcomes of interpretation of classification are as a specific cognitive system.

Chapter 3 considers the most fundamental issues concerning hierarchies: what the hierarchy in general is and what its specific manifestations are, with a focus on the hierarchies studied by biological systematics. It summarizes the basic features of hierarchies, discusses their onto-epistemic dualism, and analyzes their general structure, including topological features, leveled architecture, and overall fuzzy character. Reviewed are general principles of hierarchization of taxonomic systems in biological systematics. The currently debated controversy regarding ranked *vs.* rankless taxonomic hierarchies is regarded from a conceptualism perspective.

Chapter 4 reviews three major representation styles for the hierarchies in biological systematics, namely, text-like, tree-like, and plane-like styles.

Chapter 5 considers conceptual pyramids encompassing two problematic issues of key importance in biology in general and systematics in particular; these are the species problem and the homology problem.

Chapter 6 analyzes two issues most pertaining to the hierarchical consideration of taxonomic nomenclature, including the hierarchical structure of nomenclature regulators and its relation to the hierarchy of taxonomic classifications. Special attention is given to two major versions of taxonomic nomenclature, namely rank-dependent (rank-based) and rank-independent (rank-free).

The Bibliography section includes about 1,500 entries, which may seem too enormous for this not especially voluminous book, but in fact it is not. Such a large number of references, on the one hand, is explained by the broad scope of the book, and, on the other hand, it demonstrates the great attention paid to the hierarchy issues discussed in contemporary scientific and philosophical literature.

NOTES

1 The Russian quotation from Galich 2006 (pp. 213–214); the English quotation from Galich 1985 (p. 96).
2 An opinion that the hierarchy theme is fundamental to biological systematics is widely acknowledged, but it is not universal. For instance, in a theoretical book addressing systematic issues, the term 'hierarchy' was not even mentioned in the indexes (Williams and Ebach 2008).

1 A Historical Consideration

The history of science is science itself.
Johann Wolfgang Goethe[1]

Among the major principles of contemporary non-classical philosophy of science, including evolutionary epistemology and cognitive science (Popper 1972; Hübner 1988; Hull 1988; Radnitzky and Bartley 1993; Merkulov 1996; Abachiev 2004), the following are of prime importance for us here. First, any cognitive system can be treated as an open non-equilibrium one (see Section 2.7). Second, the historical development of such a system involves changes in the understanding of both cognitive objectives, background assumptions, and aids. Third, this development is analogous to biological evolution, and it entails differentiation due to changes in the scientific system's conceptual content, including its objectives, assumptions, and aids. Fourth, this development entails the structuration of system, including its hierarchization, which in our case implies the differentiation of concepts and ideas that constitute the content of theoretical knowledge. Lastly, such a developing system, figuratively speaking, is a 'victim of its own history', which means that at each stage of its history, its current states depend, to a greater or lesser degree, on the previous ones. Hence, in order to comprehend the conceptual contents of biological systematics at any phase of its historical development, including the contemporary, it is imperative to trace its origins and take into account entire diversity of its developing ideas. Besides, a proper awareness of history is useful for commensuration of the old and new ideas in order to assess the latter's scale and originality more correctly.

This chapter outlines the historical development of the ideas that are most relevant to the hierarchy issues addressed in biological systematics. It begins with a brief overview of the conceptual history of this discipline, followed by an analysis of the ideas of ethnobiology regarding hierarchical arrangement of folk classifications. Then, the elaboration of the rational foundations for cognitive activity by Antique natural philosophy and Medieval scholasticism is discussed, focusing on the emergence of hierarchical genus-species scheme. Two subsequent sections describe the deployment of debates over ranked *vs.* rankless inclusive taxonomic hierarchies. In addition, the history of ideas about meronomic hierarchies is cursorily sketched.

1.1 A BRIEF OVERVIEW

The historical development of biological systematics is aimed at a deeper comprehension of the structure of a particular aspect of biodiversity, which is now commonly

DOI: 10.1201/9781003597094-1

referred to as *systematic diversity*, including its hierarchical orderliness. This development follows the changing conceptual contents of cognitive activity in science, including worldviews (say, creationist or natural historical), criteria for scientificity and plausibility of knowledge, and its most appropriate cognitive means. Such a vision of the conceptual history of systematics is not so much associated with the changes in particular classifications as with the development of its theoretical contents, including taxonomic theories and concepts (Mayr 1982, 1988; Mishler 1990, 2009; Pavlinov 2018, 2021a). It allows for recognizing the following major stages in the conceptual history of this discipline.

Pre-systematics has originated from the primeval knowing activity of humans. It was and still is being developed by Indigenous people as a means for the implementation of their local pragmatic and cognitive aspirations. It is investigated by ethnobiologists, who *en mass* believe that the primordial folk classifications are ordered hierarchically and largely ranked.

Proto-systematics has been developing within a much broader context of the ideas about both the ordered universe and the ordered ways of its knowing. This meant that this very important stage of the evolution of cognitive activity has become rationally organized, and this 'key innovation' was inherited by all subsequent natural philosophy and natural science. The thinkers of those times strongly linked the specifically understood universe and knowledge about it: both were arranged hierarchically as a result of the deductive processes of generating 'ideas' (*genê*, *eidê*, and *ousiai*) of various levels of generality. The sluggish development of proto-systematics took a very long time covering Antiquity, Medieval (scholasticism), and Renaissance (Herbal epoch). The scholastic development has appeared to be of special importance for having provided early systematics with a hierarchical genus-species classificatory scheme.

Scientific systematics began with the development of natural science in the 16th century, with its primary goal being the uncovering of the all-encompassing hierarchically ordered Natural System. The elaboration of the Natural Method, whose application would ensure the uncovering of the sought System, became one of the leading ideas that guided the early conceptual development of systematics. The latter's conceptual history acquired henceforth the characteristic feature of succeeding scientific revolutions, with the emerging and diverging research programs outlining particular lineages in that history.

The first had been *scholastic* systematics, which was replaced by *post-scholastic* one in the late 18th and early 19th centuries. The initial post-scholastic development involved appearance of a number of research programs of various natural philosophical kinds (Ladder of Nature, typology, natural systematics, organicism, and numerology). Each of them was based on a rather specific worldview and provided its own substantiation of the overall structure, either hierarchical or not, of organismal diversity.

The second half of the 19th century witnessed a very significant conceptual revolution in post-scholastic systematics, which was initiated by the adoption of evolutionary idea and gave birth to the early *evolutionary* research programs (phylogenetics, etc.). They suggested their own ideas of the causes and structure of systematic diversity, with supporting its highly fragmented ranked hierarchical structure.

The next was *positivist* revolution in systematics of the early 20th century, which entailed the emergence of *biosystematics* and *numerical phenetics*, both accentuating on lower-level taxonomic hierarchies. They suggested a very simplified view of the structure of biodiversity, with ascribing a pure epistemic function to the hierarchical structure of classifications. The last to appear was *cladistic* version of phylogenetic systematics of the late 20th century, which emphasized hierarchical rankless structure of both the overall diversity of organisms and classifications reflecting it.

1.2 THE ROOTS: FOLK SYSTEMATICS

The classificatory activity of Indigenous people has evolved from a specific form of adaptive cognitive behavior of higher animals, capable of recognizing similar and different things and grouping them unconsciously on this basis (Lévi-Strauss 1966; Foucault 1970; Atran 1990, 1998; Ellen 2008). A distinctive and important human specificity of such a form of behavior is that (a) the classifying of things by their similarities is often conscious and purposeful and (b) the recognized groups thereof are designated by certain linguistic means for the communicative purposes. Such a primeval form of the human cognitive activity still persists in contemporary traditional communities and provides a crucial information concerning the prehistory of biological systematics.

These primordial classificatory efforts result in the development of the so-called *folk systematics*, which was referred to above as pre-systematics. The mental systems are developed by Indigenous people as certain images of the structure of environments they inhabit. These systems are traditionally called *folk classifications* but, keeping in mind the recently highlighted distinguishing between classifications and categorizations (Mervis and Rosch 1981; Rosch 1988; Jacob 1991, 2004), they should be more correctly denoted as *folk categorizations*. The latter are largely pragmatically motivated, they cannot be imagined outside of the local folk cultures producing them, and therefore they usually have locally distinctive characters. With this, however, the most advanced variants of folk classifications/categorizations are not as primitive and strictly utilitarian as their 'animal' predecessors: at least some of them are supposedly rational to a degree and bear certain universal features.

The terms 'category' and 'categorizations' has several special meanings, which should not be mixed. One of them is the Aristotle's understanding of category as a universal cognitive device or a specific quiddity of things. Another meaning refers to an aggregate of taxa of particular hierarchical level (rank) in taxonomic classifications, which is standard for biological systematics. Lastly, Eleanor Rosch proposed recently to apply this notion to the natural aggregates of entities of any hierarchical level that are defined with the references to their prototypical tokens (Rosch 1973, 1975, 1988). In order to avoid any ambiguity, the first meaning would better be specified as *Aristotelian*, the second as *Linnaean*, and the third as *Roschian*, respectively. With this, it seems reasonable to consider the latter categorization as a variant of classification in its broadest sense, as many contemporary empirical systematicians develop their taxonomic systems based primarily on their intuition rather than through rational justification.

Folk classifications (or categorizations, if you like) are presumed, by and large, to be framed hierarchically, with many features in common among different folk cultures, and this fact is treated in two ways (Berlin 1973, 1992; Berlin et al. 1973; Atran 1990, 1998; Medin and Atran 2004; Atran and Medin 2008; Lyubarsky 2018). When considered ontically, they are supposed to reflect a certain unified hierarchical structure of biodiversity, and such their interpretation is bolstered by certain ideas of contemporary synecology (e.g., Odum 1971; Mirkin 1985). Their epistemic consideration proceeds from a structuralist conception of a unified *cognitive matrix* or *cognitive map* which is assumed to be a part of the common intellectual abilities shared by the members of the species *Homo sapiens*. This 'matrix' makes people viewing the surrounding world as ordered hierarchically and reflecting it by the respective hierarchical classifications (Brown 1991; Winthrop 1991; Atran et al. 2002; Pumain 2006), in which certain aggregates of organisms of different levels of generality are distinguished and named. Besides, it is supposed that such a hierarchical vision of the surrounding world may result from an extrapolation of the hierarchical structure of human society onto it (Durkheim and Mauss 1963).

It should be emphasized that the supposed ranked hierarchies of the Roschian categorizations are enigmatic to a degree, as the particular hierarchical levels in them are not fixed terminologically by their Indigenous creators, even if the particular major groups of organisms are actually named (trees, grasses, beasts, fish, worms, etc.). For this reason, ethnobiologists studying such categorizations, in order to recognize and designate folk-taxonomic ranks in them, are forced to apply the standards elaborated by contemporary systematics (Berlin et al. 1973; Atran 1990, 1998; Lyubarsky 2018). One of such standards is provided by the folk nomenclature (Berlin et al. 1973; Berlin 1992; Brown 1984, 1986; Atran 1999a), which is inferred from an assumed that 'ethnobiological nomenclature is a natural naming system that reveals the way people *conceptualize* living objects in the environment' (Berlin 1992, p. 26; italics in the original). With this specific aspect of consideration in effect, folk classifications appear to their researchers as the ordered systems of folk names (Hunn and French 2000), so they can be regarded as *biolinguistic* (Chamberlain 1992).

In accordance with these standardized approaches developed by contemporary systematics, ethnobiologists usually recognize four to six major folk-taxonomic hierarchical levels or ranks (Figure 1.1) and tend to ascribe them a more or less universal cross-tribal status (Berlin 1973, 1992; Atran 1990, 1998, 2002; Atran and Medin 2008). At the highest level, the traditional 'Linnaean' kingdoms (plants and animals) are most often distinguished, and the folk taxa, identified at the next lower level, most often correspond to the life forms: these are grasses, shrubs, and trees among plants; hairy beasts, birds, snakes, fish, 'flies', and 'worms' among animals. At the far lower levels, folk genera and folk species are usually distinguished; however, if it appears impossible to correlate unambiguously these levels in folk classifications with the currently accepted standards, the fuzzy folk ranks *generic-species* and *specieme* are introduced (Bulmer and Tyler 1968; Bulmer 1974; Atran 1998, 1999a, 1999b, 2002; Tyler, 1991). The neologism 'specieme', which was introduced instead of the standard 'species', is interesting in that it is evidently derived from the linguistic term 'lexeme' (Tyler 1991).

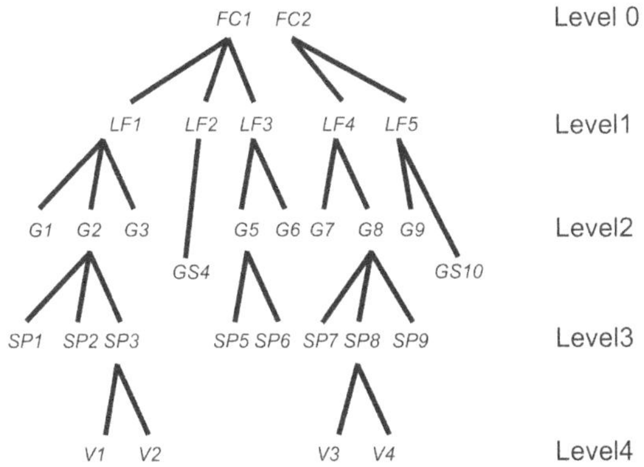

FIGURE 1.1 Graphical representation of folk-taxonomic hierarchy, as it is most usually discussed by ethnobiologists.

Abbreviations: *FC1* and *FC2*—folk classes; *LF1–LF5*—folk life forms; *G1–G9*—folk genera; *GS4* and *GS5*—folk generic-species; *SP1–SP9*—folk species/speciemes; *V1–V4*—folk varieties.

Based on: Berlin 1992, Figure 1.1; strongly modified.

With regard to the overall structure of folk classifications, experts in folk systematics pay special attention to the recognition of specific *salient* categories, which E. Rosch denotes as *basic level categories* (Rosch 1973, 1975, 1988; also Jacob 1991, 2004). They are composed of the objects that are cognitively (perceptually) or pragmatically significant for Indigenous people. However, the particular 'salient' categories are recognized differently by various researchers, depending not only on the interests of the reference native groups but also on the researchers' standpoints. Those who accentuate cognitive attitudes of Indigenous people tend to consider the higher-level Roschian categories (life forms, etc.) as the most 'salient' for their marking the major blocks of the local environments (Rosch 1973, 1988; Rosch et al. 1976; Brown 1984, 1986, 2004). To the contrary, if it is assumed that Indigenous people tend to single out before all the groups of lower ranks as most significant for the human survival 'here and now', then genera and species (speciemes) are considered most 'salient' (Greene 1909; Bartlett 1940; Boster and D'Andrade 1989; Berlin 1992; Atran 1999a; Medin and Atran 2004; Dwyer 2005).

All these facts may indicate that contemporary researchers, although aimed at uncovering the original structure of folk systematics, in fact simply impose their own 'cognitive matrices' onto the knowledge of Indigenous people and by this force them, in a sense, to speak 'an alien language'. As a result, such an uncritical transfer of the contemporary interpretative frameworks to the local folk classifications, which are generated within quite different cognitive contexts, implies their surplus standardization and can lead to the incorrect ideas about their hierarchical structure (Medin et al. 1996; Atran and Medin 2008; Pavlinov 2015, 2018, 2022a). This conclusion looks especially relevant, if the native categorization systems are treated as panarchies (on the latter term, see Section 3.3): this formalism means that such systems are supplemented with a specific highest level, shaped by the conceptual schemes of contemporary researchers. In this regard, the above-mentioned E. Rosch's idea of the recognition of natural categories by Indigenous people with the reference to certain prototypes should be especially highlighted, as she considers advanced 'logical' classifications and primordial 'prototypal' categorizations fundamentally different from a cognitive perspective (Rosch 1973, 1975, 1988; also Gardner 1985; Jacob 1991, 2004; Kleineberg 2017). This may perhaps explain the fact that, when folk categorizations (or classifications) are attempted to be investigated without applying hierarchical templates to them, they often appear to be arranged in networks or blocks (Hunn 1977; Posey, 1984; Hunn and French 2000; Ellen 2008; Alcántara-Salinas et al. 2013).

1.3 THE ANTIQUE HIERARCHICAL SYSTEMS

The most significant contribution of the Antique natural philosophy to the subsequent development of human cognitive activity includes its nearly complete rationalization. This means that both Nature itself (*φύσις*) and knowledge about it (*γνῶσις*, *επιστήμη*) are arranged in a similar regular (rational, reasonable) manner, and it is this unified regularity that supposedly makes Nature knowable by specific cognitive means that are organized in the *Method* in its most general meaning (Aristotle's *Organon*). The primary objective of the Antique cognitive activity was a comprehension of the qualitative structure of Nature, so the key part of their Method might be imagined as a specific categorization of both the observed and imaginary things following certain rules of the then elaborated logic. Accordingly, the primary tasks of thus organized cognitive process included, first, recognizing the essential features of the natural (real) objects and, second, arranging the mental objects (conceptions, notions) referring to them in a hierarchical manner that would correspond to the hierarchical order of Nature.

Despite of these general ideas unifying the whole of Antique natural philosophy, the latter was very diverse with respect to the particular understanding of the origins and structure of both Nature and Method. Coarsening this diversity, two basic worldviews and respective cognitive means are enough for us to be recognized here, namely, those elaborated by Plato and Aristotle, who, while standing on the shoulders of their great predecessors (Thales, Parmenides, Anaxagoras, Socrates, etc.), shaped the foundations of nearly the entire classical Antique philosophy (Gotthelf and Lennox 1987; Akhutin 1988; French 1994; Gaydenko 2003; Lennox 1980, 2021; Lennox and Bolton 2010; Chiaradonna and Galluzzo 2013).

The world picture of Plato (429–347 BC) was based on the general concept of *eidos* (έίδοσ, plural *eidê*, Lat. *idea*) as an unchangeable eternal supramaterial foundation of Nature; in this regard, Platonic version of natural philosophy could rightly be called *eidology*. In accordance with the latter, Nature was organized by a descending cascade of causalities, for which the *One* (ὀ ἕν) was the highest creative principle of all the being. This causal cascade began with the *One*'s emanating *eidê* of the highest order, which in turn gave rise to (emanated) the lower-order *eidê*, until the lowest of them became embodied in concrete material things to become the latter's *ousiai* (οὐσίαέ, Lat. *entia*). Of importance was that each higher-order *eidos*, when generating next lower-order ones, became a *genos* (γένοσ, plural *genê*, Lat. *genus*) with respect to the *eidê* emanated by it. Accordingly, an *eidos* became a particular *genos*, if it emanated its own lower-level *eidê*, and the latter existed only as the emanations of a particular *genos*. This meant that both *genê* and *eidê* were tightly interconnected and interchangeable in their status, so that one of them did not exist without the other, and each one can become another depending on a context. Such their specific relations implied a specific *genus-species dualism* with its concomitant *genus-species uncertainty*.

Such a process of successive emanation resulted in the emergence of an overwhelming inclusive hierarchy of the *genê* and *eidê* of different levels of generality. The sequence of the steps (levels) of emanation generated the *hierarchy of perfectness* of the emanated entities, which were ordered in a specific 'scale of perfection' according to their closeness to the complete plenitude of the initial *One*. Such linear hierarchical orderliness would subsequently be called the *Ladder of Perfection*, with the *Plenitude of Being* making it continuous *Scale of Being*, or *Scala Naturae* very popular among natural philosophers of the 16th to the early 19th centuries (O'Meara 1987). It should be noted that the *One* with all its emanated *genê* and *eidê* maintained its all-encompassing integrity and wholeness. Bearing in mind the contemporary distinction between taxonomic (aggregative) and partonomic (holistic) interpretations of the diversity of entities of any kind (see Section 2.3), Platonic emanation should be regarded as involving a specific kind of *partonomic* division of the *One*.

The natural philosophy developed by Aristotle (384–322 BC) was based on a worldview, which was largely reverse to that of Plato. In it, the concept of *ousia* took a central place and the entirety of Nature appeared to be organized by an ascending cascade of causalities, so his natural philosophical system could aptly be referred to as *ousiology*. The latter implied that the everlasting *primary* (prōtê) *ousiai* were embedded temporarily within the things thus making them what they are (*τὸ τί ἦν εἶναι*); they gave rise to the secondary, tertiary, and so forth *ousiai* of higher levels of generality, and this cascade of causalities provided the overall hierarchy of *genê* constituting the orderliness of the wholeness of Nature. The latter was superficially similar to the Platonic One in that it involved the same Ladder of Perfection but differed fundamentally from it in being ordered by a global ascending causality that began with *ousiai*'s embodiment in the things.

It should be highlighted that both Platonic and Aristotelian natural philosophies, with regard to ontology, agreed on adopting a realistic account of the intelligible categories of *genê*, *eidê*, and *ousiai*. Their reality was in that they existed outside (before and besides) the human cognitive activity; by this, they radically differed from the conceptions that arose and existed in a cognizing mind as their respective mental images. The whole of Nature was exhausted by the overwhelming hierarchy of *genê* and *eidê* of various levels of generality; this implied that there could be nothing existing in the real world that would be out of that hierarchy and would not be either *genos* or *eidos*.

With this, both of these natural philosophies presumed a twofold mode of being of these categories: they referred both to the inherent properties (quiddities) of things (including organisms) and to the things and groups thereof. Their quidditial status was thought of as primary, as these were particular quiddities of various levels of generality that appeared as the results of emanation. Accordingly, the things themselves and their aggregates were thought to be secondary by acquiring their actual existence (i.e., becoming 'what they are') due to those quiddities they embodied. In contemporary terminology, the first treatment (Aristotelian categories as quiddities of things) is referred to as *meronomic*, whereas the second one (Aristotelian categories as aggregates of things) is referred to as *taxonomic* (see Section 2.2 on their meaning). Such a twofold consideration of *genê* and *eidê* entailed a kind of their *taxonomic-meronomic dualism*, which contained no contradiction for Antique natural philosophers. In fact, it reflected a certain unity of these two modes of being of the same categories, which could be interchangeably both/either meronomic and/or taxonomic.

As for epistemology, the key idea underlying it was the recognition of the unity of Nature and Method, integrated by a shared rational principle. This unity was emphasized by the identical designation of their categories as *genê* and *eidê*, which meant their *onto-epistemic dualism*. With this, Plato and Aristotle, despite their opposing ontologies regarding the ordering of the overall cascade of causality, largely agreed on the basic principles of deductively (from the general to the particular) constructed epistemology. This meant that the foundation of the Method should be a specific cognitive hierarchy that began with the comprehension of the *greatest genos* (*megista genê*) and proceeded with the sequential division of *genê* and *eidê* up to the lowest level of *ousiai*.

In accordance to such understanding of the Method, a comprehensive knowledge of Nature was supposed to be exhausted by the hierarchy of *genê* and *eidê*. Since these categories referred primarily to the quiddities of things rather than their aggregates, the logical system underlying this Method was *intensional* rather than *extensional* (Lennox 1980; Pellegrin 1986, 1987, 1990; Gotthelf and Lennox 1987; Lansing 1995). At the same time, the same holistic vision of their hierarchy should be assumed for the relations between these two basic epistemic categories, just as in the case of their ontic prototypes. Therefore, the ancient Method could be considered a forerunner of the logical system of *mereology* (for these and other schemes and systems, see Section 2.4).

The main result of cognitive activity organized in this way was an inclusive hierarchy of *genê* and *eidê* as basic cognitive categories at different levels of generality. This hierarchical system served as a specific cognitive model of Nature; in modern terms, the latter can be considered as a kind of *classification*; accordingly, the categories in question appeared in it as *classificatory universals*. For our consideration, it is of paramount importance that the general argumentation scheme underlying that classification would be embodied by scholasticism in a classification *genus-species scheme*, which would constitute the methodological basis of the emerging systematics (see below).

The concept of hierarchically interrelated classificatory universals became fateful for their future development by neo-Platonics and scholastics (Balme 1962). As said above, the categorical status as *genê* and *eidê* in a classification were of the same kind, as in their ontically interpreted hierarchy with its genus-species dualism and uncertainty. Accordingly, they were not absolute and unchangeable but rather relative and interchangeable; with this, their pairwise subordination held strongly at each classificatory step (level), just as in the case of their ontic prototypes. Hence, an *eidos* acquired its status in a classification at its certain level as a result of the division of the respective higher-level *genos*, and the former in turn became a *genos* if subjected to the further division into more lower-level *eidê*. Such a specific interrelation between these classificatory categories could be referred to as *genê-eidê dualism*, which would become subsequently a focal point of a compound *genus-species problem* for centuries. With regard to its epistemic implementation, it would entail a *rankless sliding hierarchy* of interchangeable genera and species of various levels of generality (Pavlinov 2018, 2023; see also below).

It is of interest that Aristotle, in detailing his understanding of these categories as denoting particular groups of organisms, clearly distinguished between their ontic (natural) and epistemic (logical) accounts (Sloan 1970; Grene 1974; Pratt 1984; French 1994; Wilkins 2009a, 2010). On the one hand, when regarding them ontically as the basic elements of Nature, he treated them as continuous chains of generations and referred to them as *geneis* and *phyleis* (γενέίσ, φιλέίσ). On the other hand, when the same categories were regarded by him epistemically as the classificatory units (taxa in nowadays terminology), uniting alike things, *genos* was treated as something 'of the same essence' (*homoousios*, ὁμοούσιος), and *eidos* as something 'of the same form' (*homoeidos*, ὁμοειδῆ). Subsequently, when these categories become considered predominantly extensionally as aggregates of things, their onto-epistemic dualism would be removed. The first to do so was probably the Aristotle's disciple Theophrastus (ca. 370–ca. 288 BC), who applied in his *Enquiry into Plants* the categories of *genos* and *eidos* mostly in their extensional (taxonomic) meaning, so they would be translated in the contemporary English edition of that tractate as *class* and *kind*, respectively (Theophrastus 1916). One of the recent Aristotle's commentators suggested to designate his classificatory categories (units) as *quasigenus* and *quasispecies* to distinguish them from the ontically treated ones (Lloyd 1962; not in the most recent sense; see Section 5.1).

1.4 TOWARD THE DEDUCTIVE GENUS-SPECIES SCHEME

After Plato and Aristotle, the overall movement toward rationalization of the comprehension of both Nature and the Method of its exploration was continued by Late Antique neo-Platonics, who combined Platonic ontology with Aristotelic epistemology, and by Medieval scholastics, who placed both of them into Biblical context. From the point of view of the main theme of this book, one of their most significant achievements, resulted from such an intricate onto-epistemology, became an elaboration of the already-mentioned logical *genus-species scheme* of the sequential division of notions. It was this scheme that provided the prerequisites for the development of the first principles of classical logic. It is of special importance to us that it would give rise subsequently to the classificatory methodology of the nascent biological systematics and will lay the foundation for (a) the hierarchical style of reasoning employed by the latter's principal actors and (b) the essentialist version of early taxonomic nomenclature.

Tracing the origin of 'hierarchical thinking' in early natural philosophy and science, it should be noted that the genus-species scheme that implemented it was actually derived from the linear Ladder of Perfection and could be seen as the latter's specific

> 'extension'. It was built over the Platonic–Aristotelian ontical 'scale of perfection', according to which all entities were arranged in an order from the highest to the lowest, and the scheme in question illustrated specifically such their arrangement. For instance, the *genos* 'Living' was divided into lower-level *eidê* 'Sensitive' and 'Insensitive'.

The basic terminology of this scheme. which was being developed by scholastics, reflected the fact that the language of early European post-Antique philosophy became predominately Latin due to the transfer of the 'center of gravity' of Medieval culture from Greek to Roman (Jorge and Newton 2016). Because of this, the previous *γένοσ* became uniformly translated by Latin *genus*, whereas *έίδοσ* appeared to be either *idea* or *forma* or *species* depending on the semantic contexts it was put in. The first its meaning most closely corresponded to the original Platonic understanding of *έίδοσ*, whereas in the second and third cases, it was closer to its Aristotelian understanding as *μορφή*. Consequently, the scholastic categories of *forma* and *species* appeared in the Latin of many prominent Medieval philosophers, beginning with the Numidian theologian and philosopher Aurelius Augustine (354–430), as interchangeable synonyms (Gilson 1960; McGrade 2003). So it seems reasonable to unite them under the term *form-species* (or *species-form*) to reflect the above-mentioned taxonomic-meronomic dualism of both these universals and their hierarchies (Pavlinov 2023).

Although the classical Antique natural philosophy presumed with no doubts an objective existence of the ontically treated categories in question, a contradiction between their realistic and nominalistic accounts became one of the focal points of onto-epistemic issues in Medieval scholasticism. This was mainly because the latter acknowledged a rather strong opposition between the research objects and the cognitive instruments applied by the knowing subjects, i.e., between ontology and epistemology. Such their opposition was further reinforced by the fact that the interpretation of both genus and species as the logical (classificatory) universals became of the prime importance due to concentration of scholasticism on the principles of rational foundations of cognitive activity.

This whole issue was explicitly detailed for the first time by the Syrian neo-Platonic Porphyry (232–304) in his treatise (still in Greek) *Introduction to the 'Categories' of Aristotle* (Lat. *Introductio in Praedicamenta*; *Isagoge*) (Spade 1994; Karamanolis and Sheppard 2007). It is important to emphasize that he confessed the natural philosophy of Platonic realistic kind, and it was such a realistic ontic interpretation of *genê* and *eidê* that served for him as a basis for the reliable substantiation of their epistemic interpretation as predicates (i.e., 'those that say'). It was such an interpretation that implied a direct correspondence between the ontic Scale of Being and the epistemic genus-species scheme as a *Scale of Judgments* (Lat. *Scala praedicamentalis*). Proceeding from this, Porphyry approved the previously developed idea of both (a) deductive character of inferring epistemic hierarchy of *genê* and *eidê* and (b) general pattern of their rigid logical subordination in that hierarchy (Strange 1987; Evangeliou 1996; Karamanolis and Sheppard 2007). He enclosed the entire hierarchy of *genê* and *eidê* between the 'greatest *genos*' at the top and the 'most specific *eidê*' at the bottom. In addition, he designated all possible divisions of the respective hierarchy as subordinate (*subalterna*) *genê* and *eidê*, with each of them becoming either a *genos* or an *eidos* depending on the particular cognitive tasks (see Porphyry 2003). Within this deductively framed hierarchy, (a) *genos* always preceded *eidê* at each step of deduction, (b) they were equivalent to the extent that they were interchangeable in their predicative status, and (c) therefore the hierarchy they composed in total was rankless and sliding with no explicitly fixed steps of deduction.

> A very important part of the subsequent development of the scholastic cognitive program involved the justification of three basic worldviews, namely, realism, nominalism, and conceptualism, which still constitute the most fundamental philosophical controversy in natural science (Koyré 1973; Marenbon 1988; Gilson 1991; Klima 2017; see Section 2.1). Their disagreement served for scholastics as a general conceptual framework for the genus-species problematics, which discussions revolved mainly around the onto-epistemic status of genera and species, including their origins both in nature and in mind and their hierarchical interrelations (Gilson 1991; Spade 1994; Motroshilova 2000; McGrade 2003; Pavlinov 2023). In both cases, an examination of this problematics by scholastics concerned implicitly (in our terminology) the above-mentioned taxonomic-meronomic dualism, with the emphasis being placed on the meronomic aspect of these categories.

With this, scholasticism strengthened the old idea that the categories in question actually existed and were mentally recognized in a close conjugacy. This thesis was quite expressly highlighted by the Irish theologian Eriugena (ca. 810–ca. 870), who said in his treaties *De Divisione Naturae* that 'the genus is the genus of species and species is the species of genus' (O'Meara 1987, p. 61). He highlighted by this that (a) species was not independent of genus but merely represented its detailing, (b) a certain category became a genus, if it was divided into respective lower-level categories as its species, (c) therefore a species became a genus if further divided, and (d) hence, their particular cognitive status was not fixed. Due to this, they (let us repeat) become interconnected by the above-commented *genus-species dualism*, and this was the latter that made their overall hierarchy sliding and rankless.

* * *

A peculiar accent on the ontic interpretation of the origin and hierarchical interrelations of genera and species, rooted in Platonic natural philosophy, was brought by the Biblical worldview due to a magnificent expansion and growing influence of the Christian Church (Koyré 1973; Gaydenko 2000; Jorge and Newton 2016). This resulted in a kind of 'Christian Platonism', who had a very significant impact on the philosophical ideas of Late Antique and Early Medieval treatments of Nature and its causes (Neretina 2013; Hampton and Kenney 2021). According to one of its early versions (Pseudo-Dionysius, Eriugena, Anselm of Canterbury, Abelar, Duns Scotus, etc.), the entire hierarchy of the whole being, both ideal and material, appeared as an incarnation of divine archetypes, or 'supreme essences', comprising in a potency the wholeness of the hierarchy of multitude genera and species (Gilson 1955; Davies and Evans 1998; Erismann 2007).

One of the noticeable results of such a philosophical merger appeared to be the advent of the very term *hierarchy*. Its author was the semimythical figure

Pseudo-Dionysius (the late 5th to early 6th centuries), a purported author of the influential treatise *Corpus Dionysiacum* (originally in Greek) (Hathaway 1969; Rorem 1993). He proceeded from a biblical ontological idea of 'angelic ranks' that composed the *Heavenly* (or *Celestial*) *Ladder of Perfection* directed to the Creator (Dorofeeva 2001). This heavenly Ladder originally reflected a sequence of the celestial incarnations of divine archetypes: thus understood, the Pseudo-Dionysian neologism 'hierarchy' (*ἱερὸς ἄρχων*) etymologically referred to the *origin* (*ἀρχή*) of the order of *sacred essences* (*ἱερω*) (Lyubarsky 2018; Vasilakis 2019). Being essentially biblical in its content, this general celestial construction was evidently rooted in the antique Scale of perfection of the universe, both being arranged linearly (Lovejoy 1936).

The early variants of this celestial ordering were strictly linear to comprise a sequence of 9–11 angelic steps according to their closeness to the Creator. In its more elaborated variants (Pseudo-Clement, Aquinas, etc.), these steps were grouped into three to five fixed *orders* (Pseudo-Dionysian *τάξες*) (Dorofeeva 2001; Vasilakis 2019). Due to this grouping, a non-inclusive linear hierarchy was transformed into an inclusive several-level block one, whereby certain aggregates of the celestials were recognized, though not specifically denoted yet. This became an indirect prototype of the nested ranked hierarchy, in which several hierarchical categories (ranks) would be fixed.

* * *

With the opposing ontic and epistemic components of their cognitive situations and, at the same time, acknowledging onto-epistemic dualism of genera and species, scholastics had to pay much attention to the interconnections between their ontic and epistemic guises. Addressing this issue, the Roman philosopher and statesman Boëthius (ca. 480–524), an influential author of *Commentary on Porphyry's Isagoge* (Lat. *Isagogen Porphyrii Commentorum*), emphasized that genera and species were twofold entities, existing both in reality and in thoughts but in different incarnations. Proceeding from this idea, the German philosopher and theologian Albertus Magnus (ca. 1200–1280) divided in his treatise *De Sex Principiis* all universals into three basic categories: before things (*ante rem*, eternal divine ideas), in things (*in re*, embodiments of these ideas), and after things (*post rem*, concepts that emerge in the human mind); this triplet categorization would become famous after its refinement by the Italian philosopher and theologian Thomas Aquinas (ca. 1225–1274) (Gilson 1955; Maurer 1982; Neretina 1994; Wippel 2000; Gaydenko 2001; Tremblay 2008; Resnick 2012). The *Tractatus de intellectibus* by the French philosopher Peter Abelard (1079–1142) said that human mind looked for the common in particulars by comparing the latter with each other and likening them (Augustine's *similitudo*) to divine ideas that have generated them. It is of importance to note that Abelard seemed to be the first who explicitly included human *intention* (as *conceptio intentionalis*) into the entire cognitive situation to emphasize that each cognitive act was conscious and purposeful, so he became a recognized forerunner of conceptualism (Neretina 1994; Cameron 2007).

It is of interest that in their onto-epistemic searches, scholastics retained the Aristotelian category of genus near untouched and concentrated their intellectual efforts on species, elaborating diverse, sometimes very detailed categorizations for its manifestations. This accentuation has been initiated by Boëthius, who assured that 'if we do not realize what species is, nothing would secure us from delusion' (Boëthius 1906; quoted after the Russian edition of 1990, p. 12). Aurelius Augustine recognized in his treatise *De Trinitate* three categories of species-forms: those existing in objects (*species corporea*), in their human perception (*species impressa*, or sensible species), and lastly formalized in the human mind (*species expressa*, or intelligible species) (Gilson 1960; McGrade 2003; Nawar 2021). The *Timaeus translatus commentarioque* by the Augustinian follower, philosopher Calcidius (the 4th century) distinguished between archetypal, or main (*species archetypa*, *principalis*) and secondary (*species secunda*) species-forms (Albertson 2014). The most detailed scheme of this kind was elaborated by Thomas Aquinas in the form of a 'cognitive chain', which organized the entire cognitive processes as a causally linked sequence of the universals, namely, *species sensibiles*, *species in medio*, *species intelligibiles*, etc. (Spruit 1994; McGrade 2003). Though interesting by themselves, these categorizations of the species notion would not leave evident traces in its future development by nascent systematics, in which the categories of genus and species were taken in their original Aristotelian meaning.

The English philosopher William of Ockham (ca. 1300–1347/1349), the first great nominalist, rejected genus and species as universals at the level of ontology but paid great attention to their epistemic function of inductive kind. According to him, the process of knowing begins with a sensory perception of things, which precedes any conceptualization. The first perceptive *act of knowing* (*actus intelligendi*) is followed by a summation of particular perceptions into a certain mental *image* thereof (*habitus*), whereas every general *notion* (*conceptum*), as a universal, is formed in the 'soul' of the cognizing subject as an abstraction of the multiplicity of real things. According to Ockham, the terms referring to the universals are but units of language, and they have nothing to reflect in the real world of things (McCord Adams 1989; Panaccio 2004; Vdovina 2009; Paasch 2014). This argumentation will be repeated by the creators of the so-called species crisis in the late 19th and early 20th centuries (see Section 5.1).

* * *

All these intellectual efforts, addressing epistemology, resulted in a specific instrument of the Medieval deductive logical inferences to become subsequently known as the *genus-species scheme*. The latter, being a part of classical logic, (a) deals not with the real universals themselves, but with the mental ones, existing in mind as notions and their definitions, and (b) realizes a unified logical formula for the definition of any notion (except for the most general) by indicating its 'generic common and species particular' (*per genus proximum et differentiam specificam*) (Shapiro and Kissel 2024; Ivlev 2025). The above-mentioned Eriugena was one of the first to formalize this scheme in his *De Divisione Naturae*, in which he represented logic as 'that art which divides the genera into species and resolves the species into genera, which is called *dialectic*' (cited after: Erismann 2007, p. 79; italics in the original). Such an 'art' ensures inferring lower-level true judgments about any analyzed entities from the higher-level ones that are acknowledged *a priori* as indisputably true (such as divine archetypes). Notwithstanding the genus-species dualism in effect, the hierarchical relation between genus and species is firmly set: genus predicates of species, and the reverse route of thought (species predicating of genus) is prohibited. Respectively, species always occupies a subordinate position in relation to genus and has no independent status, being its detailing. Therefore, the late scholasticism, unlike its earlier predecessor, became paying main attention to genus and its divisions, and this logical approach was adopted by the earliest systematicians for the elaboration of classifications of living beings.

The basic terminology for the scheme in question had already been developed by Porphyry and subsequently specified by the English philosopher Roger Bacon (ca. 1214–1292), who refined the above-outlined Porphyrian ideas by setting the initial (*genus generalissimum*) and intermediate (*genera subalterna*) logical generic divisions and finalized it by the recognition of the last species, which was not further divided (*species specialissimum*, or *species individuum*). The latter designated in this hierarchical scheme merely the last step of cognition, which might pertain to any level of generality, depending on a particular cognitive task. All these steps, except for the terminal one, were designated as genera, just because they were subject to further divisions.[2] It was suggested later to designate the initial point of division as *Genus summum*, its intermediate steps as *Genera intermedia*, to recognize the last step of division as *Genus proximum*, and to call its final point denoted as *Species infima*. The latter was not further divided (otherwise it would be a genus) but can be exemplified by *tokens*; these were Plato and Socrates of the human species in the original Porphyrian scheme.

Thus, the final terminological version of the inclusive hierarchy of genus-species scheme elaborated by scholasticism, with which mastering the earliest systematics emerged, looked like this:

```
Genus summum
  Genus intermedium
...............
    Genus intermedium
      Genus proximum
        Species infima
```

In this multi-level inclusive hierarchy, the division steps corresponding to the *Genera intermedia* were rather arbitrary, they were not designated by specific terms, and their number was not fixed, depending on the particular exploratory tasks. The same held for the terminal step that could easily become intermediate: for instance, as would be suggested much later by the English botanist George Bentham (1800–1884), who began his scientific career as a logician, 'generally speaking, … there is no such thing as an infima species, as any one, how ever low, is susceptible of still further subdivision' (Bentham 1827, p. 117). It was this specific feature of genus-species scheme that, as said above, made its hierarchy rankless, non-fixed, and sliding.

It is to be noted, however, that this rankless scheme in fact contained implicitly a certain germ of a possibility of its subsequent ranked interpretation. This was due to an explicit recognition of two basic hierarchical categories, *genus* and *species*, and more so, of several major steps in the sequential generic division (*summum*, *intermedium*, and *proximum*). These steps would became later the fixed ranks of the first versions of ranked hierarchy of biological systematics in the 18th century. With this, recognition of the interrelated pair of Aristotelian categories of genus and species would entail a very peculiar version of taxonomic nomenclature by making taxonomic designations of the concrete species first *binary* (consisting of two parts) and then *binomial* (consisting of two words) (Pavlinov 2015, 2022a; see Section 6.1).

* * *

Considered from the perspective of future biological systematics, an important advance in the development of the logical genus-species scheme was made by the elaboration of its tree-like diagrammatic representation. Its first versions appeared in the early re-editions of and commentaries on Boëthius' *Isagogen* (Barsanti 1992; Hacking 2007; Rosalinde 2010; Archibald 2014; Verboon 2014; Richards 2016; Dardağan 2018). This visual representation gained its commonly acknowledged denotations as *Tree of Porphyry* (Lat. *Arbor Porphyrii*, after neo-Platonic Porphyry) in the *Tractatus* by the Spanish theologian Peter of Spain (Lat. *Petrus Hispanus*; fl. 13th century) (de Rijk 1972). The later copies of *Tractatus* contained drawings of *Arbor Porphyrii* that differed in their formats (Barsanti 1992; Rosalinde 2010; Dardağan 2018), and one of them, rather picturesque, would become most frequently reproduced (Figure 1.2).

The basic idea of hierarchy of both the genus-species scheme and Tree of Porphyry was inferred from Platonic natural philosophical concept of successive emanation of the *One* into particular *genê* and *eidê*. Proceeding from this background, two important conjectures could be made concerning the tree in question: it illustrated (a) more meronomic rather than taxonomic interpretation of the respective Aristotelian categories and (b) both the steps of their sequential categorical divisions and connections (links) between them. Both divisions and connections were shown by the same lines linking the interrelated genera and eventually species with each other, which implied a characteristic simplified *line format* of Tree of Porphyry in all its modifications (see Section 4.2).

Its subsequent development has led to the splitting of such its 'complex unity'. It so happened that the 'connection' interpretation of this tree was retained by logicians, who would apply it most actively just in the line format up to the most recent times (e.g., Huth and Ryan 2004). Unlike this, the 'categorization' tree was developed by philosophers and naturalists as a pure classificatory one, so that the vertical line format of the Medieval Tree of Porphyry was transformed into a horizontal *bracket format* of the Renaissance *Tree of Ramus*. It was appeared probably for the first time in the *De Inventione Dialectica Lucubrationes* (1479) by the Dutch humanist Rudolphus Agricola (1444–1485) and was refined afterward by his follower, the French logician Pierre de la Ramée (Lat. *Petrus Ramus*, 1515–1572) in his *Dialecticae Institutiones* (1556) to become named subsequently after him (Ong 1958; Archibald 2014; Dardağan 2018). The hierarchically arranged bracket structure of this tree-like diagram (Figure 1.3) was especially designed to indicate logical relations between the notions as an inclusive hierarchy of their aggregates of various levels of generality; its horizontal orientation was probably due to the fact that it more fitted the

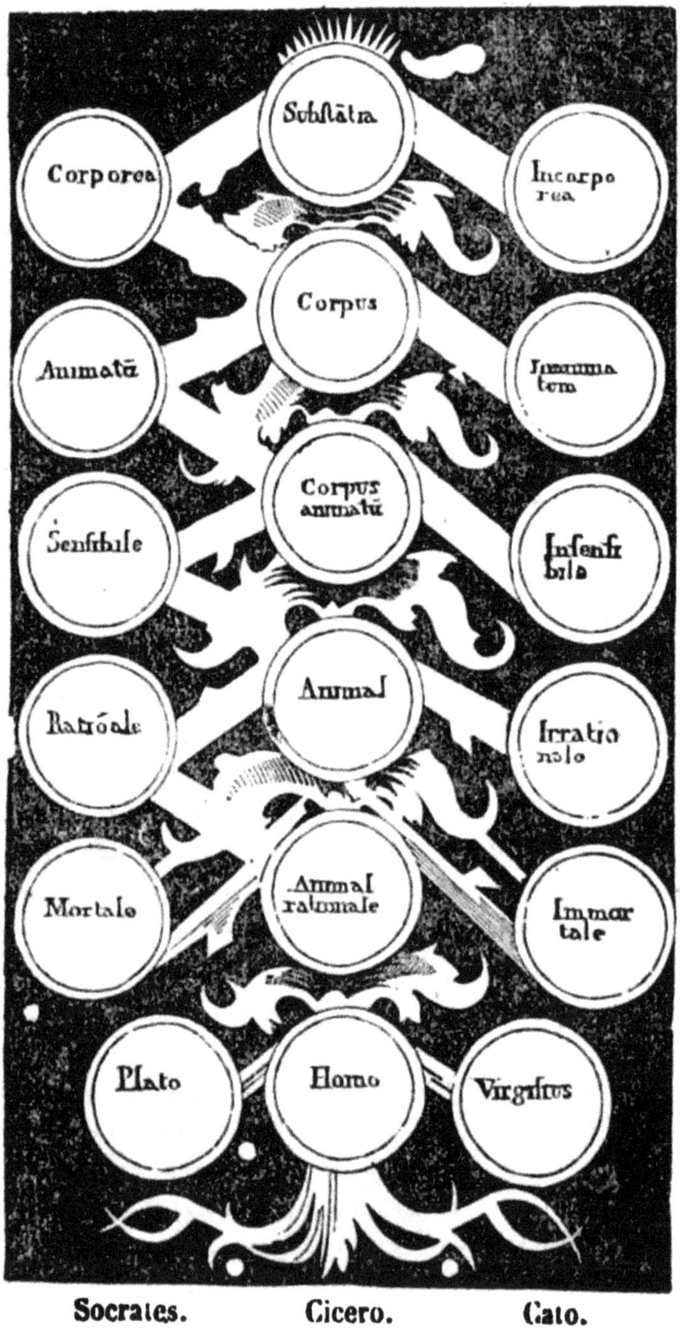

FIGURE 1.2 One of the earliest drawings of Tree of Porphyry, as it appeared in *Isagoge* by Porphyry. The open Internet.

Source: The Logic Museum (https://www.logicmuseum.com/wiki/File:Porphyrian_tree_PL_64.41.jpg); modified (side comments are not reproduced).

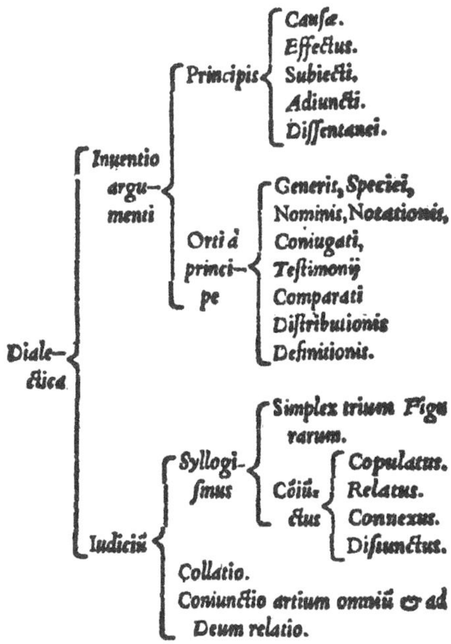

FIGURE 1.3 The Tree of Ramus drawn in its typical bracket format.

(Ramus 1547, p. 100).

standard book format, which had become widely accepted by that time. All this made Ramean tree soon very popular among scholars in both natural philosophy and natural history. Of the former, the most interesting (to us) were those of the Englishman John Wilkins (1614–1672) with his 'philosophical language' illustrated diagrammatically by numerous trees of Ramean kind (Wilkins 1668; see below). It was this tree-like representation format that would become subsequently most popular in biological systematics.

Considering the latter, a tree-like style representation of classifications appeared for the first time in some herbals, bestiaries, and the earliest rational systematizations of organisms—and the trees in them were those of Ramus, not of Porphyry (see Barsanti 1992; Pietsch 2012 for reviews).[3] Of them, the most advanced in this regard were *Nomenclator aquatilium animantium* by the Swiss naturalist Conrad Gesner (1516–1565), *Methodi Herbariae* by the Bohemian naturalist Adam Zalužanský (1555–1613), and *De animalibus insectis* by the Italian naturalist Ulisse Aldrovandi (1522–1605) (Gesner 1560; Zalužanský 1592; Aldrovandi 1618). With the advance of the natural method of scholastic systematics, such a diagrammatic representation became widely applied to illustrate hierarchical arrangement of not only synoptic classifications of plants and animals (R. Morison, J. Ray, P. Magnol, etc.) but also some other important information pools. For instance, Carl Linné illustrated by such trees both his idea of plant 'sexual system' and relations between different taxonomic systems of his predecessors (Linnaeus 1735, 1738).

* * *

An important movement toward the future incorporation of scholastic ideas of genus and species, including the genus-species logical (and by this, classificatory) scheme, in systematics involved a more or less clear differentiation of their meronomic and taxonomic accounts. Just to emphasize the importance of this endeavor, it is to be reminded that the former refers to the quiddities (essences) of things, whereas the latter does so regarding aggregates of the same things. These two accounts are united by taxonomic-meronomic dualism, with Medieval scholars paying most attention to the quiddities by considering them (keeping in mind Platonic eidology) primary with respect to the things and their aggregates. Meanwhile, systematics will become more interested in the aggregates of things and will take their features just as means for defining those aggregates. This position is evident from the early classificatory synopses of plants and animals, which were arranged as 'checklists' of taxa with their diagnoses.

The first quite evident step in this direction was made by already mentioned Albertus Magnus in his fundamental multivolume encyclopedia *De Animalibus* (1479), based on the Aristotelian zoological treaties and called by its first commentators *Summa zoologica*, by analogy with the Aquinas' *Summa theologica* (Wallace 1999). Albertus declared in Book I of his *Summa* that he intended to represent descriptions of all animals 'according to [their] genera and species' and added that he used '"genus" in the sense of "genus of birds" or "genus of fish", since each of them contains many smaller genera and species' (see Albertus Magnus 1999, Vol. I, pp. 47, 52). Based on this premise, he summarized the diversity of animals as rather detailed accounts of their 'genera' and 'species', sometimes indicating their ranks in the titles of their accounts, e.g., *De calais duobus generibus falconum*. The same basically 'taxonomic' approach was characteristic of the *Synopsis* of his botanical series: it included section *Divisiones genera species plantarum*, in which the higher genera (*Genera summa*) were trees, shrubs, herbs, algae, fungi, with one of its sections being titled *Genera et species secundariae* (see Albertus Magnus 1867). And yet, his Aristotelian categories of genus and species were often mixed taxonomic-meronomic units of various levels of generality.

It was especially noteworthy that Albertus comprehended ontology of the animal species in a sense that was akin the modern one by referring to the so-called *generative idea*. So he wrote in Book I of the *De Animalibus* that 'species <...> participate in an eternal existence by *reproducing one individual from another*, and this can only be so if the force forming the animal is transmitted in the seed from the one who reproduces' (cited after Albertus Magnus 1999, Vol. I, p. 72; italics added). Although *De Animalibus* was in fact just a compilation of zoological works of Aristotle, so Albertus only iterated the idea of his great predecessor, he deserved to be counted as a real forerunner of what is now known as the *reproductive species conception*.

* * *

The general idea of the hierarchical deductive genus-species scheme was willingly adopted by the dawn systematics of the 16th–17th centuries as a basis for the ordering plants and animals into classifications with rankless hierarchies. For example, the

English naturalist John Ray (originally Wray, Lat. *Joannis Raius*, 1627–1705), emphasized in several of his treatises (*De variis Plantarum Methodus*, *Methodus plantarum emendate*) that the plants should be classified 'according to the rules of logic [and] a complete definition of [species] consists of proximal genus and essential difference' (Ray 1696, p. 23; Ray 1733, p. 6). In this 'formula', the 'proximal genus' had evidently meronomic and not taxonomic meaning. In his *The wisdom of God*, Ray mentioned four 'major genera, or orders' of beasts, birds, fish, and insects (Ray 1714, p. 21). A similar classificatory approach was evident in the treatise *ΠΙΝΑΞ theatri botanici* by the Swiss botanist Kaspar (Gaspard) Bauhin (Lat. *Casparus Bauhinus*; 1560–1624), which subtitle read *Methodicè secundum earum & genera & species proponens* (Bauhin 1623).

A commitment of early systematicians to the genus-species classificatory scheme of late scholastics could be seen in their focus on genera, which was evident in *De plantis libri XVI* by the Italian philosopher and phytograph Andrea Cesalpino (1524–1603) (Cesalpino 1583) and in *Opuscula Botanico-Physica* by the German philosopher and mathematician Joachim Jung (1587–1657) (Jung 1662, 1747). Another manifestation of the scholastic character of the works of that time could be seen in the fragmentation of the categories of both genus and species due to their sequential division into subcategories *prima*, *secunda*, and *tertia*. The resulting classifications of this kind were expressly exemplified by those in *De historia stirpium* by the German herbalist Leonhart Fuchs (1501–1566) (Fuchs 1542), as well as in the just-mentioned Ray's *Methodus plantarum*.

In this regard, it should be emphasized that the genera and species, which naturalists of that time used to recognize on the scholastic deductive basis, were understood by them as onto-epistemic (classificatory) universals without discernible particular biological meaning. Such an apparent non-biological comprehension of Aristotelian categories was rather typical for the systematicians of that epoch. For instance, J. Ray in his *The wisdom of God*, when characterizing the results of divine creation, referred to species of all material things and attributed 'various species of inanimate bodies to the divers figures of the minute particles of which they are made up' (Ray 1714, p. 61).[4] Therefore, the English naturalist Arthur Cain (1921–1999), in his analysis of treatises, denotes Bauhin's categories not with standard but with modified terms: *generoid* ('as if genus') and *speciate* ('as if species') (Cain 1994). This is reminiscent to an extent, how Aristotle's applications of these categories were suggested to be termed *quasigenus* and *quasispecies* (Lloyd 1962; see above).

> To conclude this section, it may be worth noting that although inclusive hierarchy has been widely applied by both scholastics and early naturalists, the very word 'hierarchy' was never met in their treatises (Verdier 2006). The most likely motive for this was that this term has been being predominately used at those times in its original theological meaning by referring to the linear celestial hierarchy. The first and quite occasional instances of its application in the scientific contexts with just the 'inclusive' meaning appeared in the works of the early 19th century: it was mentioned by J.-B. Lamarck in his *Philosophie zoologique* (see Lamarck 1963, p. 24), as well as by some sociologists (A. de Tocqueville, A. Comte, R. Virchow, etc.) (Whyte 1969; Verdier 2006).

1.5 THE DEVELOPMENT OF RANKED TAXONOMIC HIERARCHY

The general trend of the conceptual development of early systematics involved replacement of the rankless hierarchy, inherited from Medieval scholasticism, with the ranked one, which represented another, non-scholastic version of rationalization of our discipline. The significance of this 'new rationality' was that the inclusive ranked hierarchy provided a more solid and robust framework for biological classifications. Therefore, there are sufficient grounds to believe that this replacement was of revolutionary importance for the subsequent development of systematics (Lyubarsky 2018; Pavlinov 2021a). The elaboration of a ranking system for classifications can be compared in significance to the elaboration of the Cartesian coordinate system for a graphical spatial-like representation of certain relations between objects. It is noteworthy that this innovation has strengthened discrete categorization analytics and moved rationalization of systematics, being a part of natural history, in a direction opposite to that of natural philosophy (physics, chemistry, etc.), in which numerical continualization of the analytical method (differential calculus, etc.) was dominated.

Notwithstanding its revolutionary character, this hallmark shift preserved two important provisions of the general design of the scholastic rankless hierarchy. One of them was that the earliest four-rank version of taxonomic hierarchy (Tournefort, Linné) was based on the principal four-step structuration of scholastic genus-species scheme (see below). Another 'legacy' became the fundamental *principle of rank equality*, whereby each division of a higher-level classificatory unit resulted in two or more next lower-level equiranked ones.

The first signs of fixed ranks in classifications of the living began to appear in some of the advanced treatises of the Herbal epoch of the 16th and early 17th centuries (Sachs 1906; Arber 1938; Pavlinov 2015, 2018, 2021a). Although the ranks themselves were not fixed yet in their positions in the hierarchies of particular classifications, they became designated by new, non-scholastic classificatory terms. For example, the classificatory units of *Herbarum vivae eicones* by the German theologian and naturalist Otto Brunfels (1488–1534) were entitled by the terms borrowed from the music: 'generic' chapters were denoted as *Rhapsodia*, while he termed 'species' sections *Sententia* or *Paraphrasis* (Brunfels 1530–1536). The final historical phase of that epoch was marked by the introducing certain hierarchical categories that would become subsequently basic in systematics. For instance, the above-mentioned Conrad Gesner designated chapters in his *Nomenclator aquatilium animantium* as Orders: *Ordo III. De Pisciculus*, *Ordo XII. De Cetis*, etc. (Gesner 1560). The Dutch (by birth) theologian and naturalist Charles de l'Écluse (Lat. *Carolus Clusius*; 1526–1609), the author of *Rariorum Plantarum Historia*, applied standard terms 'genus' and 'species' throughout but recognized meronomic 'classes' in some of the chapters to individuate variants of the leaves and flowers (Clusius 1601).

A very peculiar version of classification with fixed ranks appeared in the treatise *An essay towards a real character* by the above-mentioned philosopher John Wilkins in the mid-17th century (Wilkins 1668). When developing a universal 'philosophical language' for a unified description of the universe based on a specific deductive classificatory approach (Slaughter 1982; Maat 2004), he elaborated a strictly ordered hierarchy of the overwhelming System of Nature in the form of the classificatory

Tree of Ramus (incorrectly called Tree of Porphyry by J. Maat). Wilkins recognized in it three basic levels, denoted by him (in a descending order) 'Genera' 'Differentiae', and 'Species', and placed the notions referring to particular things, including biological organisms, at the tips of the tree branches. A noteworthy linguistic (nomenclatural) part of Wilkins' approach included a special codification of the designations of those things, which indicated their positions in his enormous universal classification. It is to be added that J. Wilkins tried to involve the above-mentioned naturalist J. Ray in his enterprise, but this phytographer found the latter quite unsatisfactory (Sloan 1972; Slaughter 1982).

* * *

An epochal shift toward the development of the ranked hierarchical classifications for plants and animals, with the ranks being strictly fixed both in their position in hierarchy and terminologically, began with the works of several systematicians of the late 17th and the 18th centuries.

The first of them to be mentioned here was the French botanist Pierre Magnol (1638–1715), who fixed in a specific manner supra-generic hierarchy of botanical classification by using hierarchical categories (ranks) *Sectio* and *Familia* in his *Prodromus historiae generalis plantarum* (Magnol 1689). The category of *Familia* was borrowed from the social organization of humans to emphasize its not logical but evidently natural status. It would become basic in classifications by Magnol's compatriot and follower Michel Adanson (1727–1806), as well as in the most of subsequent classifications both in botany and zoology since the early 19th century.

The most significant contribution to the future development of the inclusive ranked hierarchy in biological systematics was made by the Magnol's compatriot and disciple Joseph Pitton de Tournefort (1656–1708), who became the first among naturalists to modify completely the sliding rankless hierarchy of genus-species scheme of scholastics. In his *Élémens de botanique*, he introduced and consistently applied a simplified taxonomic hierarchy of four ranks, which were firmly fixed both by their position and mutual subordination in the hierarchy, as well as by specific terms, namely, *Classis*, *Sectio*, *Genus*, and *Species* (de Tournefort 1694). There is every reason to assume that such a four-level hierarchy appeared not accidentally on a Tournefort's whim but was derived from four major steps of the rankless hierarchy of scholastic classificatory scheme: *Genus summum*, *Genus intermedium*, *Genus proximum*, and *Species infima* (see also below on Linné). This suggestion was most probably motivated by an aspiration toward a more solid framework for the representation of hierarchical structure of *Systema Naturae*, and the basic generic structure of scholastics already in hands appeared quite appropriate for this purpose. As a result, the former genus-species dualism/uncertainty became resolved by the ranked hierarchy with a few specifically fixed and denoted categories.

Two very important features of this innovative hierarchy should be especially highlighted. First, Tournefort definitely separated genus from higher-level categories, so the latter ceased to be just 'other genera' of the sliding hierarchy, and each of them acquired its own status and designation. An importance of this idea will be subsequently acknowledged by calling Tournefort the 'father of concept of genus' in botany (Carr 1923; Bartlett 1940; Stuessy 2009). Second, the same was true for species, which was deprived by Tournefort of the functions of (a) becoming a genus, if subjected to further detailing, and (b) terminating a particular deduction inference at any arbitrary hierarchical level.

These Tournefortian ideas regarding the ranked taxonomic hierarchy were refined in several decades by the Swedish naturalist Carl Linné (Lat. *Carolus Linnaeus*, 1707–1778) in his multiple treatises. He adopted four main ranks suggested by Tournefort with a minor alteration (*Sectio* was replaced with *Ordo*) and emphasized their significance by indicating them in the title of his *Systema Naturae*, beginning from its first edition (Linnaeus 1735). Discussing these basic ranks in *Critica botanica*, Linnaeus approved their scholastic source by noting that '*Classis* corresponds to *Genus Summum*, and *Order* to *Genus intermedium*', and defined their subordination accordingly: 'Genera consist of Species, Orders of Genera, and Classes of Orders' (Linnaeus 1736, § 251). To these four Tournefortian ranks (categories), to be subsequently called *principal*, he added more two: *Empire* above and *Varietas* below. The latter infraspecific category was discussed by Linnaeus in the special section of *Philosophia botanica* (Linnaeus 1751), and it appeared to be of paramount significance: due to this innovation, the species could from now on be divided into varieties without losing its categorical status as exactly the *Species*.

With this, it is to be noted that the Linnaean ranking system was not definitely settled by him; e.g., the major groups of plants (fungi, algae, etc.) were called 'Species' in his *Fundamenta botanica* (Linnaeus 1736), 'Families' in *Philosophia botanica*, and 'Tribes' in the second volume of the 12th edition of *Systema Naturae* (Linnaeus 1767). Moreover, in a brief outline of the *Delineatio Plantae* in the latter treatise, he indicated only two basic categories, *Genus* and *Species*, with *Class* and *Order* being listed as particular cases of *Genus* in its broad, evidently scholastic understanding.

It so happened that both this primordial ranked taxonomic hierarchy and species as its lower-level taxonomic category would be denoted in the early 19th century as *Linnaean*, and both these terms are now commonly adopted in biology. However, it should be emphasized that all these categories (ranks) in their Linnaean interpretation had no specific biological meaning: in fact, they still remained classificatory universals equally applied to the Kingdoms of *Plantae*, *Animale*, and *Lapideum* (minerals), and he was followed by some of his devotees in such their unified interpretation (e.g., Scopoli 1777; Fischer 1811).

Introducing these principal categories in his *Critica botanica* and *Philosophia botanica*, Linné seemingly ascribed different ontic status to them; this can be supposed by an analysis of some of his rules for the formation of the names of taxa allocated to them (Pavlinov 2015, 2022a). First, the one-word generic name should be a noun in the singular, reflecting by this an idea of the genus being a fundamental unit of the System of Nature. Unlike this, second, the one-word names of suprageneric taxa were usually plural nouns, which indicated that they were not so much real natural units of their own but rather aggregates of genera. This implied, third, that the generic part of the complete species name was very important in that it indicated the generic allocation of the respective species, whereas there was no particular need to indicate their belonging to either classes or orders by specific lexical means. However, this rule for naming species followed not so much an ontic consideration as the standard scholastic formula 'generic common and species particular'. It should be noted that such a standpoint of 'early' Linné concerning ontic status of taxa radically differed from his later idea of the divine creation of these categories (see below).

* * *

The very idea of the ranked higher-level inclusive hierarchy became very popular among those systematicians of the second half of the 18th century who were adherent to the fundamental concept of *Systema Naturae*, as opposed to that of rankless linear *Scala Naturae* (see Section 1.6 on the latter). With this, a new trend appeared in the development of the ranking system, which involved its significant fragmentation and usually another terminology as suggested by Tournefort and Linnaeus. Such a trend was most probably explained by the widening horizon of systematic diversity that made their four principal ranks deficient. This particular hierarchy problem was resolving in a rather pragmatic and half-measured manner: when a well established group became too large due to newly described genera, it was not split in several other groups of the same rank, but instead several subgroups were differentiated within it to make its diversity more manageable (Stevens 1994, 2002). The newly introduced hierarchical categories gained specific denotations borrowed from social, military, juridical, and other sources, which quickly made the diversified system of multi-ranked hierarchy in systematics of that time largely non-Linnaean (Needham 1911; de Queiroz 2005; Pavlinov 2015, 2018). However, it became a 'historical curiosity' that the essentially renovated taxonomic hierarchy, as it was shaped in the early 19th century, was called 'Linnaean' by A.P. de Candólle (de Candólle, 1813, 1818), and this designation would be adopted nearly unanimously by systematicians.

A significant multiplication of both ranks and their designations was evident from the comparison of different treatises of the 'Linnaeans', as Frans Stafleu (1971) called them. One of such variants of non-canonical ranked hierarchies could be found in *Introductio ad historiam naturalem systems* by the Italian natural philosopher Giovanni Scopoli (1723–1788), a student of Linnaeus, who applied the following higher-level taxonomic categories (ranks): *Tribus* ⊃ *Genus* ⊃ *Divisio* ⊃ *Order*, whereas the genera proper were listed without specifying their rank (Scopoli 1777). The German naturalist Gottlieb Storr (1749–1821) devised in his dissertation *Prodromus methodi mammalium* a quite different, much more fractional system of classificatory categories: *Phalanx* ⊃ *Cohort* ⊃ *Ordo* ⊃ *Missus* ⊃ *Sectio* ⊃ *Coetus* ⊃ *Genus* ⊃ *Family* (Storr, 1780). The books of the French zoologist Pierre Latreille (1762–1833) provided a special case of instability of early versions of higher-level ranked hierarchies in systematics: he applied different ranking systems in his different studies. These looked like that (in French): Classe ⊃ Sous-classe ⊃ Division ⊃ Sous-division ⊃ Ordre in *Histoire naturelle des Crustacés et des Insectes* (Latreille 1801); Classe ⊃ Legion ⊃ Centuria ⊃ Cohors ⊃ Ordre in *Genera Crustaceorum et Insectorum* (Latreille 1806); Classe ⊃ Ordre ⊃ Section ⊃ Tribus (not always) ⊃ Famille in *Considérations générales* (Latreille 1810). Besides, the same category (rank), denoted by the same name, could be placed in different positions by various authors: e.g., *Divisio* was placed either between *Order* and *Genus* (Scopoli 1777;

Lamarck 1809), who applied their opposite subordination, or between *Classis* and *Order* (Latreille 1801).

This terminological mess soon became more ordered in a specific manner: the fragmentation frequently entailed not completely new categories of principal kind, but rather as *auxiliary* that were linked terminologically to the *primary* ones. It was noteworthy that such 'auxiliaries' were introduced initially as the *subordinate* categories (suborder, subfamily, etc.), whereas the *superordinate* ones (superorder, superfamily, etc.) would appear much later, and this seemed to be a specific legacy of the old scholastic divisive manner of the hierarchization of the genus-species scheme (Pavlinov 2018). Besides the just-mentioned Latreille's *Histoire naturelle*, this emerging trend was clearly illustrated by a short treatise *Prodrome d'une nouvelle distribution* by the French zoologist Henri de Blainville (1777–1850), who applied the following ranks to hierarchize Règne Animaux: Sous-règne ⊃ Type ⊃ Sous-type ⊃ Classe ⊃ Sous-classe ⊃ Ordre (de Blainville 1816). Pragmatic and nominalistic character of these auxiliary ranks was evident from the fact that they were frequently designated not by specific terms but by figures or formal symbols (such as §). The old 'Linnaean' and some of new principal (primary) ranks were denoted as *mandatory*, and the 'auxiliaries' as *non-mandatory* (Kwok 2011). All these suggestions resulted in new terminology, and this seemingly unstoppable growth of rank fragmentation would be denoted as *rank inflation* (Stuessy 2009).

* * *

Those naturalists and systematicians of the late 17th and entire 18th centuries, who have been inclined to a natural philosophical thinking, when discussing taxonomic ranks, usually intended to recognize natural groups in classifications. The consideration of their ontic status was based on the scholastic (ascending to neo-Platonics) opposition of realistic *vs.* nominalistic treatments: the basic criterion was, whether the groups in question could be attributed to Nature or were they the products of human mind. Accordingly, the main issue regarding the reality (naturalness) of ranks rested ultimately on the appeal to the natural causes of the groups allocated to them. The respective questions were initially answered based on the biblical natural philosophy with the reference to divine creation; later, such causes were sought based on other natural philosophical principles of Nature that structure and order it.

An appeal to creation was characteristic, e.g., of J. Ray, who in his *Historia plantarum* asserted that 'the number of species in nature is strictly determined: God interrupted on the sixth day his great work, the creation of new species' (Ray 1686, p. 40). This evidently indicated that for Ray, species was a natural category, whereas genus was a result of the application of the *Methodus generalis* as a 'convenient means of brief presenting botany' (*ibidem*, Prefatio, p. 2). A similar substantiation of the naturalness of lower categories was provided by 'early' Linné in his *Genera plantarum* and *Philosophia botanica*, who followed Ray in believing that 'there are as many *species* as there are different *forms* created by an initially Infinite Being' (Linnaeus 1737, *Ratio operis*, § 5; Linnaeus 1751, § 157; italics added). This quotation allows for supposing that Linnaeus seemed to believe that created have not been the 'species' as particular groups of organisms but rather the 'forms' (most probably in their

Aristotelian sense) as the above-mentioned incarnations of divine archetypes (Pavlinov 2023). With regard to genus, Linnaean position was not so definite: in § 159 of his *Philosophia botanica*, he considered it either secondary relative to species or of the same ontic status, being 'made in the first place such as it is'. One may consider the following passages in his *Philosophia* as summarizing his earlier ideas on this subject as follows: 'Species and Genera are the work of Nature', Classes and Orders are 'the work of Nature and Art', whereas 'Variety is quite often the work of cultivation' (Linnaeus 1751, § 162).

The 'founding fathers' of early post-scholastic systematics concurred with this view of the naturalness of species but accentuated other natural causes. For them, the species category was natural because all the real species of plants and animals were represented actually by the chains of generations produced by the natural process of breading.[5] In accordance with this, the species category became regarded as the basis for natural classifications they elaborated, whereas the other categories were recognized as species groupings. This idea was more than clearly expressed by the English botanist John Lindley (1799–1865) by saying that 'the Natural System *consists of species* disposed in genera, genera in orders' (Lindley 1835, p. 41; italics added). An opinion of unnaturalness was especially true for the auxiliary (secondary) categories, which were introduced for a convenience and, thus, certainly were (in the Linnaean words) 'the work of <...> Art'.

However, quite a distinct idea of the natural groups of living beings was developed by Linné in his older age, based on a quite different conception of divine creation that was more consistent with Platonic eidology. His novel perspective was at first stated in the dissertation *Fundamentum fructificationis*, which was formally authorized by his student Johannes Gråberg, though C. Linné was known to prepare his students' dissertations mostly by himself (Pulteney 1805; von Hofsten, 1958; Larson 1968; Bobrov 1970). Anyway, that dissertation read that 'in the beginning the Creator created only one plant of each *natural Order*, differing from others by habitus and fructification'; then 'He inseminated them with each other, so that from their offspring <...> there turned out as many natural *Genera* as there were different variants of parents'; and lastly, 'some genera produced many species, while others did not' (Linnaeus 1762, pp. 20–21; italics in the original). As is evident, this version of divine creation implied that the higher categories were primary with respect to the lower ones (including species) by having been initially created by God as particular *forms* (again, most probably in their Aristotelian meaning) (Pavlinov 2023). This viewpoint was subsequently reproduced and, hence, approved by the Swedish Linné's disciple Johan Murray (1740–1791) in *Caroli a Linné equitis Systema Vegitabilium*, where he wrote that in the beginning God created 'as many different plants as there are *Natural orders*' (Murray 1784, p. 8; italics in original).

Fundamentally different interpretations of the general idea of naturalness of the higher-level groups and their ranks to be uncovered by naturalists were advanced by several natural philosophical ideas of non-biblical kind in the first half of the 19th century. A very popular doctrine was developed by the members of the Schellingian school of organicism natural philosophy,[6] most active of which was the German naturalist Lorenz Oken (Okenfuss; 1779–1851). As is seen from the name of this concept, it likens the entire Nature to a biological organism, and this implies that Nature is partonomically partitioned into compartments of various levels of generality, just as an organism is partitioned into organs, tissues, and cells. Accordingly, if naturalists intended to identify natural groups of living creatures, they had to discover a certain fundamental *principle of hierarchization* of wildlife, with the naturalness of the groups implying also the naturalness of their ranks. Oken attempted to implement this idea in his *Grundzeichnung des natürlichen Systems* (Oken 1809 and other editions; also Oken 1847); however, multiplicity of his classifications in its different editions indicated that he seemed to appear unlucky in discovering the sought natural arrangement of wildlife.

An indirect impact of the organicism natural philosophy on the ideas about the causes of the natural ranked hierarchy of the living could be found in the phylogenetic concept by E. Haeckel, who equated the history of life with the development of a 'genealogical individuum' (Panchen 1992; Pavlinov 2005). The evolutionary substantiation of both natural hierarchical categories (ranks) and groups (taxa) allocated to them will be in more detail considered below.

Another natural philosophy was implemented by the so-called *quinarian* system of the English-Australian zoologist William MacLeay (1792–1865), who exposed its principles in the first volume of his *Horae Entomologicae* (MacLeay 1819). His neo-Pyphagorean numerical idea implied that at each level of taxonomic hierarchy, five subgroups should be recognized within a given group to reflect the numerical quinarian harmony of Nature. Accordingly, he paid much attention to finding the properly subordinated true 'fives' of some invertebrate animals but did not discuss substantiation of their ranking. The English naturalist William Swainson (1789–1855) adds trinarity to quinarity by dividing five groups of the same series based on their closest and distant affinities (Swainson 1837).

* * *

In the first half and middle of the 19th century, a certain integrative trend in the development of the ranked taxonomic hierarchy became evident, which combined fragmentation of ranks with their standardization and codification. The Swiss botanist Augustín Pyrámus de Candólle (1778–1841) became the first to initiate the movement in this direction by his fundamental treatises *Théorie élémentaire de la botanique*, *Regni vegetabilis*, and *Prodromus systematis naturalis*. He proposed in them the most seriously substantiated, for his time, account of post-Linnaean higher-level taxonomic nomenclature, which contained also a codified ranked hierarchy for botanical classifications. This hierarchy, in its somewhat simplified version, included the following ranks: *Class* ⊃ *Subclass* ⊃ *Cohors* ⊃ *Ordo* ⊃ *Tribus* ⊃ *Genus* (de Candólle 1818), whereas its more detailed version implied also fragmentation of genera into sections and species into varieties (de Candólle 1813, 1818, 1824). This nomenclature system of A.P. de Candólle perpetuated an important innovation, which has been offered by M. Adanson several decades before in his *Familles des plantes*. He suggested that the names of plant families (orders of de Candólle) should be based on the names of their most typical genera (Adanson 1763), and this became the first idea regarding the future *principle of typification* in systematics (Pavlinov 2015, 2022a; see Section 6.2).

The development of ranking system of inclusive hierarchy in such a manner led ultimately to its 'official' fixation by the Codes of Nomenclature beginning from their earliest editions. Botanists appeared to be especially successful in this regard: the *Lois de la nomenclature botanique* (de Candólle 1867; also Dall 1877) announced the system of 15 ranks above the species level, starting from the *Regnum* (Figure 1.4). Zoologists were noticeably lagging behind botanists in the codifying taxonomic hierarchy: the *Code of the British Association* (also known as the *Strickland Rules*) did not present any kind of comprehensive taxonomic hierarchy for animals, though reviewed the rules for naming families, subfamilies, genera, and species (Jardine 1866).

* * *

Biology of the 19th century experienced a very significant shift in the worldview initiated by a global evolutionary idea about the causes of the diversity and structure of wildlife. This idea was enthusiastically adopted by systematics as a more realistic (materialistic) metaphysical foundation of the original idea of natural system. It was initially mastering by biological systematics in several versions, which could be

Regnum vegetabile.
Divisio.
Subdivisio.
Classis.
Subclassis.
Cohors.
Subcohors.
Ordo (gallice : *Famille*).
Subordo (gall. *Sous-famille*).
Tribus.
Subtribus.
Genus.
Subgenus.
Sectio.
Subsectio.

FIGURE 1.4 The lower-level ranked hierarchy developed in systematic botany in the mid-19th century and published in A. de Candólle's *Lois de la nomenclature botanique* (de Candólle 1867, p. 15).

grouped in two major conceptions, *macro*- and *micro*evolutionary. They suggested different explanations for the origin and nature of the diversity of organisms at various levels of generality and therefore became crucial to the issue of taxonomic hierarchy.

The first of them was the macroevolutionary conception developed by the French naturalist Jean-Baptiste de Lamarck (1744–1829) in his *Philosophie zoologique* (1809; see Lamarck 1963), who based it on the idea of Scale of Being and therefore rejected the naturalness of any inclusive taxonomic hierarchy (see below). His evolutionary idea has appeared to be ahead of that time and therefore did not enjoy much popularity. The next became two major macroevolutionary conceptions of the second half of the 19th century, authorized by the German morphologist Ernst Haeckel (1834–1919) and American paleontologist Edward Cope (1840–1897). They both considered anagenetic macroevolutionary trends, which were coherently defined by a presumed 'Scale of perfection', now interpreted as an evolutionary progression. With this, their evolutionary models differed markedly in paying much attention to either divergent (Haeckel 1866) or parallel (Cope 1868, 1887) aspect of evolution. This general approach was united under the Haeckelian term *phylogenetics*, and it became concentrated on the revealing and taxonomic implication of the genealogical (kinship) relations referred to as genealogical or phylogenetic. Their graphical representations were denoted as *genealogical* (*phylogenetic*) *trees*, which overall appearance (topology) varied according to the particular phylogenetic versions. The Haeckelian divergent scheme was represented by a typical tree-like branching pattern (Figure 1.5; see also Section 4.2), whereas the Copean one depicted parallel lineages and more resembled a 'bush' or a 'lawn', very characteristically illustrated by Cope's disciple Henry Osborn (1857–1935) in his monograph of Proboscidea (Osborn 1936; Figure 1.6).

Taxonomic implication of phylogenetic conceptions gave rise to *phylogenetic systematics* in the same two major versions. Its basic tenet implied such a transformation of phylogenetic trees into phylogenetic classifications, in which the tree branches (lineages) of various orders were represented by taxa of various ranks. The 'vertical' orderliness of phylogenetic tree or bush defined a similar orderliness of phylogenetic classification in its standard text-style representation: this was formalized by the *principle of taxon sequencing*, which implied that within a particular taxon, the sequence of its subtaxa began with the least and ended with the most advanced ones. Within this general orderliness, the Haeckelian version was guided by the *principle of monophyly* emphasizing the unity of origin, whereas the Copean one was based on the *principle of polyphyly* accentuating the unity of anagenetic steps in parallel lineages. As far as both phylogenetic process and resulting phylogenetic relations were deemed objective, the same was assumed both for the phylogenetically interpreted taxa and their ranked hierarchy.

Quite different microevolutionary conceptions of Darwinism and neo-Lamarckism focused on the evolutionary mechanisms that acted at the infraspecies levels of the systematic diversity (Darwin 1859; Bailey 1894). These evolutionary explanations, indirectly supported by the then becoming popular positivist philosophy of science, offered rather radical 'non-Linnaean' resolution to the principal issue considered here: only the lower units of diversity and their ranks were real (natural), while the higher ones were largely nominal (artificial). Such a standpoint equated

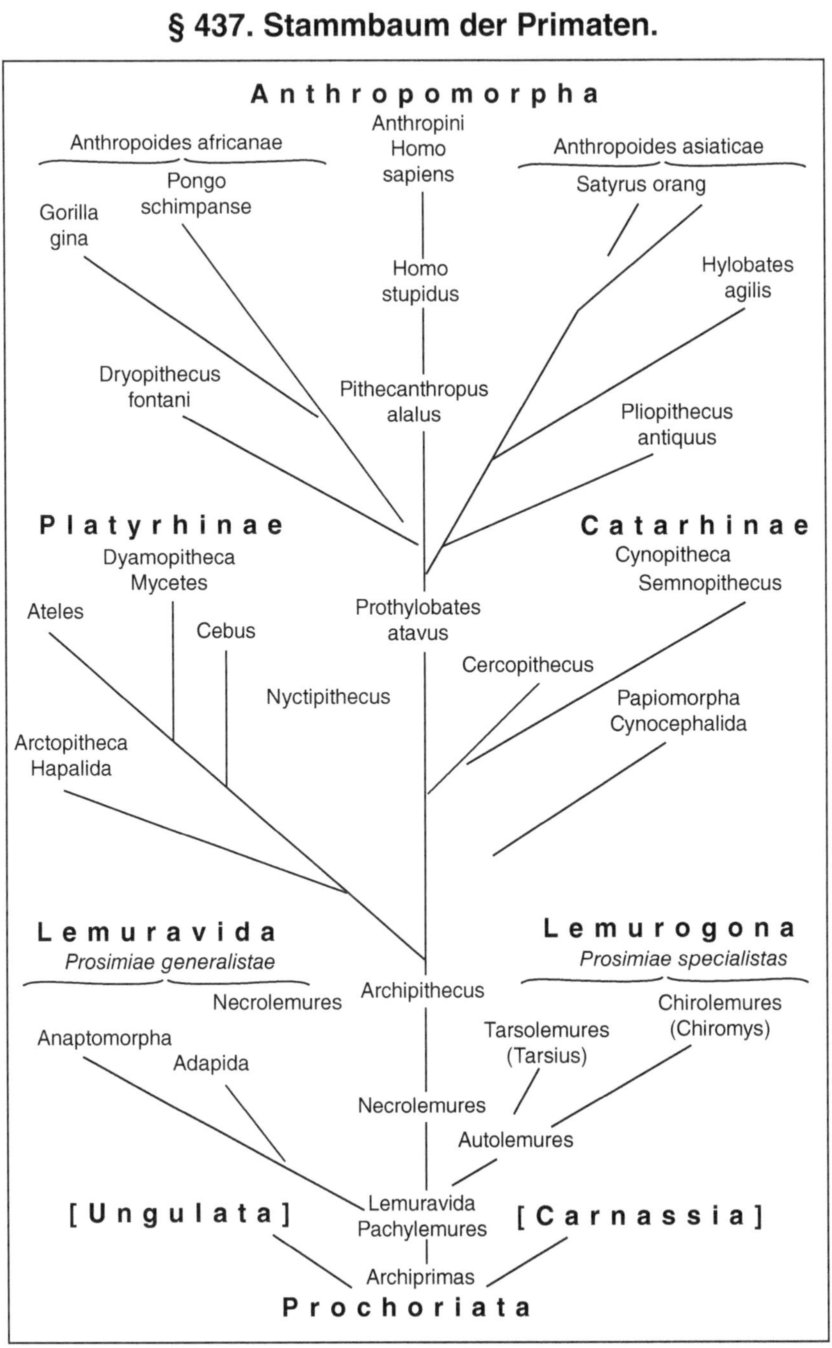

FIGURE 1.5 A typical simplified branching tree illustrating divergent primate phylogeny in E. Haeckel's *Systematische Phylogenie* (Haeckel 1895, p. 601).

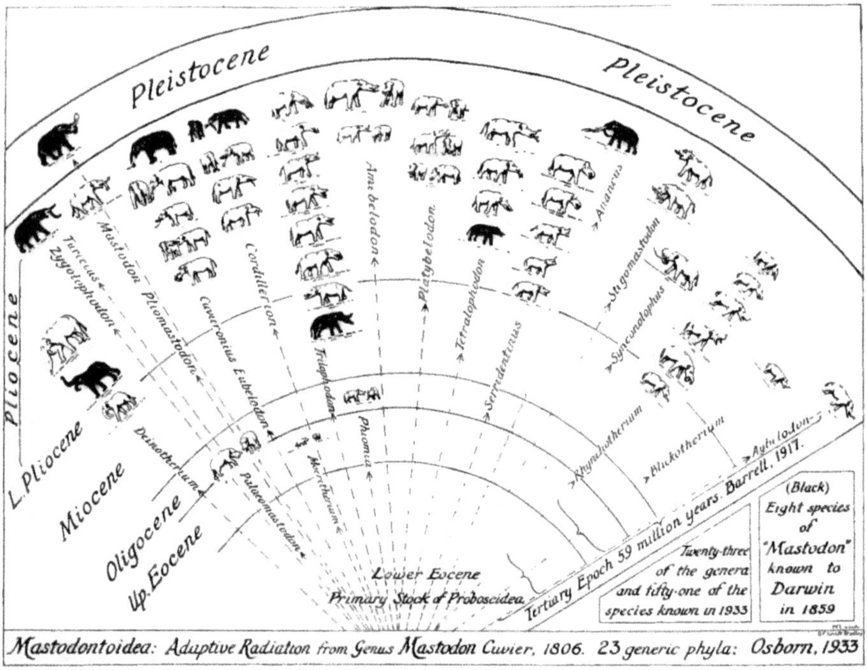

FIGURE 1.6 A graphical 'lawn-like' scheme illustrating parallel phylogeny in H. Osborn's *Proboscidea* (Osborn 1936, p. xiv).

Species.
Subspecies (vel Proles, gall. *Race*).
Varietas.
Subvarietas.
Variatio.
Subvariatio.

FIGURE 1.7 The lower-level ranked hierarchy developed in systematic botany in the mid-19th century and published in A. de Candólle's *Lois de la nomenclature botanique* (de Candólle 1867, p. 15).

species and infraspecies taxonomic categories in their status and by this annihilated the salient status of species in its classical account (see Section 5.1). All this inevitably attracted the primary attention of many systematicians of the late 19th and early 20th centuries to the infraspecies diversity and strongly motivated them to elaborate the latter's fairly fragmented hierarchical systems (Figure 1.7).

As this trend was progressing, it entailed recognition of various systems of within-species classification units, recognized by specific criteria and arranged in various ranked hierarchies. Its coarsened variant was expressed by the suggestion to recognize *microspecies* within the classical *macrospecies*: the former were denoted as *jordanons* (after the French botanist Claude Jordan; 1814–1897) and the latter were referred to as *linneons* (Lotsy 1916, 1931; Thorp 1940). A more fragmented variant implied a complicated system of 'through' lower-level hierarchy with two major levels: the higher one included fairly stable *subspecies*, *race*, and *natio*, and the lower one included *morph*, *aberration*, and *phase* (Semenov-Tyan-Shansky 1910; Bianchi 1916). With an emphasis on the ecological criterion, the basic infraspecific elements were defined as *ecotypes* (Turesson 1922; Valentine 1949); according to the genetic criterion, two within-species hierarchies were proposed, one composed of *komparium*, *kommiskuum*, and *konvivium* (Danser 1929) and another of *biotype*, *pure line*, and *clone* (Du Rietz 1930). Combining phenetic and genetic criteria resulted in the recognition of *homogenon*, *alloploidon*, *mycton*, *phenogen*, *phenon*, etc. (Camp and Gilly 1943). Such a chaotic situation was rightfully qualified as an 'experimentation' with infraspecies categories (Dean 1979), which ultimately resulted in several dozens thereof (Sylvester-Bradley 1952; Valentine and Löve 1958).

* * *

The subsequent conceptual development of phylogenetic systematics continued mostly its Haeckelian version, which resulted in two its recent conceptions emerged around the mid-20th century (Löther 1972; Mayr 1982; Rieppel 2016; Pavlinov 2018, 2021a). One of these was *evolutionary taxonomy* developed by the American zoologist George Simpson (1902–1984). It was enriched by some additional taxonomically significant assumptions about evolutionary process, including the suggestion to delineate and rank higher-level taxa according to the scales of *adaptive zones* occupied by the respective groups of organisms (Simpson 1961; Van Valen 1971; Bock 1977). Another version was suggested by two Germans biologists, first by the botanist Walter Zimmermann (1892–1980) and second by the zoologist Willie Hennig (1913–1976) (Zimmermann 1931, 1934; Hennig 1950, 1966). It was based on a very simplified evolutionary scenario, in which the whole evolutionary process appeared to be reduced to cladogenesis, therefore this approach was suggested to be called *cladistics* (Mayr 1965; Nelson and Platnick 1981; Pavlinov 1990a, 2018, 2021a; Williams and Ebach 2008; Rieppel 2016). The resulted phylogenetic tree was reduced to *cladogram*, which displayed only branching pattern of phylogeny. The ultimate objective of *cladistic systematics* was defined as the elaboration of *cladistic classification*, in which the hierarchy of taxa would reflect only that branching pattern. To reach this end, the broadly understood classical (Haeckelian) monophyly became decomposed into two components, *paraphyly* and *holophyly* (strict monophyly), and only holophyletic groups were acknowledged as meeting the condition presumed by the principle of monophyly (Nelson and Platnick 1981; Wiley 1981; Pavlinov 1990a, 2018, 2021a; Wägele 2005; Williams and Ebach 2008).[7]

A consistent implication of the latter guiding provisions, enthusiastically advocated by W. Hennig and his most devoted followers, resulted in a very fragmented higher-level *monotone* (i.e., without dividing into primary and secondary) ranked hierarchy. The basic principle of rank equality was specified to the one of *rank equality of sister groups*, which contributed to the ranked hierarchy over-fragmentation, especially due to its implying rank equality of ancestral species and their descendant groups of whatever ranks. This caused a specific term-creation, as each taxonomic rank should be designated by a distinct term (Hennig 1950, 1966; Wiley 1981).

For example, in a cladistic classification of mammals developed by the American paleontologists, in the rank group of order, along with the already established auxiliary ranks (superorder, suborder), there were also infraorder, parvorder, mirorder, etc. (McKenna and Bell 1997). As a consequence, cladistic systematics, while winning in the rigor of its initial presumptions and classificatory procedures, lost in the certainty of hierarchy of its classifications. In order to minimize this specific uncertainty, several versions of non-strict (degenerate) hierarchy was elaborated (Wiley 1979, 1981, 2009; Schoch 1986; Gauthier et al., 1988; Pavlinov 1990a, 2005; Stuessy and König 2008).

An evident natural reaction to such a rigor oversimplified taxonomic theory becomes an idea of '*post-phylogenetic*' systematics, which in fact is *post-cladstic* one (Wheeler 2008; Hörandl 2010; Zander 2010a, 2013; Wheeler and Hamilton 2014; Minelli 2015; Pavlinov 2018, 2021a). It presumes the revival of a more substantively rich version of evolutionarily interpreted taxonomic theory with a noticeable addition of the ideas of evo–devo and with substantiating the general ranking approach in a specific manner.

* * *

The development of the systems of inclusive ranking of taxonomic classifications during the 20th century was marked by several ideas, some of which were fairly conceptual and others more technical.

The first to be highlighted here was an explicit distinction between the notions of *taxon*, *rank*, and *category* (Woodger 1937). It has been stimulated by the introduction of the notion of taxon for the designation of classificatory units in the traditional (basically typological) taxonomic classifications (Meyer 1926).[8] Prior to this important separation, the notion of category was of very wide meaning by referring either to the groups of organisms or their levels of generality or occasionally to both (not to mention its philosophical connotation going back to Aristotle). The distinction in question, now widely accepted, implies the following interpretation of the respective notions: *taxon* is a particular group of organisms recognized in a taxonomic classification, *rank* is a particular level of generality in that classification, and *category* is a set of taxa allocated to a particular taxonomic rank (Beckner 1959; Simpson 1961; Van Valen 1964; Blackwelder 1967; Mayr 1969, 1978; Pavlinov 2018, 2021a). In accordance to this specific terminology, taxa are *allocated* to ranks and the latter are *assigned* to them, whereas taxa are *members* of categories which *encompass* them.

Of importance became an explicit introduction of the specific *ranking criteria*, which made the taxa, hierarchized based on their provisions, more or less comparable, at least in principle, by some of their inherent properties. Typologists suggested inferring the ranked hierarchy of taxonomic classifications from that of the hierarchical arrangement of archetypes (bauplans), provided that the latter's elements (merons) of various levels of generality define the respective ranked taxa (Naef 1919; Troll 1951; Skarlato and Starobogatov 1974; Starobogatov 1989; Vasilieva 1989, 1992, 1999; Lyubarsky 1996, 2018); this idea was subsequently formalized as *taxonomic-meronomic correspondence* (see Section 2.2). Phylogenetic systematics introduced a chronological ranking scale, which was based on the time of origin (evolutionary age) of the groups and implied that the older the groups, the higher the rank ascribed to the respective taxa. The evolutionary age of a group was suggested to be determined with the reference either to geochronology of paleontological data or 'molecular watch' inferences (Hennig 1950, 1966; Simpson 1961; Avise and Johns 1999; Avise and Mitchell 2007; Avise and Liu 2011). Evolutionary systematics proposed to take into account also diversity of a taxon (the number of its subtaxa, etc.), its morpho-biological specificity (correlated with the specificity of the adaptive zone occupied by the respective group), etc. (Simpson 1961; Mayr 1969).

Some more technical subject-specific suggestions were made by systematic botany and zoology for the purposes of taxonomic nomenclature. The *Code of Botanical Nomenclature* preserved the entire highly fragmented ranked hierarchy elaborated since the mid-19th century, covering the levels from *Divisio* to *Subvario* without their grouping (Anonymous 1934). Unlike this, the *Code of Zoological Nomenclature* limited itself by a narrower range of ranks and structured their hierarchy by establishing several *rank-groups* (Anonymous 1964). Three such groups are now codified: species-group, genus-group, and family-group, each including the respective principal rank and all its secondary (also occasionally *intercolated*) sub- and super-ranks as its specific auxiliaries. Besides, infrasubspecies ranks (*natio*, etc.) were ruled out by Zoological Code, but this concerns only nomenclature and does not imply that zoologists have no right to recognize taxa of those lowest ranks in their research. The idea of rank-groups was also adopted by the draft *BioCode* (Hawksworth et al. 1994; Greuter et al. 1998) and the draft *Linz ZooCode* (Dubois et al., 2019, 2022): in both the ranked hierarchy is extended up to the highest levels, and the second draft suggested to denote rank-groups as *series*.

As far as nomenclature is concerned, it should be mentioned that the contemporary Codes specify several other hierarchies of non-taxonomic kind, which are relevant to the accomplishing certain classification tasks. These include, e.g., ranking of nomenclature regulators and nomenclatural types (Pavlinov 2015, 2022a; see Section 6.2).

1.6 AGAINST TAXONOMIC HIERARCHY AND/OR ITS RANKING

As shown above, systematics has commenced and was initially developing with the implementation of the inclusive hierarchy of a rankless (with certain reservations) kind due to having borrowed genus-species classificatory scheme from scholasticism. Its ranked versions began to be elaborated in the 18th century by the adherents of the natural philosophical conception of the hierarchically arranged *Systema Naturae*. However, both the very idea of inclusive hierarchy and especially its ranked arrangement encountered serious criticisms that came from various sources.

The first and most fundamental of the latter was the natural philosophical conception of the linear Ladder of Nature (*Scala Naturae*), or Chain of Being, according to which the entire universe was thought of as a continuous progression of forms from the lowest to the highest (Lovejoy 1936; Archibald 2014). This conception went back to Aristotle's ideas of the Ladder of Perfection and assumed the following fundamental properties of the orderliness of the universe: its linearity, polarity, plenitude, certainty, and continuity. It was postulated that such a ladder was a wholeness generated by a specific creative principle (prototype, etc.), so that all beings with their essences took certain unchangeable (fixed) positions in it, and all of them were united by an all-encompassing network of the affinities with no gaps in it. It should be clearly understood that this natural philosophy opposed not hierarchy at all but only an inclusive one, whereas the linear hierarchy (from the lowest to the highest) formed its ideological core. Such a linear hierarchy is sometimes called *scalar* (from Lat. *scalae*, steps) (Romme 2021).

This worldview became quite popular among European natural philosophers and naturalists in the 18th and early 19th centuries. Its proponents rejected the ranked inclusive hierarchy of *Systema naturalis* in favor of the rankless linear *Order naturalis*, which was supposed to reflect the Chain of Being most accurately. One of its most devoted adherents was the French naturalist Georges-Louis Leclerc de Buffon (1707–1788), who followed the Leibnitzian natural philosophical maxim '*Natura non facit saltus*' and argued in the introductory section of the first volume of his *Histoire naturelle* (1749) that 'only individuals really exist in nature, while genera, orders, classes exist only in our imagination' (cited after: Buffon 1835, p. 44). Buffon's compatriot and colleague J.-B. de Lamarck, who was most actively involved in the systematic issues, emphasized in his *Philosophie zoologique* (1809) that 'Nature has made nothing of this kind [meant inclusive ranked hierarchy]; and instead of deceiving ourselves into confusing our works with hers, we should recognize that classes, orders, families, genera and nomenclatures are weapons of our own invention' (cited after: Lamarck 1963, p. 20). However, all contemplations of such 'ladder' kind in fact implied not organismal but some macro-level consideration of Nature, which is evident from an analysis of the works of another devoted 'ladderist', the Swiss natural philosopher Charles Bonnet (1720–1793). Indeed, when considering an ordered structure of the diversity of animals in his *Traite d'Insectologie* (1745), he individuated rather large 'links' of that Chain, such as quadrupeds, bats, aquatic birds, and worms (Figure 1.8), which presumed certain elements of a primitive two-level hierarchy (Anderson 1976). Such a representation of the structure of diversity presupposed (a) the presence of certain elements of a two-level hierarchy and (b) certain discrete linearly arranged scales (steps) in it.

This ideal natural philosophical scheme did not fit well with the real diversity of organisms, so Lamarck, solving his practical hierarchy problems of classifying the Animal Kingdom, recognized about a dozen classes and many orders or sections in it (see Lamarck 1963).

In the early 20th century, an objection to taxonomic hierarchy stemmed from the natural philosophical idea of one of the versions of *onto-rational* systematics (on the latter, see Pavlinov 2011c, 2018). One of its early adherents, the German natural philosopher Hans Driesch (1867–1941), criticized in his *The Science and Philosophy of the Organism* inclusive hierarchical classifications (he called them 'Linnaean') for their inability to predict about properties of organisms, whereas periodic system in chemistry did this excellently about chemical elements. So he insisted upon the looking for a specific orderliness in the organismal diversity that would enable predictions about those properties, and he considered this really natural (Driesch 1908). Driesch's reference to the periodic system as a 'role model' found a number

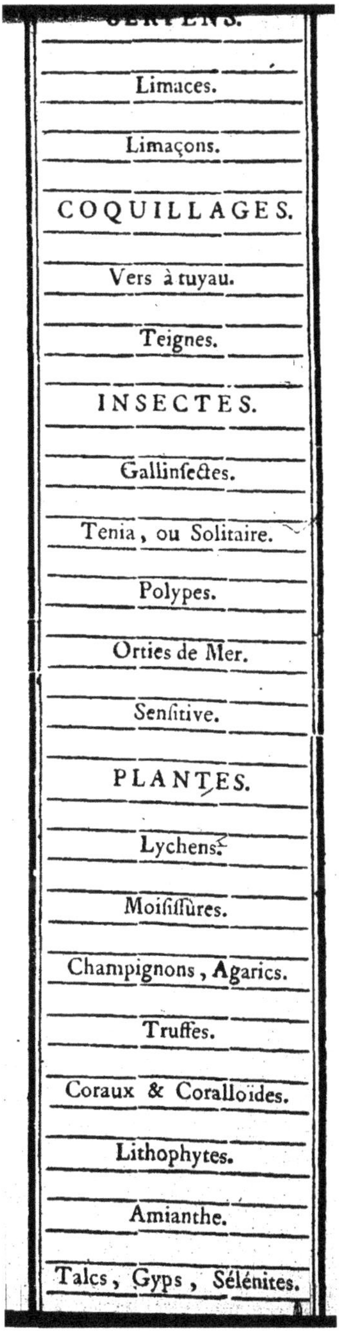

Limaces.

Limaçons.

COQUILLAGES.

Vers à tuyau.

Teignes.

INSECTES.

Gallinſectes.

Tenia, ou Solitaire.

Polypes.

Orties de Mer.

Senſitive.

PLANTES.

Lychens.

Moiſiſſures.

Champignons, Agarics.

Truffes.

Coraux & Coralloïdes.

Lithophytes.

Amianthe.

Talcs, Gyps, Sélénites.

FIGURE 1.8 A fragment of the Chain of Being in C. Bonnet's *Traite d'insectologie* (Bonnet 1745, unnumbered page).

of followers, who focused on the search for a rational system of periodical kind in the diversity of organisms (Lubischev 1923; Tchaikovsky 1990; Lyubarsky 1996; Popov 2008; etc.). There was at least one real attempt to implement this idea into a particular graphical form for a group of organisms (Arkhangelsky 1996). It is to be mentioned that Arkhangelsky's table was suggested not instead of, but as an addendum to the ranked phylogenetic classification. A taxonomic system in such a table representation was proposed to be denoted as *combinative-gradative* (*reticular*) (Preobrazhensky 2010).

The serious attack on the ranked higher-level inclusive hierarchy of taxonomic classifications came from the positivist-oriented schools of biological systematics in the early to mid-20th century. The most noticeable actors of this new movement were biosystematics and numerical phenetics (both mentioned above). They echoed the general arguments of Ockham and his earlier followers, as well as those forwarded by 'laddersits', and all of them concurred in that hierarchically arranged classifications were just specific devices for a compact and manageable representation of the immense diversity of organisms.

In the second half of the 20th century, cladists and their proponents picked up the baton of attacking taxonomic ranks, now suggesting that an assumingly monotonous character of cladogenesis made any discrete steps in the taxonomic hierarchy 'subjective' (e.g., Eldredge and Cracraft 1980; Eldredge 1985; de Queiroz and Gauthier 1992, 1994; Smith 1994; de Queiroz 1997, 2005; Ereshefsky 1997, 2002, 2004; Laurin 2010, 2024; Zachos 2011). Among other rationales, cladists substantiate their negative position regarding ranks by refusing the objective status of paraphyletic groups. It was their insistent promotion of these ideas culminated in the declaration of an 'anti-ranked' revolution in taxonomic nomenclature that sparked the current heated debates over ranked *vs.* rankless taxonomic hierarchy (see Section 3.7). Many of contemporary evolutionary systematics, while acknowledging higher taxonomic hierarchy as being ontically justified, criticizes its ranking as largely artificial and biased. One of objections points at the substantive incomparability of taxonomic categories designated by the same terms (Van Valen 1973; Rasnitsyn 1992, 1996; Minelli 1993).

In order to make hierarchical classifications more convenient pragmatically, it was proposed to substitute the traditional terminology of ranked hierarchy with a numerical codification of the levels in the rankless hierarchy, which resulted in the so-called *numericlature* (Jahn 1961; Michener 1963; Little 1964; Hull 1966, 1968; Yochelson 1966; Griffiths 1981). Cladistic systematics most recently developed its own version of rankless nomenclature, based on their interpretation of phylogenetic (actually, cladogenetic) pattern (de Queiroz and Gauthier 1992, 1994; de Queiroz 1997, 2005; Ereshefsky 1997, 2002, 2004; etc.). This campaign against ranked taxonomic hierarchy appeared successful, as its initiators and most active actors brought their idea to the practical realization in the form of *PhyloCode* (Cantino and de Queiroz 2010, 2020; Laurin 2024; see Section 6.4).

1.7 A BRIEF HISTORY OF MERONOMIC HIERARCHIES

All of the issues, discussed above in this chapter, concern mainly taxonomic hierarchies, which involve groups of organisms. With this, many of these issues seem to be equally relevant to both inclusive and non-inclusive hierarchies of the meronomic aspect of systematic diversity (see Section 2.2 on this and other of its aspects).

A trivial idea of the dismemberment of organisms into distinct parts was evidently born within the folk biology, which is evident from its rich vocabulary of the terms denoting those parts in animals and plants. The terminology developed by Indigenous people in this regard indicates that their meronomic views of the 'body plans' are implicitly hierarchical, which is clearly instantiated by recognition of the whole human body, limbs as the body's parts, elbow/knee and fingers as the limbs' parts, nails as the fingers' parts, etc. (McClure 1975; Brown 1976; Majid et al. 2006).

The Antique and Medieval natural philosophies did not explicitly distinguish between taxonomic and meronomic aspects of the diversity of things, including organisms, which were tightly connected by the taxonomic-meronomic dualism (see above). So it was not until the late scholastic systematics of the 17th to 18th centuries that these aspects were cognitively separated and their structures, including hierarchies, became more or less distinct research objects.

The prehistory and early history of biological systematics witnessed several attempts to develop more or less standardized meronomic terminology by some herbalists, systematicians, and anatomists (e.g., Bock 1546; Linnaeus 1751; Barclay 1803). Typologists of the 19th century actually developed an idea of meronomic 'near-ranked' hierarchy based on the body plans (bauplans) or archetypes surmised for major groups of animals and plants (Cuvier, Baer, Goethe, Owen). One of the most significant outcomes of their considerations became the foundation of the general concept of homology defined through the relations between parts within a whole (Owen 1848; Naef 1919; Kälin 1945; Woodger 1942; Remane 1956; etc.; see Section 5.2). In the 20th century, the so-called *new typology* elaborated an interesting idea of the usage of the meronomic 'near-ranked' inclusive hierarchies for inferring taxonomic ones of the same kind (Naef 1919; Troll 1951; Lyubarsky 1991, 1996; Beklemishev 1994; etc.).

The meronomic hierarchies seemed to have never been subjected to a standardized ranking, including terminological fixation of the unified levels of generality of merons. With regard to non-inclusive meronomic hierarchies and their ranking, this issue was being ceaselessly discussed since Aristotle, who has distinguished between *essential* and *accidental* properties (features) of organisms. In the subsequent development of essentialism by scholastics, this Aristotelian ranking of properties was refined by the introduction of the notions of *principal essence* by Anselm of Canterbury (Holopainen 1996), and it became one of the cornerstones in the fundament of scholastic systematics.

Different schools of systematics developed different criteria to estimate significance of characters with respect to their contributions to the resulting classifications. Those systematicians who followed (explicitly or implicitly) Aristotle in this regard, accentuated the functional importance for organisms of their anatomical (and eventually other) structures represented by the diagnostic traits (Cesalpino 1583; Jussieu 1789; Cuvier 1801; de Candólle 1818; Lindley 1835, 1836; Strickland 1841; etc.). Others paid special attention to the diagnostic value of characters: thus, C. Linnaeus (1751) suggested distinguishing between *natural*, *essential*, and *artificial* characters, of which the first defined genera (as *definitio*) and the second delineated them (as *differentia*). However, there was no agreement among systematicians as to which particular organs should be considered the 'most important' in the taxonomic respect: for instance, the earliest actors of systematic botany were either the so-called corollists or fructists. Typologists ascribed the greatest weight to the characters, which were most correlated with the archetypes (ground plans, etc.) (Cuvier 1801; Owen 1848; Naef 1919; Kälin 1945; Beklemishev 1994; Vasilieva 1989, 1999; Lyubarsky 1996; etc.). Evolutionarily interpreted systematics 'weighted' characters according to their contribution to the phylogenetic reconstructions: C. Darwin (1859) and E. Haeckel (1866) were the first to have suggested this criterion, and they were subsequently followed by next to all phylogeneticians.

Currently, an evaluation of the classificatory significance of characters is commonly referred in systematics to as their *differential weighting*, and the respective values ascribed to the characters are termed as their *weights*. This approach is based on specific non-inclusive characters ranking/scaling systems, and it is formalized by the *principle of differential weighting of characters* (Zarenkov 1988; Pavlinov 2018, 2021a), which constitutes a significant part of methodology of systematics.

* * *

Along with the denial of taxonomic hierarchy, the same trend involved characters ranking/scaling based on their taxonomic significance (weight). The just-mentioned principle of differential weighting is opposed by the alternative *principle of taxonomic equivalence of characters* that implies their *equivalent weighting*. Its prototype was developed by M. Adanson as a part of his original version of the natural method (Adanson 1763). However, it was not until the middle to late 20th century that this principle became actively mastered and implemented by systematicians, first by the adherents of phenetic idea (Gilmour 1940, 1961; Cain and Harrison 1958; Sokal and Sneath 1963; Colless 1967; Sneath and Sokal 1973; etc.) and later by some cladists (Eernisse and Kluge 1993; Kluge 1998a, 2004), who substantiated this approach with the reference to the epistemic *principle of total evidence*.

NOTES

1 Cited after: Wise (2016, p. 347).
2 However, this general rule was not always followed literally: for instance, *Patrologiae Cursus Completus* by the French priest Jacques-Paul Migne (1800–1875) in one of the tree-like graphical representations of the generic-species scheme applies *genera subalterna* and *species subalterna* as synonyms (Migne 1847, p. 41). Such terminology could be found also in the treaties of early systematicians.
3 It should be noted that these trees of Ramean kind are incorrectly denoted as Porphyrean in some recent literature (e.g., Maat 2004; Humphries, 2005; Pavlinov 2015, 2018). See Section 4.2 on their difference.
4 Therefore, an accepted view that J. Ray was a 'founding father' of the truly biological understanding of species (e.g., Zavadsky 1968; Löther 1972; Mayr 1969, 1982; Wilkins 2009a; Zachos 2016) seems to be incorrect (Pavlinov 2023).
5 It is to be reminded that this generative conception of species was rooted in the Aristotelian natural philosophy and confirmed by scholastics like Albertus Magnus (Wilkins 2009a; Pavlinov 2023).

6 Not in the sense of contemporary organicism theories (on them, see McGuire 2007; Wolfe 2024).

7 It should be noted that cladists prefer to apply traditional 'labels' (phylogenetic, monophyletic, etc.) to denote their approach and its results, which is by no means correct.

8 As said above, this Meyer's term was initially introduced to separate typologically (as taxa) and phylogenetically (as *phylons*, Haeckel's term) defined groups. Currently, 'taxon' is applied to designate any classificatory unit.

2 Some Philosophical Considerations

> Science is what you know. Philosophy is what you don't know.
>
> *Bertrand Russell*

The philosophy of science addresses the most fundamental problems that pertain to all issues regarding what and how science studies and what is scientific knowledge. All these problems are concentrated around certain ontic and epistemic ideas, about which no consensus is observed among philosophers and scientists. These challenges of philosophical kind are definitely relevant to biological systematics, which explores a specific aspect of biodiversity, referred to as *systematic diversity*, including its hierarchical pattern. The primary objective of such a philosophical drafting in systematics is to elaborate its specific *conceptual framework* (in the sense of Botha 1989; Gärdenfors 2000). The latter's basic function is to organize all cognitive activity within this discipline in a certain philosophically (and hence scientifically) meaningful manner, including the development of general foundations for its own *taxonomic theory*. Accordingly, any serious in-depth discussions of the theoretical issues concerning hierarchies explored and inferred by biological systematics should begin with the consideration of its onto-epistemic fundament.

The chapter provides a brief overview of some general ideas that have been developed by contemporary conceptualism in this area, to the extent that they concern biological systematics and specifically the hierarchical arrangement of classifications it elaborates.[1] The following issues will be addressed: what constitutes a cognitive situation in systematics and how it is structured; how theoretical knowledge can be developed in systematics; what is the systemic worldview and how it is relevant to the hierarchical vision of both biodiversity and the knowledge about it; and what are the principal outcomes of interpretation of classification as a specific cognitive system.

2.1 THE HIERARCHICAL STRUCTURE OF COGNITIVE SITUATION

One of the tenets of contemporary conceptualism coupled with the ideas of cognitive science asserts that any knowing activity is carried out in a certain *cognitive situation* (Stepin 2005; Swoyer 2006; Velichkovsky 2006; Thagard 2023). The latter refers to the very basic properties of the whole of cognitive process and the objects it addresses, so it has no formalized rigorous definition, as is typical for the most general ideas and notions in science. Rather, it exists as an informal and largely intuitive grasping of what are the general initial and boundary conditions of that activity.

DOI: 10.1201/9781003597094-2

A focal point of such a grasping is the comprehension of any cognitive situation as being comprised of three basic interrelated components. These are (a) the *ontic* component defining (delineating) particular objects to be cognized, (b) the *epistemic* component with specific cognitive means that are elaborated to achieve the desired end, and (c) the *subjective* component being identified with a cognizing subject operating with these means in order to cognize the objects in question. All three components interplay with each other in an intricate manner, thus making cognitive situation an integrated hierarchically arranged whole, which is convenient to be thought of metaphorically as a *cognitive triangle* (Pavlinov 2018, 2021a).

* * *

Regarding the ontic component from a conceptual perspective, it is crucial to emphasize first and foremost that in the case of directly unobservable natural entities (structures, processes, etc.), the latter are represented in cognitive situations not by themselves but by more or less formalized *cognitive* (*representation*) *models* (Bertalanffy 1969; Wartofsky 1979; Hacking 1983; Ahl and Allen 1996; Van Fraassen 2008; Frigg and Nguyen 2016). Each such model can be understood as a specific conceptual *knowledge unit* (Junkkari and Niinimäki 1999), which is designed to capture those properties of the studied entity that are deemed most essential for solving the specific cognitive task set by a knowing subject. Accordingly, a cognitive model is a kind of *ideal type* in the sense of Max Weber (1949). This general idea is formalized by the *principle of representation* (*modeling*), and these are the particular properly construed cognitive models that constitute the actually analyzed objects.[2] It is assumed that any directly unobservable natural entity can be comprehended to the extent that its cognitive model represents its natural prototype (not in the senses of Rosch 1983 nor of Rosen 1983) as adequately as possible, which may be referred to as their *isomorphism*. With this, however, it is acknowledged that the model's isomorphism cannot be 'absolute' even in theory, as there are many factors that make it incomplete and biased (see below).

In general, any representation of a research object, which meets certain criteria, is eligible to be treated as a specific kind of its cognitive model, which may be verbal, mathematical, graphical, physical (material), etc. In systematics, tree-like or plane-like representations are quite familiar graphical models of the hierarchical structure of both systematic diversity and classifications representing it (see Chapter 4). As far as elaboration of cognitive situation is meant, the model in question is largely a particular *conceptual* construction.[3] It is designed to outline the research object, which in our case is systematic diversity and any of its manifestations (aspects, fragment, etc.), in a specific manner presumed by the respective cognitive task (Wartofsky 1979; Langacker 1990, 2009; Górska 2001). A properly organized array of such interrelated cognitive models of various levels of generality, each referring to a particular manifestation of the studied object, constitutes a specific *conceptual reality* that represents its natural prototype in the given cognitive situation. In systematics, the lowest level of generality in this hierarchy corresponds to specific classifications as *empirical models* of the corresponding fragments of systematic diversity.

Any representation model is simpler than its prototype entity and therefore can only partly reflect it. This is because it results from a selective perceiving of the natural object: as said above, only those features are included in its model which are deemed essential for solving a particular cognitive task (Weber 1949; Bertalanffy 1969; Wartofsky 1979). For instance, when considering the diversity of organisms, they can be characterized from ecological, phylogenetic, genetic, ethological, physiological, and other perspectives. Respectively, any complexly structured natural entity may be represented by several models, each capturing its distinctive manifestation of interest. This cognitive *principle of model multiplicity* (Black et al. 2022; Heljakka et al. 2023; Jiang et al. 2024), especially when it is extended to that of *conceptualization multiplicity* (Massey 1999), is indispensable for understanding general approaches to construing cognitive situations in systematics, including definitions of its research objects and shaping its reduction cascades (see below on this).

An important part of the principle of representation is the development of specific *correspondence rules* that make different models representing the same natural prototype mutually interpretable (Rosen 1983). In systematics this provides a possibility for different classifications, which represent particular manifestations of systematic diversity (typological, phylogenetic, biomorphic, etc.) of the same group of organisms, to be integrated by a certain 'super-classification' of *faceted* kind (in the sense of Broughton 2006; Chan and Salaba 2016).

* * *

The natural entity that biology deals with is *biota*. It is a complexly structured entity with the organisms being its elementary units. They are diverse, and the basic research task of any biological discipline includes exploration and explanation of specific manifestations of their diversity. The latter is now commonly referred to as *biodiversity*, which is (incorrectly, see Pavlinov 2018, 2021a) usually regarded as a subject domain of biological systematics. Unfortunately, this buzzword notion, due to its frequent uncritical common-sense usage, especially in non-scientific contexts, lacks its conceptual clarity by referring to something obscure and not especially meaningful (Ghilarov 1996, 2001; Ivanter 2012). Therefore, it needs a proper conceptualization, and one of the latter's possible versions implies that it should be understood as a systemically arranged totality of organisms, i.e., not just a sum of their variability (as it is defined in: https://en.wikipedia.org/wiki/Biodiversity), but a specific manifestation of complexly structured biota (Pavlinov 2018, 2021a). Such systemic comprehension serves as a prerequisite for its conceptual individuation as a distinct natural phenomenon that can be subject to scientific explorations.

The next step of conceptualization, crucial from the standpoint of biological systematics, involves *near-decomposition* (the term of Simon 1962; see Section 3.1) of biodiversity into its lower-level particular manifestations, with two approaches for their identification being most significant: *object-based* and *aspect-based* (Pavlinov 2018, 2021a). In the first case, certain natural objects themselves are assumed to exist as inherent components of biota: an organism, an organismal constituent (organ, cell, etc.), a population, an ecosystem are proper examples. In the second case, the studied entities are delimitated by specific consideration aspects fixed by the cognizing subjects guided by specific cognitive (or other) tasks within the particular cognitive situations. With this, it is assumed that such aspect-based manifestations of biodiversity may be treated as fixed more or less naturally (non-arbitrarily), if certain natural causes underlying them can be supposed.

That latter provision implies that the *causation* seems to be the most significant condition of a correct ontic consideration of organismal diversity and its hierarchy. This major theme of ontology begins with two basic questions, (a) whether a studied aspect of the structure of this diversity is caused by any particular factors or not, and (b) if it is, what are the concrete causal mechanisms most responsible for it. With regard to the particular mechanisms of whatever kinds (emanation, evolution, epigenesis, systemic structuration, else), the fundamental question addresses their predominantly either *downward* or *upward* direction (Campbell 1974; Dessalles and Phan 2005; Santos 2015, 2024). The former is characteristic of a holistic worldview and implies that higher-level natural phenomena largely determine lower-level ones (the most prominent example is Platonic emanation); the latter implies an opposite direction of causation, which is characteristic of a reductionist worldview (such as Darwinian microevolution). Of course, these two basic *causation flows* co-occur and interact in nature; however, this dichotomy is of great relevance, as it may affect an understanding of the ontic status of some entities studied by systematics. For instance, species can be treated either as a phenomenon in the case of downward causation or as epiphenomenon in the case of upward causation (Cracraft 1989; see below).

In the evolutionary-oriented systematic schools, one of the basic constituents of the ontic component of their cognitive situations is referred to as *evolutionary scenario* or *evolutionary model* (Eldredge and Cracraft 1980; Felsenstein 2004; Pavlinov 2005; Tëmkin and Eldredge 2015). It defines the key parameters of the evolutionary process deemed most significant from a certain cognitive perspective. The parameters in question are *adaptatiogenesis* (adaptations of organisms to certain environments), *cladogenesis* (divergent evolution), *anagenesis* and *parallelisms* (directed evolution following certain unified trends), *termination* (extinction), *timing*, *dispersion*, etc. (Huxley 1942; Simpson 1953; Rensch 1960; Gould 2002). Particular scenarios/models may vary significantly depending on the inclusion or exclusion of certain parameters, which make them richer or poorer with respect to their contents. For instance, the classical model of evolution as an adaptatiogenesis involves all of the parameters just listed and assumes its largely ordered character; to the contrary, the currently popular parsimonious model of evolution discards most of these parameters and implies its largely stochastic character. Keeping in mind the problem of ontic reduction (see below), one can argue that the more complex scenarios are more realistic, i.e., the corresponding evolutionary models are more adequate to the real evolutionary processes.

However, the content-rich evolutionary conceptions face a principal impossibility of a strong isomorphism between evolutionary models and classifications, because they relate to two fundamentally different consideration aspects of biota, *dynamic* and *stationary*, respectively (Skarlato and Starobogatov 1974). Consequently, each particular complex scenario can conceptually substantiate an array of near-equally plausible hypotheses of phylogenetic relations, and each of the latter can be transformed into a number of as near-equally plausible phylogenetically consistent classifications (Simpson 1961; Felsenstein 2004; Pavlinov 2018). This controversy is formalized by the methodological *principle of NP-completeness* (Garey and Johnson 1979; Semple and Steel 2003; Felsenstein 2004), according to which, the more complex an ontic model of systematic diversity, the less attainable may appear a classification that adequately represents it. This problem is evidently most relevant to those research programs of systematics, whose ontic models are very rich in content, such as classical evolutionary or typological approaches.

* * *

The epistemic component includes an array of normative prescriptions that regulate all cognitive operations on the ontic component. Before all, it provides criteria, according to which particular natural entities are assigned a certain onto-epistemic status by acknowledging their (a) actual (objective) existence and (b) accessibility for exploration by the available scientific means. Three well-known classical worldviews—*realism*, *nominalism*, and *conceptualism*—differ drastically in this regard. Realism considers all the feasible things and processes, addressed by natural sciences, to be unequivocally existing objectively and cognizable at least in principle. From a nominalist perspective, only physically perceived objects are eligible to be treated as actually existing and rationally cognizable. Conceptualism acknowledges objective existence of the complexly structured universe, which various manifestations are fixed as the research objects by the respective conceptions and notions elaborated by knowing subjects. The hierarchical structure of biodiversity provides an evident instance of this fundamental trichotomy: it is considered either indisputably existing (realism) or lacking (nominalism) in nature or outlined properly by the respective conceptions (conceptualism).

Another very significant function of epistemology, connecting it with ontology, includes formulation of general principles of construing cognitive models, conceptions, respective notions, etc. Considered epistemically, hierarchization of such models can be treated as a specific, universally applicable cognitive device designed especially for the information compression ('convolution') (Solomatin and Menyaev 1982; Wolff 2019; Chen et al. 2014). For instance, specific *programming* models in computer science can be considered in this capacity (Labrosse et al. 2007); in systematics, these are models of biodiversity and its distinct manifestations (fragments, levels, etc.), down to taxonomic and meronomic classifications. Of paramount importance is the elaboration of specific descriptive (analytical) languages as cognitive tools most adequate to the structure of specific manifestations of the universe studied by particular natural science branches (Pattee 1970; Langacker 1990; Croft and Cruse 2004; Geeraerts 2006; Barrett and LaCroix 2020). The resulting linguistic models are worthy mentioning in this regard for their underlying the so-called *linguistic world pictures* (Gumperz and Levinson 1996; Talmy 2000). At last, epistemic component is responsible for the elaboration of exploratory methodologies and criteria for the evaluation of their scientific status. Its important part includes substantiation of research sample as a kind of *operational model* of the studied natural phenomenon, including biodiversity; in particular, classificatory characters are cognitive 'mini-models' of distinct organismal features. Of special significance is the correct formulation of sampling protocols corresponding to the particular exploratory tasks (Ahl and Allen 1996; Pavlinov 2018).

* * *

The subjective component of cognitive situation, just as the two others, is organized hierarchically, encompassing individual researchers, research groups (schools, labs, etc.), and eventually certain higher-level human communities. At a personal level, it manifests itself in the so-called *personal knowledge* (in the sense of Quine 1969; Polanyi 1998), including various *thinking styles* (holistic or elementaristic, intuitive or rational, etc.). For our consideration, systemic and hierarchy thinking styles (Ahl and Allen 1996; Mella 2009; Hester and Adams 2014) and phylohenetic (tree, lineage) thinking (O'Hara 1988; Pavlinov 2005; Jenner 2024) are of special significance; an opposition of extensional and intensional thinking styles with regard to the hierarchies (Vasilyeva and Stephenson 2008) should also be mentioned. At a higher level, this component involves particular research groups (schools), whose members adhere specific research programs (in systematics, say, phylogenetic or typological); the members of each such group share a certain thinking style that 'dictates' them how to mentally look at a research object and manipulate with its cognitive models (Hester and Adams 2014). At last, it seems reasonable to include here, at a far higher level, some decision-making bodies (such as grant allocators), as far as their resolutions about supporting particular directions of the exploratory activity in systematics (such as molecular phylogenetics) affect to a degree the overall structure of the latter's cognitive situation.

The role of subject component, understood in the broadest sense, in cognitive situation is deliberately downplayed by both classical and positivist philosophies of science, whereas it is emphasized by conceptualism (Quine 1969; Toulmin 1972; Maturana and Varela 1980). Its most significant role is determined by the fact that cognition is not a passive 'reflection' of nature by human mind but rather an active interaction of both object and subject, which shapes the entire cognitive activity. Accordingly, the Swiss psychologist Jean Piaget (1896–1980) argued that the knowledge about external world came not from it as such but rather from a subject's (observer's) interaction with it (Piaget 1971).

An indispensable presence of a cognizing mind in a cognitive situation is evident before all in that it is this active mind that frames the entire contents of the respective situation. Its decisive influence begins with the setting of specific cognitive objectives to be achieved within that situation and continues with outlining both ontic and epistemic components, together with their interrelations. After all, an answer to a question depends on the latter's formulation: a proper asking implies a possibility of a proper answer, whereas a wrong asking would hardly promise a correct answer—but it is no other than the subject who both asks and searches for answers. Indeed, regarding the main theme of this book, a hierarchical cognitive model of a complexly structured entity may be treated as a result of the latter's specific 'hierarchical' vision by the so-called hard-line hierarchists (Mesarovic and Macko 1969; Pattee 1973a; Ahl and Allen 1996). No less significant impact of subjective component comes from the *pragmatics* of cognitive activity; it makes exploratory and other tasks dependent on the particular subjective motivations and intentions. This is quite evident in the case of classification systems, which can be elaborated as either general purpose or special purpose ones (Gilmour 1940; Sokal and Sneath 1963; Meyen and Shreyder 1976; McKelvey 1982; Mai 2004; Shatalkin 2012; Pavlinov 2018). In particular, pragmatic accentuation is obviously present in the contemporary discussion of the problematic hierarchical issues in systematics from a nomenclature perspective (e.g., de Queiroz 1992, 1997; de Queiroz and Gauthier 1992; Lee and Skinner 2008; Laurin 2024).

* * *

The ontic and epistemic components are related to each other in a complex manner. The Argentine philosopher of science Mario Bunge (1919–2020) confines their relation by the dependence of epistemic premises on ontic ones (Bunge 1969). However, they are in fact mutually interrelated, so that not only ontic premises affect to a degree epistemic ones but the reciprocal affect also occurs, which entails specific *onto-epistemic correspondence* (Pavlinov 2011a, 2018, 2021a). The latter implies that (a) both ontic and epistemic parameters of a particular cognitive situation are shaped interdependently and (b) they configure coordinately a *conceptual framework* or *conceptual space* of the entire cognitive situation (Botha 1989; Gärdenfors 2000; Efremov 2009).

The onto-epistemic correspondence is evidently *incomplete*, which poses the most serious challenge to the cognitive activity that addresses natural entities. Its most fundamental and irremovable cause is the significantly simpler structure of the descriptive language of natural science compared to the described natural entities (Nalimov 1979; Chomsky 1987; Taylor and MacLaury 1995; Devitt and Sterelny 1999). With regard to biological systematics, this implies a principal impossibility for a classification system to be fully adequate to any given fragment of systematic diversity. This entails specific *taxonomic uncertainty* and necessitates acknowledging the fuzzy character of classifications representing systematic diversity (see Section 3.5).

The correspondence in question makes its two components interconnected by a specific *complementarity principle*, which presumes the acknowledgment of a kind of *onto-epistemic dualism* of the cognitive models representing the research objects in cognitive situations. Such a dualism means that a prototype object in question can be comprehended only through its conjoint ontic and epistemic consideration (Pattee 1973a, 1978); in particular, regarding classifications as specific cognitive systems (see Section 2.7), they 'comprise a *synthesis of metaphysics and epistemology*' (Bryant 2000, p. 17; italics in the original). If a subject is included in this consideration, the onto-epistemic dualism may be treated as an *object-subject* one (Pattee 1978), which is sometimes called an *observed-observer dualism* (Allen and Starr 1982; Wood 1994). This complex dualism (or trinity) entails an *onto-epistemic relativism*, which implies a kind of *propositional relativism*: the propositions about research objects are context-dependent, so they may have different truth-values presumed by the particular onto-epistemic premises (Feyerabend 1999; García-Carpintero and Kölbel 2008; Baghramian and Carter 2025). This important point shapes in a specific manner the dilemma of objective *vs.* subjective status of the judgments about hierarchies studied by systematics (see Section 2.6).

The just-mentioned onto-epistemic relativism has two evident hypostases, ontic and epistemic. The *principle of ontic (ontological) relativism* (Quine 1969; Dupré, 1993; Feyerabend 1999) implies, in our case, that the recognition of complexly structured biodiversity as a scientifically meaningful research object depends on certain epistemic premises. For instance, according to the physicalist *principle of observability*, taken in its extreme form (e.g., Contessa 2006), only concrete observable organisms can be studied, while the so-called biodiversity cannot, as it is conjectured and not actually observable. The *principle of epistemic relativism* (Feyerabend 1999; Einheuser 2005; Carter 2025) implies that there are no 'absolutely' true and universally applied epistemic instruments, but instead the latter's eligibility depends largely on the implied ontic premises. The most evident example is provided by two basic argumentation schemes, deductive and inductive, which are elaborated based on certain ontic assumptions regarding downward or upward causalities that order the universe. In a much more particular case of our concern, the particular algorithms for inferring phylogenetic hypotheses depend on the assumed evolutionary models (Wiley 1981; Pavlinov 1990a, 2005; Wägele 2005).

The recognition of onto-epistemic dualism has a characteristic impact on the defining ontic status of biodiversity constituents studied by systematics, to the degree that they are not observed directly but rather conceptually conjectured. When considered traditionally, they are divided into two basic categories, either objective (real) or subjective (artificial) (see below). However, the onto-epistemic relativism makes this dichotomy too primitive and oversimplifying a possible spectrum of ontic treatments. An evident illustration of the latter's variety is provided by the contemporary discussion of the versions of ontic status of species (see Pavlinov 2023 for a review; also see Section 5.1).

Another intriguing consequence of the onto-epistemic relativism includes the multiplicity of treatments of the phenomenological nature of studied objects (Pavlinov 2007b, 2018, 2021a; Loreau 2010; not in the sense of Husserl). From a classical perspective, the latter, if they are recognized as 'objective', are typically treated as the *natural phenomena* supposed to be inherent elements of objective reality that actually exist independently of a cognizing subject. Unlike this, from a conceptualism perspective, at least two kinds of ontic status should be distinguished: the research objects can be treated as either phenomena or *epiphenomena* (Shapiro and Sober 2007; William 2019). Partly adapting the arguments of the just-cited authors to the central theme of this book, it seems justifiable to discriminate these two statuses as follows. In considering a complexly structured natural system, which is 'by default' a natural phenomenon, its inherent constituents can be regarded as specific system *microparameters* that can rightfully be regarded as natural phenomena. On the other hand, there are certain system properties and/or relations, which are assumed to have no distinct objective existence of their own but rather are surmised by the cognizing subjects based on the analyses of the system's constituents. They should be more correctly thought of as specific system *macroparameters* and treated ontically as epiphenomena. An impressive example of such an offbeat addressing of the issues directly related to biodiversity is provided by the above-mentioned interpretation of species as 'epiphenomena, developed or evolved <...> from lower-level processes' (Cracraft 1989, p. 48).

> The metaphor of 'colored' universe may help to understand, what this distinction is about (Jackson 1982). In brief, what does really (objectively) exist in the physical world are electromagnetic waves of different frequency, being universe microparameters, whereas all those 'reds', 'blues', 'greens', etc. appear as specific universe macroparameters resulting from a certain physiological and mental activity of an observer. This activity includes perceptions of the reflected waves of certain frequencies by special cells in the eye retina and the subsequent processing of the results of these perceptions by special neurons in the brain (Zeki 2022). So, the universe is really colorless, whereas the allegedly 'visually evident' colors are epiphenomenal *qualia* (in the sense of Jackson 1982).

Considered in such a vein, a seemingly trivial issue of the observability of *similarity* between organisms, upon which systematics bases all its inferences, appears in an unexpectedly intricate perspective. Contrary to the common belief in its 'obviousness' and therefore 'objectivity' (Gilmour 1940; Simpson 1961; Sokal and Sneath 1963; Colless 1967; Meyen 1977, 1978; Lyubarsky 1996), conceptualism implies that a cognizing subject actually observes not a similarity as such but rather particular organisms with their particular features. As a matter of fact, what is routinely denoted as an 'objective similarity' is an observer's personal judgment about the commonality of these organisms by certain selectively perceived features, which results from a fairly sophisticated cognitive act (Tversky 1977). Thus, according to the basics of cognitive psychology, 'similarity' seems to be not a real natural phenomenon but rather another epiphenomenon, so it can also be regarded as a special kind of qualia (Jackson 1982; Dessalles 2001) resulting from a selective cognitive activity of particular subjects addressing the particular objects (Hjørland 2008; McCabe 2021). Moreover, from the conceptual perspective, 'philosophers have long realized [that] similarity without theory is empty' (Sober 1984, p. 336), i.e., it is not a 'pure objectivity' but rather conceptually dependent and context-relative.

Of paramount importance is that this consideration is pertinent to the very notion of biodiversity, which is central to biological systematics. If biota is taken for a complexly structured natural phenomenon, then its constituents (organisms, local ecosystem, etc.) with their physical interrelations and inherent features can be justifiably treatable as its microparameters deserving status of natural phenomena. Unlike this, concerning what we denote and study as bio- (or systematic) diversity with its hierarchical structure, it seems to be more plausible to be regarded as a biota's macroparameter and ascribe to it an epiphenomenal status. In other words, the hierarchically structured systematic diversity should be thought of as a kind of 'intellectual qualia', being a specific conceptual image of a certain manifestation of the presumably real structure of biota (Hoffman 1979; Torday 2016).

* * *

In accordance to the onto-epistemic dualism, any research object can be approached predominantly either from an ontic or epistemic standpoint, depending on a subject's cognitive attitude. This 'either-or' may imply a rigid demarcation between ontically or epistemically defined hierarchies, which are regarded either as attributes of nature or cognition (e.g., Wilson 1969; Allen and Starr 1982; Ahl and Allen 1996; Wu 2013).

Considered from the perspective of organization of cognitive activity in systematics, such a subjectively motivated predominance entails a kind of *exploratory precedence*, so that these components turn out to be connected metaphorically by a 'horse and cart' link. For example, in 'pre-numerical' phylogenetics, epistemology follows ontology: specific methodologies of phylogenetic reconstructions are inferred from certain assumed regularities of phylogenesis (Hennig 1966; Wiley 1981; Pavlinov 1990a, 2005). In contrast, in recent numerical phyletics, ontology follows epistemology: phylogeny is regarded as a stochastic process, which is substantiated by the reference to certain statistical provisions of numerical methods for inferring phylogenetic trees (Felsenstein 2004).

In the case of epistemic precedence, a hidden 'epistemic pitfall' appears in a cognitive situation (Williamson 2000), if the onto-epistemic correspondence is arranged in such a manner that the epistemic (methodological) consideration not only precedes the ontic one but even dominates over it. This means that, accordingly to the famous Zen Buddhism aphorisms, 'a person is inclined to mistake his finger pointing at the moon for the moon itself'. In our case, when biologists analyze hierarchical pattern of systematic diversity, they often fail to realize that it is a particular concept of the latter (a pointing 'finger') that makes them 'seeing' a certain pattern in nature they call *the* hierarchy (the 'moon') (Bartlett 2015; Pavlinov 2023). V. Ahl and T. Allen (1996, p. 78) discuss in this regard a challenging relation between 'an ideal ordering according to definitional criteria [and] a description of a particular material system that can be empirically uncovered'.

> A similar specific 'pitfall' is also built into the so-called *mythological thinking*, to which many practical systematicians are inclined. They forget their conceptual or personal knowledge that underlies their empirical findings and therefore sincerely take the latter for what actually occurs in nature. This is even more true for those who rely heavily on modern technical means to solve specific research tasks and, in particular, 'fetishize' the hierarchically branching trees inferred using numerical molecular phylogenetic methods. This point can be illustrated by the following curious fact. During the theoretical seminar at the Moscow Zoological Museum, a prominent molecular phylogeneticist has been asked what, in his opinion, was phylogenesis. He at first did not seem to understand the proper meaning of the question, but then pointed with his finger to the tree graph on the slide and declared: 'This is *the* phylogenesis!'. So, to him, phylogenesis was not a natural multi-aspectual process of the historical development of a group of organisms but merely a graphical scheme produced by his computer. Just look at the captions to the drawings in the molecular phylogenetic articles to see that such a primitive 'technical' vision of phylogeny is a rule. Isn't it both funny and sorrowfully?

* * *

As seen from the above considerations, the cognitive situation in systematics represents a specific, hierarchically arranged, multifaceted conceptual construct designed to organize and carry out all cognitive activities that address biodiversity. This is apparently the first conceptually construed hierarchy, with whose specific formatting biological systematics began each of its conceptual developments.

At the highest level of conceptualization of systematics, its onto-epistemic framework encompasses the broadest presumptions regarding systematic diversity (its existence, overall structure, origin, etc.) and the valid approaches to its exploration (representation by the pertinent cognitive models, elaboration of the most appropriate methodologies, etc.). At this level, the resulting conceptualizations of cognitive situation can by no means serve as a kind of operational device for the practical taxonomic research. To ensure the latter, specific interpretations of both its ontic and epistemic components are to be developed, and such a multi-aspectual reduction operation entails an array of lower-level, more particular cognitive (sub)situations. Their underlying conceptual frameworks become the 'ideological cores' of the respective research programs implemented by various taxonomic schools.

2.2 THE HIERARCHIES OF REDUCTION CASCADES

One of the most fundamental cognitive generalizations known as the *principle of abstraction* (or individuation, separation, etc.) says that any manifestation (fragment, structural or functional unit, process, etc.) of the universe, at which cognitive activity is targeted, becomes a research object, if it is properly individuated by its 'carving' from its environment. It is supplemented by the *principle of limited complexity*, according to which any such object is represented in a cognitive situation by a reasonably simplified model making it actually accessible for comprehension and description at a desired level of detail (Levins 1973). These fundamental principles are a specific aftermath of the interrelations of three components of cognitive situation, so they unavoidably underlie any cognitive activity. This is because the latter is always guided by a specific predefined task and therefore intentionally selective: a subject (observer) applies a kind of '*cognitive filter*' to the universe and considers only (a) an individuated object and (b) those of its features, which are deemed to be essential for solving the task in question.

Hence, abstraction means nothing else but an interpretative *ontic reduction* of the object, which results in its *representation incompleteness* and entails a certain *ontic gap* between a research prototype (object itself) and its simplified model (Wartofsky 1979; Salthe 1985).

The subject-driven multistage procedure of ontic reduction shapes a specific *ontic reduction cascade*, which begins with a concept of the broadest (within the given cognitive situation) meaning and ends with the operational conceptions designed to solve specific research tasks. This cascade is characterized by the following major parameters. Its basic component is a *reduction path*, comprised of successive steps of particular reduction operations. By its logical design, it is similar in its construction to the hierarchical genus-species scheme of scholasticism. At each *reduction step*, several simpler models are elaborated, each representing in a specific manner the respective complex prototype entity individuated at this step. Therefore, the entire ontic reduction cascade consists of an array of branching reduction paths.

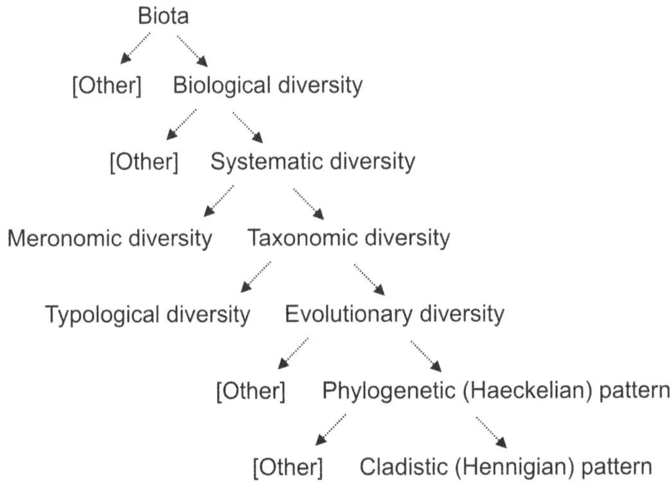

FIGURE 2.1 The hierarchy of ontic reduction cascade leading from concept of biota to that of cladistic pattern.

(Based on: Pavlinov 2024, p. S4; significantly modified). See text for explanation.

In systematics, such a reduction cascade (Figure 2.1) begins with the 'carving' of biodiversity from the overall structure of biota. At the next step, the biodiversity is near-decomposed into its aspect-based major constituents, which are most usually defined with the reference to particular features (morphological, ecological, ethological, etc.) and/or specific relations (similarity, kinship, spatial contingency, etc.) of organisms. These aspect-based manifestations of the diversity outline distinct ontic components of particular cognitive situations of the respective biological disciplines. One of these manifestations of the overall organismal diversity provides *systematic* (taxonomic *s.l.*) *diversity*,[4] while other aspects define its biogeographic, synecological, and other components. As is evident from the preceding consideration, all of these should be treated as specific conceptual realities constructed for the particular cognitive demands; the one studied by biological systematics is a specific *systematic* (taxonomic *s.l.*) *reality* (Zuev 2002, 2015; Pavlinov 2011a, 2018, 2021a; Shatalkin 2012).

Regarding systematic diversity, considered in its general sense, its initial aspect-based near-decomposition in turn entails the recognition of its two basic aspects, namely, *taxonomic* (*s.s.*) and *meronomic*, which are interrelated in a complex manner (Pavlinov 2018, 2021a). Their aspectual status is evident from the facts that they do not exist separately from each other but are individuated by a cognizing subject: indeed, there cannot be things (including organisms) without their attributes, just as there cannot be attributes without things as their 'bearers'. Accordingly, their interrelation, considered at this basic ontic level, can be understood as a specific manifestation of the above-introduced *taxonomic-meronomic dualism*, which can be treated as a peculiar analogue to the wave-particle dualism assumed for the quantum objects in microphysics (see, e.g., Chester 1987 on it). Because of this, taxonomic and meronomic aspects are equipotent to each other and to the entire systematic diversity, and they complement each other in the wholeness of the latter. Their unity is cognitively disrupted and entails two partial hierarchies of the same name, taxonomic and meronomic, each being individuated as an array of taxa or merons, respectively, to become relatively distinct research objects set by specific exploratory tasks. Their equipotence and dualism provide an opportunity to link them conceptually by the *taxonomic-meronomic correspondence*, which plays an important role in the hierarchization of classification systems (Pavlinov 2018, 2021a; see Section 3.6).

It is to be reminded that these two fundamental aspects of the diversity of things were not segregated by Antique and Medieval natural philosophers, and their unity seemed to be most consistent with the taxonomic-meronomic dualism in its primary understanding.[5] The systematicians began to regard them separately because they became interested mostly in the recognition of taxa, whereas any meronomic considerations served for their diagnosing. These two aspects are currently distinguished terminologically: the taxonomic one is called *diversity* proper, whereas the meronomic one is called *disparity* (e.g., McGhee 1999; Kaplan 2004; Pavlinov 2011b); accordingly, taxonomic and meronomic hierarchies are referred to as diversity and disparity patterns, respectively (Tëmkin and Eldredge 2015).

* * *

Before proceeding with the discussion of this important issue, a special clarification is to be given for the notions of *taxonomic*, *meronomic*, and *partonomic*, as they are applied in this book.

Their recognition was initiated by two conceptual novelties. On the one hand, the meronomic partition, as based on the whole-part relations, has been specifically formalized by the Polish mathematician Stanisław Leśniewski (1886–1939), who in his *Podstawy ogolnej teorii mnogosei* (*Foundations of a General Theory of Sets*, 1916) elaborated a logical system of *mereology* for this case (Pietruszczak 2005; Calosi and Graziani 2014; see also Section 2.4). On the other hand, the German theoretical morphologist Adolf Meyer-Abich (Meyer before 1938; 1893–1971) suggested the notion of *taxon* to designate a typologically defined classificatory unit as opposed to Haecleian *phylon* as a phylogenetically defined unit (Meyer 1926).

Following these innovations, the British theoretical biologist Joseph Woodger (1894–1981) suggested to recognize two major kinds of hierarchies in systematic biology, *aggregative* 'Linnaean' and phylogenetically defined *divisive* (*D*-hierarchy) (Woodger 1937, 1952). They were based on two fundamentally different kinds of partitioning, one based on traditional extensional and another on mereological systems, respectively. The first of them was subsequently denoted as *taxonomic* partition (Gregg 1954), it was based on an 'aggregative' standpoint and implied division of composite aggregates (sets, classes, etc.) into kinds (subsets, subclasses, etc.). The second approach was denoted as *meronomic* partition (Meyen and Shreyder 1976; Meyen 1977, 1978; see also Lyubarsky 1996, 2018; Pozdnyakov 2015; Chebanov 2017; Kull 2019), which was supported by a holistic viewpoint and implied partition of integrated wholes into their parts; it is occasionally referred to as *mereology hierarchy* (Jansen and Schulz 2014). E. Mayr denoted these two kinds of partitions as *aggregational* and *constitutive* hierarchies, respectively (Mayr 1982; Rieppel 2009a). The term *partonomy* was proposed in the 1980s with a meaning similar to meronomy, and it also gained a certain popularity since then (Winston et al. 1987; Tversky 1989, 2013; Mahner and Bunge 1997; Beghtol 2000; Rieppel 2006a; etc.).

In this book, these three notions are suggested to be specified as follows.

The terms 'partonomic' and 'meronomic' are both implied by a holistic treatment of the compound entities. The *partonomic* ones is adopted in a broad meaning to be applied to the wholes of any kind and any levels of generality, from the *One* in Platonic cosmology down to the lower-level collective units in wildlife (species, colonies, ecosystems, etc.) and human society (tribes, organizations, etc.). In systematics, such is the treatment of biodiversity adopted by organicism, typology, and partly phylogenetics. The *meronomic* consideration is understood in much narrower senses to refer to: (a) partitions of the solid physical bodies into their attributes (features, parts, functions, etc.), (b) one of specific aspects of systematic diversity, as they were defined above, and (c) to classification systems representing this particular aspect of the diversity of organisms. In the case of biological organisms, these attributes are particular organs, cells, biomolecules, and they are designated uniformly as *merons* (Meyen 1977, 1978; Lyubarsky 1991, 1996, 2018) or *morphemes* (Göpel and Richter 2016; Göpel 2024). The just-cited authors referred to the anatomical structures, but it seems reasonable to extend the notion of meron to encompass all organismal attributes, including physiological functions, coloration, behavior patterns, etc. (Pavlinov 2018, 2021a).

With regard to the term 'taxonomic', its application here in the sense of Barbara Tversky and Co seems to be inappropriate, because it traditionally has quite another meaning in systematic, which is evidently followed in this book. Accordingly, it will be preserved to designate both (a) another specific aspect of systematic diversity, alternative to the meronomic, and (b) specific classification systems as its cognitive models: they will all be uniformly referred to as *taxonomic*. When partitions of sets/classes into subsets/subclasses are implied, as opposed to the partonomic ones, the respective hierarchies are referred to as *aggregative*.

However, an attention should be called to the fact that, as is usual for the ontology-related complex structurations, the accepted dichotomy of aggregative (taxonomic *sensu* Tversky) *vs*. partonomic partitions is too rough for an adequate analysis of the issue. As Elźbieta Górska pointed out proceeding from the ideas of cognitive grammar (in the sense of Langacker 2007), the inclusive hierarchy of military units, which is usually regarded as a typical example of aggregative partition, is in fact endowed with certain partonomic features (Górska 2001, 2002). This is because each such unit of whatever rank is of systemic nature, with its elements (soldiers, commanders, etc.) being strongly interrelated by carrying out specific functions, and therefore each unit is clearly more than just a sum thereof. Hence, *systemic* partition should be placed between aggregative and partonomic ones (Shatalkin 2012), which seems to be quite relevant to the consideration of near-decomposition of the overall hierarchical pattern of biodiversity into its particular aspect-based manifestations.

* * *

The aspect-based near-decomposition of taxonomic (*s.l.*) diversity results in the recognition of its specific manifestations that are studied by various schools of systematics, namely, phylogenetic, typological, biomorphic, population, etc., formalized by the respective conceptions. Each of them may in turn be reduced to the next lower-level conceptions elaborated for more specific manifestations of taxonomic diversity. For instance, typological diversity may be fragmented into Cuvierian, Goethian, or Baerian typologies, whereas phylogenetic diversity may be fragmented by Haeckelian, Darwinian, Simpsonian, or Hennigian conceptions (see Pavlinov 2018, 2020, 2021a for their review). The object-based near-decomposition of taxonomic diversity (*s.s.*) involves the recognition of its particular elements, which are specific groups of organisms customarily referred to as *taxa* (species, families, biomorphs, etc.); besides, levels (ranks) of taxonomic hierarchy may also be considered as another class of taxonomic objects.

The near-decomposition of meronomic diversity can also be either aspect-or object-based, and it results in the individuation of particular constituents of a generalized organism (archetype), these are its particular attributes (features, properties, parts, etc.) referred to collectively as *merons*. The aspect-based merons are coloration, behavior patterns, physiological functions, etc., and the object-based ones are organismal parts. Homologization plays a key role in the recognition of merons of whatever kind and level; it serves as a procedural framework for an elaboration of the meronomic classifications. At an operational level of reduction cascade, merons serve as a basis for the individuation of *classificatory characters* (traits) to be used in the elaboration of taxonomic classifications.

* * *

The reduction cascade is arranged asymmetrically by consisting predominately of an array of descending reduction paths between conceptions of various levels of generality. Hence, the entire cascade can be interpreted as a hierarchically ordered *deductive-nomological* cognitive system (in the sense of Hempel 1966; Woodward 2003), in which the lower-level conceptions are derived from the higher-level ones. However, there may also be feedback loops in it, resulting from specifying the certain higher-level conceptions by the contents of respective lower-level ones. Besides, some 'anastomoses' within this cascade may appear due to partial fusions of the distinct paths due to exchanges of different conceptions of (nearly) the same levels of generality by specific ideas. This point is exemplified by evolutionary typology, which emerged as a fusion of Haeckelian phylogenetics with typology of Platonic kind (Vasilieva 1998, 2007); another example is provided by pattern cladistics that couples Hennigian cladistics and Baerian typology (Nelson 1978; Platnick 1979; Nelson and Platnick 1981; Patterson 1983).

In evaluating cognitive status of the resulting conceptions that take different positions within a reduction cascade, it is important to bear in mind the above-mentioned ontic gap that tend to widen with the degree of reduction. It is reasonable to suppose accordingly that the longer the reduction path, leading from the large-scale higher-level conceptions to particular lower-level ones, the wider the ontic gap separating them.

The following important features of the inclusive hierarchy of ontic reduction cascade are to be taken into account when regarding the basic ideas most relevant to the subject of this book. First, there are many ways of shaping particular cascades depending on the specific contents of the conceptual frameworks of particular cognitive situations. Second, each such cascade may begin deliberately with a specific highest-level concept and end with as specific lowest-level ones, with different degrees of fragmentation of the paths connecting them. Third, a reduction cascade presumes deductive structuring of respective conceptual space so that each lower-level conception is inferred at a certain step of reduction not by itself (*ad hoc*) but rather as a particular interpretation of the including higher-level one (Quine 1969; Carrier 1994). Consequently, each reduction path encompasses a specific sequence of the semantically interconnected conceptions of various levels of generality. This means, fourth, that any conception of a certain level of generality gains its substantive meaning only within the contexts of the respective inclusive higher-level ones. This is especially true for the operational conceptions, which may appear to be biologically meaningless if elaborated and/or applied formally without any reference to some biologically sound conceptual models. Fifth, as the natural entities systematics deals with are substantively manifold, there is a possibility for each of them to be reduced to (represented by) several particular cognitive models (conceptions), each emphasizing some of its specific features. Accordingly, an overall reduction cascade involves a number of 'parallel' sequentially branching reduction paths, which results in the multiplication of lower-level conceptions. The above-mentioned particular interpretations of typological and phylogenetic conceptions exemplify this idea; another well-known demonstration of the latter's relevance is the general species concept, for which about dozens of particular interpretations have been suggested (see Section 5.1).

* * *

As is seen, the overall ontic reduction cascade results in an inclusive (nested) hierarchy of cognitive models (conceptions) of different levels of generality, with these models multiplying from top to bottom of the cascade, so the latter's totality may be thought of as a kind of the so-called *conceptual pyramid* (Pavlinov 2018, 2021a). Biological systematics has been addressing this most fundamental hierarchical construction from the very beginning of its historical development. Of course, it has not been its own 'invention'; actually, the historic predecessor of this conceptual construct had been a *pyramid of notions* of Medieval scholasticism (Makovelsky 2004).

Although our pyramid refers to the ontic component of cognitive situation, it should not be forgotten that it is actually generated by the epistemic operation of reduction. The latter comment allows for paying special attention to the fact that the entire ontic reduction cascade can be considered from a broader onto-epistemic perspective as a specific representation of the so-called *decision tree* (in the sense of von Winterfeldt and Ward 1986; Moshkov 2019; Azad et al. 2022). Indeed, within each of the reduction paths, the transition from one step to another is not carried out by itself but rather result from a certain decision made by an explorer, who choose among several possible lower-level reduction paths. Such a hierarchy was suggested to be termed a *hierarchy of decision layers*, or *analytic hierarchy* (Mesarovic and Macko 1969; Saaty and Vargas 2012; Brunelli 2015). This interpretation is important in that it obliges highlighting several important points that are quite pertinent to our consideration.

With analogizing this cascade with a decision tree, it becomes more than apt to identify its reduction paths with the sequences of particular decision-making sequences. It makes more evident that the running of each of such reduction cascades constitutes a specific part of cognitive activity and cannot be imaged out of the latter. The decision tree metaphor makes it reasonable to alike the ontic reduction cascade (and the respective conceptual pyramid) to a hierarchically arranged *information system* and each of the reduction paths to the particular *information flows* in it (see also Section 2.7). This means that at each step of a reduction path, the decisions involve certain changes in the information about systematic diversity as a research object. Of paramount importance is that the reductional character of the paths implies that such changes within them result in a stepwise decrease in the information content about this object while passing through the information flows. With this, each of the respective decisions regarding the choice among the next lower-level paths is probabilistic and contains some uncertainty. Therefore, the entire hierarchies of both reduction cascades and the respective conceptual pyramids are probabilistic and therefore fuzzy constructions full of uncertainties.

All these considerations are relevant to the main topic of this book, i.e., to the nature of hierarchies biological systematics deals with. Indeed, as emphasized above, a conceptual near-decomposition of the total biodiversity into particular hierarchical patterns (ecological, systematic, etc.) is aspect-based, with the particular aspects being fixed by the subjects in accordance to their respective exploratory or other tasks. The same holds true for the particular hierarchies, into which systematic diversity can be conceptually near-decomposed, namely, typological, phylogenetic, biomorphic, etc. Being internested, they all constitute together the wholeness of the overall hierarchical arrangement of biota, and this is the cognizing subjects that, first, substantiate the very possibility of 'near-decomposing' the wholeness in question into particular components and, second, actually individuate them as particular research objects. One may reasonably question, given certain onto-epistemic premises, whether or not all these partial hierarchies are real ('objective') attributes of the total biodiversity. A possible answer to this fundamental question is not as obvious as one would like it be (see below on this).

* * *

The ideas regarding ontic reduction also apply to the epistemic component of cognitive situation, where it takes the form of *epistemic reduction* resulting in a similar *epistemic reduction cascade*. Accordingly, the latter entails a pyramidal construction of epistemic operations of various levels of generality, which can properly be termed as an *epistemic pyramid*. One of the highest levels in the latter is occupied by the above-mentioned representation principle, which implementation substitutes a research object with its cognitive model and transforms by this the respective fragment of objective reality into a specific conceptual one. An important part of epistemic reduction cascade (and respective pyramid) involves *instrumentalist reduction*, implying that any cognitive activity of an analytical kind may be ultimately reduced to the particular technological operations by applying specific algorithmized procedures as cognitive instruments ("shut up and calculate", as N. Mermin 1989 metaphorized this point on another occasion). This specific form of reduction seems to be of great relevance for the studies of systematic diversity.

Such an algorithmized methodological approach, if absolutized, entails an epistemic problem of *instrumentalism* (Stanford 2016), which represents one of the extreme versions of pragmatism (John 2009). This standpoint results eventually in the substitution of the thinking about a complex object with the conducting certain standard technical operations on its simplified cognitive models, so that the creative personality of scientists is replaced by an artificial intellect. This is perhaps most apparent in quantum physics, where the respective substantive cognitive models of the elementary particles are reduced to the concrete decisions of the Schrödinger equation (e.g., Chester 1987). In systematics, this is most evident in the case of the above-mentioned reductionist 'substitution' of the real phylogeny by its surrogates in the form of graphical tree-like representation schemes. For this reason, one of the most intriguing questions regarding the knowledge, inferred in such an 'instrumentalist' manner, is as follows: what is a ratio in it between the properties of the research object as such and those of the applied particular method? In systematics this kind of uncertainty involves, e.g., the hierarchies of classifications: to what extent do they reflect a real property of systematic diversity, and to what extent are they merely a specific by-product of the hierarchical classification algorithms applied (Sneath and Sokal 1973; Rieppel 2007; Pavlinov 2018, 2021a)?

Another kind of epistemic reduction, which is opposite in a sense to the just-considered instrumentalist one, involves *phenomenological* reasoning, according to which the focus is not on the cognizable object itself but on the method of its intuitive comprehension: 'the subject <...> of consideration is not a taxon, but a way of 'dealing' with a taxon' (Oskolski 2007, p. 224). Respectively, 'the reality of a taxon <...> has nothing to do <...> with its correspondence to an "objective reality"' (Stekolnikov 2007, p. 114). It is needless to say that such an intuitivist reduction yields the same cognitive problem as the instrumentalist one: the adequacy of taxonomic systems to the structure of taxonomic diversity appears uncertain.

2.3 SOME COMMENTS ON THE NOTIONS AND DEFINITIONS

One of the basic functions of epistemology includes elaboration of the notions that shape the entire onto-epistemic framework of cognitive situation in systematics by composing its basic thesaurus. With regard to ontology (in its philosophical sense), one of the most fundamental provisions of conceptualism states that these are the properly defined notions, being the 'markers' of respective conceptions, that allow explorers to individuate the particular manifestations of the structure of biodiversity as the distinctive research objects. This is achieved by specifying the latter's essential features, which are necessary and sufficient for their recognition and inclusion in the respective thesaurus. In a sense, as said above, a proper definition can be considered as a cognitive model of an object representing it in a cognitive situation. With this, it is implied that if an entity is not explicitly defined in a proper manner, it is not represented in a cognitive situation conceptually by the respective notion and, accordingly, is absent in a conceptual reality being construed. This means that the entire conceptualization of systematics includes the proper definitions of its notions that constitute in total its basic thesaurus, and the latter can be interpreted, given certain assumptions, as a compressed representation of its ontology (Smith 2004).

In the developing such a thesaurus, it is essential to distinguish between *theoretical* (conceptual) and *operational* (working) definitions. With regard to biodiversity, the former refer to its constituents in their general sense, whereas the latter are designed to facilitate recognition of these constituents in wildlife, so they are identified rather than defined. For instance, in the case of species (see Section 5.1), a theoretical definition of the respective notion (higher-level concept) is elaborated in order to make it possible to distinguish 'species' from some 'non-species' phenomena (such as ecomorphs and guilds). The same holds true for the lower-level theoretical species conceptions: they allow for outlining, say, ecological and phylogenetic species based on the definitions of respective notions. These lower-level theoretical conceptions underlie practical explorations of biodiversity and predetermine, to a significant extent, their results. They constitute the bases for the elaboration of operational definitions that specify how *the* concrete species of organisms are empirically identified (recognized) by using available and relevant raw data. Therefore, any empirical accounts of species diversity will be biologically meaningful to the extent that the respective theoretical species conceptions underlying them are clearly formulated, which implies that the referential notions are properly defined.

With regard to theoretical definitions, it is important to distinguish between *absolute* and *relative* ones. The former implies that the object of interest can be defined by indication of its own properties, whereas the latter refers to any kind of its relations to other objects. For instance, a taxonomic rank can be defined either absolutely with the reference to a certain exterior hierarchization parameter or relatively with the reference to other ranks in a classification hierarchy (see Section 3.1). However, generally speaking, all definitions are relative/relational in a sense, just because any definition is given for an object in order to distinguish it from other objects 'of the same kind', and all of them should be related to each other in some way within a meaningful context, in which they are compared. The relative character of definitions is particularly evident from their context-dependence, with the respective contexts being multifold. On the one hand, a meaningful context for a particular definition is provided by the whole hierarchy of higher-level definitions within a particular conceptual pyramid. On the other hand, a specific context can be posed by a cognitive task set by a subject. In general, this means that a studied entity is not given 'objectively' as such but rather is shaped conceptually by a particular theoretical definition based on the relevant onto-epistemic provisions (Ahl and Allen 1996; Kearney and Rieppel 2006). In systematics, this is true, for instance, for the context- and task-dependent definitions of homologues (see Section 5.2).

In systematics, the objects to be defined before all are systematic diversity with all its manifestations, classification as its cognitive model, and structural features of the both, including their hierarchies and basic constituents (taxa and characters). An *intensional* theoretical definition is characteristic of essentialist and typological approaches, it implies that the notions are defined by indicating the intrinsic properties of the research objects and their particular manifestations and structure. Unlike this, an *extensional* definition, which is largely characteristic of phenetic and cladistic approaches, implies either listing of the membership of the object in question (say, genera within a family) or its relations (such as kinship and similarity) to other respective objects. Combination of these two kinds of definitions entails a two-parametric conception embracing both intensional and extensional characteristics of the research object (Stumme 1999). At last, an *ostensional* (ostensive) definition implies a direct indication of the tokens presumably belonging to the objects being defined, which makes it more empirical rather then theoretical. A possibility of such definition of a taxon is sometimes supposed by indication of its nomenclatural type (Heise and Starr, 1968; Dubois 2008) or its hypothetical ancestral species (Dupuis 1984; de Queiroz 1992).

Regarding hierarchies and their levels (ranks), considered as specific research objects of systematics, they can probably be given intensional and extensional definitions but not ostensional ones. Their intensional theoretical definitions may refer, among other things, to the processes supposedly generating hierarchies and their essential properties, and the extensional ones refer to the structure and position of particular levels (ranks) in the overall hierarchy.

* * *

There are several fundamental epistemic generalizations that are relevant to the issue of correct definitions of systematic notions (Kuraev and Lazarev 1988; Voyshvillo 1989; Hjørland 2009; Margolis and Laurence 2011).

The *constructivity principle* asserts that both a research object itself and any of its features may be counted as embraced in a cognitive situation, only if each of them is given a certain definition, which should be elaborated within the given cognitive situation or borrowed from outside. Such a definition of whatever origin should meet certain provisions, before all, *generality*, *completeness*, *strictness*, and *explicitness*, formalized by the respective principles. When properly defined, a studied 'empirical entity' becomes a 'definitional entity' represented in a cognitive situation by the respective notion as its cognitive model (Ahl and Allen 1996).

The observations of all rational provisions for the proper definitions of the studies objects are essentially limited by the most general multifold *uncertainty principle*. It implies that the meaningful definitions of each of the empirical entities, including those studied by biological systematics, contain a certain portion of vagueness. The latter has many causes and guises, and the most significant of them are as follows.

A very specific cognitive principle suggests an *inverse relation between the rigor and meaningfulness of a notion*: the more rigidly a notion is defined, the less likely there is something in nature, to which it may exactly correspond (Voyshvillo 1989). This principle implies that it is impossible for the just-listed and other provisions to be fulfilled literally and simultaneously, so the scientists actually have to deal with the approximate compromise formulations (Levins 1966; Weisberg 2006; Matthewson and Weisberg 2009; Brooks and Eronen 2018).

In systematics, a good example of such an inverse relation is provided by the notion of holophyly (strict monophyly) in cladistics, which becomes less meaningful biologically, when it is attempted to be defined more rigidly. The notion of *minimally monophyletic group* provides additional constraints to the correct operational definition of monophyly (Zander 2025), which makes it not a simple conception to be easily applied in practice. Besides, a strict definition of holophyletic group (ancestor and all its descendants) yields the so-called *Platnick's paradox*: a monotypic genus cannot be considered holophyletic, since it includes only one species, which is assumingly a descendant (Nelson and Platnick 1981; Pavlinov 1990a, 2018, 2021a; Wägele 2005; Kwok 2011).

The *incompleteness principle*, which is an extension of the above-mentioned representation incompleteness, says that it is logically impossible to exhaustively define any notion referring to a specific natural entity (Kuraev and Lazarev 1988). One of its provisions (see below about another one, no less important) says that any rational knowledge about biodiversity, embodied in the respective notions, is inevitably incomplete, and it underpins the above-discussed ontic reduction. To make definitions substantively more complete, the deficiency of their rational 'precise realism' is compensated by an intuitive grasping, which makes them fuzzy and metaphoric to an extent (Kittay 1987; Murphy 2002; Madsen 2016). As a consequence, any notion that refers to a complex natural (or any other) entity may be thought of as a 'sphere' consisting of two parts, more or less formalized 'core' rational definition and its metaphoric 'envelope'(see also Kittay 1987).

Accordingly, the semantics of taxonomic thesaurus appears to be most adequately described based on *fuzzy logic* (Zadeh 1965, 1971, 1992; Dubois and Prade 1980; Kosko 1993), which is specifically implemented by a *probabilistic model* of the natural science language (Nalimov 1979). This means that it is possible for each notion to be defined more or less strictly only in its conceptual 'core' using the logical relation '*A* is *B*', while its 'periphery' remains fuzzy, presuming contextual particular definitions of the type '*A* can be *B1*, *B2*, *B3*...' (Smith and Medin 1981; Bryant 2000). Therefore, the notions composing basic thesaurus of systematics, as a natural science branch, cannot be treated as the full-fledged *rigid designators* (in the sense of Kripke 1980; LaPorte 2022). Such fuzzily defined notions typically differ in their 'cores' but may overlap by their 'peripheries', and they are mutually interpretable to the extent that their 'cores' are substantively comparable (Kuhn 1962; Hacking 1983).

The above observations imply that each taxonomic notion (conception) occurs in general cognitive situation of systematics as an array of interrelated context-dependent particular definitions that constitute a kind of 'flock of meanings'. This means that a certain fuzzily given notion (*B*) is endowed with a 'confidence interval' of its possible meanings with realization probabilities ascribed to them. In each particular cognitive situation, the notion in question becomes specified in a characteristic way by the selection of one of its particular definitions (either *B1* or *B2* or *B3*...), which becomes a particular exploratory tool (Cartwright 2016; Brooks and Eronen 2018). The notion of taxon provides a typical instance of this kind: its 'core' definition refers uniformly to a certain unity, whereas its 'peripheries' are specified differently (context-dependently) in typology, phenetics, phylogenetics, etc., by specific interpretations of the presumed unity.

All this makes the entire conceptual framework of our discipline fuzzily delineated and structured (Pavlinov 2018, 2021a). This is especially characteristic of strictly empirical (including primordial) cognitive activity, which is unframed by any *a priori* elaborated rational schemes. This results in the recognition of what Eleanor Rosch denoted as *natural categories* (not in the sense of taxonomic categories, see Section 3.4), which are perceived and distinguished intuitively by grasping their non-apparent prototypes as their cognitively salient 'cores' (Rosch 1973, 1983, 1988; Mervis and Rosch 1981; Bryant 2000).

* * *

The above-mentioned pyramidal construct, understood generally as a pyramid of conceptions, represents at the same time a deductive inclusive hierarchy of notions and their definitions of various levels of generality. The resulting *definitional reduction cascade* implies that each definition is formulated not by itself but as a specification of the definition of a certain higher-level inclusive notion. Such a deductive approach is most consistent with the classical theory of definitions, according to which any 'logical species' are possible to be defined properly by the logical division of their inclusive 'logical genus' (Smith and Medin 1981; Voyshvillo 1989; Makovelsky 2004). The same conclusion, but in a more regular conceptual sense, is achieved, if the pyramid in question is supplemented with the above-mentioned incompleteness principle. The latter, as it was developed toward a general epistemic meaning, says that a notion cannot be correctly defined within its own basic thesaurus but rather within a wider cognitive context (Antipenko 1986). This means that in order to formulate a proper definition of a particular notion, a more general notion is required that would include the one being defined and provide a meaningful context for this. Considered from such a standpoint, an inductive (down–top, ascending) scheme of definitions from the simplest to the most complex ones, which is unanimously advocated for by empiricism and especially positivism, is not correct.

For instance, when considering the structure of basic taxonomic notions, an initial one to be defined are 'classification' and 'hierarchy', then a 'unit of classification' and a 'hierarchical level', followed by the definition of 'taxon' as a certain unit of classification, and finally, say, a notion of 'family' as a taxon of a certain hierarchical level. A quite impressive example is provided by the species notion: its currently most popular definition by the reference to certain interplays of the organisms within it is inductive and therefore seems to be incorrect. Instead, with considered species as a certain biotic constituent, it can be conceptually defined as one of the structural units of biota, which necessitates a prior definition of the biota, its structure, and its structural units (Pavlinov 2013, 2023; see Section 5.1). Such a deductive approach implies that for a hierarchy of any kind to be defined consistently, it is required first to comprehend and define a certain phenomenon, to which the hierarchy can be attributed as its inherent property. Looking ahead, it is worth noting that such a phenomenon is a *system* in its most general sense, be it physical or mental, natural or artificial (see below).

* * *

The most recent computer-oriented studies suggest a specific approach to the general problem of definitions by seeing its focal point in the elaboration of some *full-blown ontology* as a part of the bioinformatically outlined cognitive situation (e.g., Page 2004; Schulz et al. 2008; Franz and Thau 2010). Considered in a broad historical perspective, this approach continues the elaboration of standard anatomic terminology by naturalists of the 16th–18th centuries (Pavlinov 2018). Its central concept of ontology is understood in a quite technical sense as a 'computer-readable, semantically structured knowledge base used as representation of terms and their definitions' (Göpel 2024, p. 4). Such ontologies address mostly meronomic aspect of systematic diversity, dealing with the accurate definitions of the structural elements of organisms. Examples of this approach include the biomedical, gene, and cell ontologies (Ashburner et al. 2000; Bard et al. 2005; Schulz et al. 2008); more special are 'phenotype ontologies' for some groups of organisms (e.g., Mabee et al. 2007; Mungall et al. 2010; Cooper et al. 2013; Vogt et al. 2013; Dahdul et al. 2014; Wirkner et al. 2017; Göpel 2024). These meronomic ontologies are designed to provide standardized templates for making descriptions of organisms more suitable for their unified comparisons. In particular, this approach is promised to facilitate a more detailed and operationalized refinement of the concept of homology, which is essential for the progress in both taxonomic and meronomic branches of systematics (Sereno 2007; Roux and Robinson-Rechavi 2010; Dececchi et al. 2015; Göpel 2024; see Section 5.2).

However, an excessive optimism, with which the proponents of technically standardized ontologies are charged, faces the same problems, which was discussed above regarding philosophically understood ontology. In the case under consideration, such a 'technical ontology' becomes relativistic due to the fact that any elementary entities as merons can be naturally identified only within the context of hierarchies of within- and between-organismal relations. The problem arises from the multiplicity of such hierarchies, which tends to load this entire issue with a significant portion of epistemology, and due to this any particular entity becomes endowed with a fuzzily given 'flock of definitions', which versions depend on particular onto-epistemically defined contexts (Lyubarsky 1996, 2018; Minelli 2023a). A good example of such an 'epistemic pitfall' is provided by a multiplicity of near-equally eligible definitions of a seemingly clear notion of the tetrapod limb, which is caused by the variety of possible contexts of its consideration within the overall hierarchy of a vertebrate organism (Kuznetsov 1999).

2.4 THE PYRAMIDS OF REASONING: LOGICS, METHODOLOGIES, AND METHODS

All natural science branches are governed from their origin by the idea of rationality, which descents to Antiquity; its contemporary classical version is usually referred to as *Cartesian* after the French philosopher René Descartes (Lat. *Cartesius*; 1596–1650), who has made the most notable contribution to the development of its foundations (Gaydenko 2003). Its high relevance certainly holds true for biological systematics, which emerged and was initially developing based on the mastering of rational genus-species logical scheme (Pavlinov 2018, 2021a; see Section 1.4).

Therefore, an explicit elaboration of the rational foundations of the epistemic component of cognitive situation of our discipline constitutes one of the key objectives of its conceptual development. In order to approach successfully this end, the just-named idea is formalized by the general *principle of rationality*, which presumes that cognitive activity, in order to be scientifically consistent and meaningful, must be carried out based on certain rationally substantiated, explicitly formulated, and properly applied tenets, rules, and precepts. This most general principle implies that the scientific character of both knowing and knowledge of systematic diversity is largely provided for by being based on certain epistemic propositions, which constitute the multifold *principle of scientificity*.

* * *

The implementation of the principle of rationality entails a complexly structured and hierarchically arranged epistemic construct, which can aptly be termed a *pyramid of reasoning*. Its overall design is determined by the above-considered reduction cascade and is formalized by the *pyramid principle*, according to which the reasoning procedure follows the general deductive scheme by beginning with the most general and ending with the most simple ones (Minto 2008). In the case of substantive considerations, which biological systematics deals with, all reasoning begins with the worldviews, which take the top position within this pyramid and make it ontically constrained. Each of them serves as a source, first, for the formulation of specific provisions of cognizability and, second, for the substantiation of certain exploratory approaches (in their general sense of Aristotelian *Organon*) that agree with those provisions and lead to the rationally inferred and therefore considered scientifically meaningful knowledge.

The next level of implementation of the principles in question, now pure epistemic, implies that, for an exploratory methodology to be acknowledged as scientifically consistent, it must be based on a solid logical background. In regard to biological systematics, this general *principle of logic* in its extreme embodiment demands that any classifying is primarily a logical procedure and therefore must be accomplished based on a certain universal logical system equally applied to the diversities of any kind (Thompson 1952; Lubischev 1969, 1972; Zarenkov 1983, 1988; Pokrovsky 2014). However, contemporary logic is quite diverse in its basic syllogistics, so there is in fact a number of logical systems that are more or less local in their meaning and applications, which makes an idea of some 'general' and allegedly universal logic evidently untenable. These particular logics are elaborated as distinct axiomatic systems, which differ in the arrays of 'internally' consistent basic axiom and inference rules (Ivlev 2025). Proceeding from this, a critical question arises as to which particular logical system can (or should) be treated as preferable according to the just-mentioned principle.

The first of such systems has been developed by early Medieval scholasticism to become known as *two-state* (binary) or *classical* (Aristotelian) logic. It is this system that was and still is usually considered 'general' or 'universal' logic, but it exists in several versions, of which two pairs of opposites are most relevant to systematics. One of these pairs includes *intensional* logic of classes and *extensional* logic of sets

(Pellegrin 1987): the first implies definition of both taxa and merons by referring to their essential features (attributes, etc.), whereas the second defines them by referring to their composition. These logical systems are implemented in taxonomic nomenclature: intensional logic underpins the principle of diagnosing, and extensional logic underlies the principle of circumscription of taxa (Pavlinov 2015, 2022a). Another pair includes *deductive* and *inductive* logical systems, which imply, respectively, either top–down (from general to particular) or down–top (from particular to general) substantiation of the true status of logical inferences. In systematics they are implemented by the respective classification algorithms, which are well-known today from the practice of numerical taxonomy (Sokal and Sneath 1963; Moss and Hendrickson 1973; Sneath and Sokal 1973; Abbot et al. 1985).

The classical deductive logic was the first to have been mastered by nascent biological systematics, and it persisted in it for several centuries. It is based on the *principle of excluded third*, which is supplemented with several others, its original version can be summarized in a simplified form, as it is most pertinent to systematics, as follows. The sequential stepwise division of taxa is carried out on a *single basis* (combined use of several uncorrelated characters is prohibited), must be *exhaustive* (a taxon is divided into subtaxa without any remainder), provides *discrete* taxa (those of both the same and different ranks must be mutually exclusive), and their hierarchy is *continuous* (all fixed hierarchical categories are to be uniformly applied with no gaps in all parts of a classification). In the inductive version of classical logic, the last three basic axioms are similarly applied.

Since the mid-19th century, several systems of *non-classical* logic have been being developed to shape the basis of non-classical rationality (Mikhailov 1983; Shuman 2001; Burgess 2009). Of them, the *multi-state* logic became the most significant for natural science branches, including systematics. Its *three-state* variant adds *indefinite* judgments about research objects, whereas its *probabilistic* variant considers the probabilities of each of those judgments being true. In systematics, the three-state logic formally substantiates a usual practice of the violation of the principle of exhaustive delineation of taxa by identifying 'taxonomic remnants', which include taxa with unclear allocations (*incertae sedis*).

The classical principle of a single basis of division was originally implied by the essentialist interpretation of classifications. It makes taxon definition *monothetic*, and it was actively used by scholastic systematics, though with certain reservations. However, it is rejected by post-scholastic systematics in favor of the *principle of multiple basis* of division presuming a *polythetic* taxon definition.[6] The latter principle gains a specific ontic substantiation in the evolutionarily interpreted taxonomic theories, though it is considered pure methodological in phenetic systematics.

The requirement for the continuity of taxonomic hierarchy, which presumes the ordering of all taxa within a given classification according to a unified hierarchization parameter (ruler), is one of the most significant in classical systematics. Its application to the classifications based on extensional logic entails the so-called *Gregg's paradox* (Gregg 1954), which sparked a lively discussion (Buck and Hull 1966; Löther 1972; Griffiths 1974; Shatalkin 1988, 1995, 2012; Mavrodiev 2002; Pavlinov 2018). This condition is largely relaxed by the *principle of non-strict* (degenerate) *hierarchy*, which is specifically implemented by cladistic systematics (Wiley 1981; Pavlinov 1990a, 2018).

The already-mentioned *fuzzy* logic implies that all judgments and their truth assessments in science are probabilistic and context-dependent, so they cannot be strictly defined in a unified manner (Zadeh 1992; Dompere 2009a, b). This very flexible logical system substantiates the context-dependent character of all judgments about taxa, including their features, boundaries, and ranks, depending on the contents of partial taxonomic theories (see Section 2.5 on them). Of similar non-rigid kind is the *Bayesian logic*, which operates with *conditional* probabilities of the true status of judgments (Lin 2024); it is sometimes rated as the 'Bayesian philosophy of science' (Sprenger and Hartmann 2019) and is widely applied in contemporary numerical phyletics (Felsenstein 2004; Nylander et al. 2004; Bolstad 2007; Chen et al. 2014; Nascimento et al. 2017).

A conceptually important version of non-classical logic is *modal logic*, which accentuates an active role of a cognizing subject in the reasoning by making particular judgments as true or false personally dependent (Blackburn et al. 2001).

A very specific version of non-binary logic is suggested by a *one-state* logical system (Vasiliev 1989; Bazhanov 2009). It negates the universally applied classical *principle of equivalence* of both positive *A* and negative *non-A* judgments and asserts instead that only positive judgments are meaningful, whereas negative judgments are not. In systematics, this concerns taxonomic significance of the presence *vs.* absence of classificatory characters, which are considered equivalent in classical systematics and nonequivalent in some of its non-classical schools (Pavlinov 2018, 2021a). This logical system is adopted in cladistics, which postulates diagnosing taxa by synapomorphies only (positive *A*), whereas symplesiomorphies (negative *non-A*) are excluded as meaningless (Hennig 1966; Wiley 1981; etc.).

Of two already discussed general accounts of compound entities, referred to as either *aggregative* (elementaristic) or *partonomic* (holistic), the classical logical systems seem to be applied to the former. Their 'formal' difference can be imagined as implying either 'is-a-kind-of' or 'is-a-part-of' relations, respectively (Salthe 2012). The latter is considered by the already mentioned logical system of *mereology* formalizing the whole-part relations (Luschei 1962; Calosi and Graziani 2014). In systematics, it is relevant to those of its branches that consider biodiversity holistically (organicism, typology, and partly phylogenetics). An opposition between elementaristic and holistic considerations of taxa/merons concerns such issues as the distinction between their definitions as classes *vs.* descriptions as quasi-individuals, the significance *vs.* irrelevance of similarity of organisms as a *sine qua non* condition for their taxon membership, etc. (Buck and Hull 1966; Hull 1978; Mahner 1993; Mayr and Bock 2002; Rieppel 2006b).

Of particular importance for non-classical systematics is *quantum* logic, which substantiates consistent procedures for uniting formally incompatible statements about the same object by their *complementarity relation* (Vasyukov 2005; Armand 2008). This specific logical system was initially substantiated in quantum physics (hence its name); in systematics it implicitly underpins the elaboration of the so-called *faceted classifications*, which integrate different conceptions related to the same entity, such as those about species, without eliminating their specific characters (Pavlinov 2018, 2023).

This cursory overview of the logical foundations of biological systematics should be concluded with the following important considerations. Any logical system is just a specific technical instrument designed to ensure the logical consistency ('logical truth') of inferred judgments with respect to certain initial ones accepted *a priori* as unconditionally true within a certain axiomatically given logical system. Therefore, an accentuation on the logical foundations of systematics provides another version of the above-touched *instrumentalism*, with the logical syllogistics being specific inference instrument. The problem here is that the latter's 'logical truth' does not say anything definite about the relations of the respective judgments to the studied natural entities, i.e., about their 'objective truth'. This is all the more correct in light of the existence of diverse logical systems, whose judgments are 'internally' consistent with respect to their own axiomatics, regardless of whether they are 'externally' consistent with respect to those entities. This important point obliges us to remind the mismatch between intersubjectivity and objectivity of scientific knowledge (see also below on this): now one may say that the former appeals to the 'logical truth', whereas the latter implies the 'objective truth'. Of special importance is that, contrary to the classical logicism, a non-classical conceptual rationality implies that different ontologies may presume specific epistemic tools, including logical ones, that are proved to be most consistent with them (Berkov and Yaskevich 2001; Shuman 2001). In terms of the above-discussed onto-epistemic correspondence and 'precedence', this conclusion means that, at least as far as biological systematics is concerned, these are the logical principles of elaboration of both taxonomic and meronomic classifications that should follow ontological premises, and not *vice versa*.

* * *

The next lower levels of pyramid of reasoning are occupied by numerous methodological principles and prescriptions, which in systematics are summarized by a specific *classification methodology*. The principal objective of this part of epistemic component of cognitive situation is to substantiate and develop, first, argumentation schemes and, second, concrete methods as specific research tools to implement these schemes in systematic research.

Argumentation schemes represent specific 'templates', according to which the entire research activity in systematics is organized and carried out. In this capacity, they underlie more particular methodological principles. Three basic schemes of such kind are commonly recognized, which are differently implemented by particular research programs in systematics.

The *inductive* scheme is based on the idea that scientific knowledge, in order to be objective and thus true, should be inferred directly from the observed facts, and it should be both theory- and subject-neutral. Its natural philosophical background was based on the Aristotelian ontology of an ascending cascade of causal relations in nature, which is complemented with the inductive logic. The latter's implementation results in the inductive classification algorithms. In its contemporary guise, this scheme has begun with the empirical ideas of Cartesian rationalism and became most popular subsequently due to positivist philosophy of science. The inferred generalizations about natural phenomena are treated as hypotheses, which truth status is to be *corroborated* by facts. The latter in systematics are observations on organisms with their characters, and the primary generalizations involve the judgments about their similarity relations. One of the main methodological tools for inductive inferences is the *principle of total evidence* developed by positivism and widely used in phenetic and partly cladistic systematics (Gilmour 1940; Sokal and Sneath 1963; Lines and Mertens 1970; Sneath and Sokal 1973; Eernisse and Kluge 1993; Kluge 1998a). However, in post-positivist philosophy, strictly empirical inductive inferences are acknowledged to be unfeasible.

The *deductive* argumentation scheme presupposes the substantiation of lower-order (particular) judgments about natural phenomena by referring to higher-order (general) ones, with the latter constituting the ontic background of cognitive situation. It has been originally founded on the natural philosophy of Platonic kind, which implied a descending cascade of causal relations between natural phenomena. All in all, the ultimate knowledge thus inferred is theory-dependent and subject-neutral; the respective judgments about particular entities have no hypothetical status, as they are considered true to the extent that the judgments about higher-order phenomena are assumed to be true and deductive logic of their inference is applied properly. Biological systematics emerged with the mastering of this scheme, and the latter dominated in it until the early 19th century, when it was implemented by several natural philosophical schools (super-organismal character of nature, Pythagorean harmony of numbers, etc.).

The *hypothetico-deductive* argumentation scheme includes a certain combination of both inductive and deductive reasoning. It implies quite a rich basic ontology, in which nature is represented as a complexly organized hierarchical system, ordered by both ascending and descending cascades of causal relations. The judgments about natural entities are inferred from the observations of empirical data, but they become scientifically sound only within the contexts of background knowledge of ontic kind, which makes them initially theory-dependent. With this, the subject's active occurrence in cognitive situation is also acknowledged, which makes scientific knowledge subject-dependent. The hypotheses about natural phenomena are tested according to the Popperian *principle of falsification* (Popper 1972). A possibility of application of this scheme in systematics was actively discussed over the last several decades initiated by the development of contemporary phylogenetics (Platnick and Gaffney 1977; Ruse 1979; Wiley 1981; Pesenko 1991; Kluge 1997a,b, 2009; Hull 1999a; Rieppel 2003; Rieppel et al. 2006; Stamos 2007; Fitzhugh 2016).

Of a more local meaning is the *abductive* argumentation scheme, which includes a non-formalized personal knowledge as a significant instrument of reasoning and therefore makes the latter rather fuzzy (Douven 2021). It is responsible for a special *flexible* kind of cognitive rationality (Masalova 2013) and is considered relevant to the inferences in systematics (Fitzhugh 2006).

With regard to the methodologies proper, they are usually summarized as a *scientific method* in its most general connotation; in classical systematics, the so-called *natural method* is usually discussed in this capacity. The most intriguing issue being addressed here is as follows: what exactly makes a method scientific; the possible answers to this question are given by the above-mentioned principle of scientificity with its concrete *criteria of scientificity*. It is evident that both this principle and respective criteria it implies are not universal, being dependent on particular philosophies of science.

Following the basic provisions of conceptualism, it seems reasonable to ground consideration of this section of pyramid of reasoning on the principle of onto-epistemic correspondence, which specific explication turns it here into the *principle of methodological* (*methodical*) *correspondence*. The latter implies that for a method to be recognized as scientifically sound, it must comply with the provisions of specific onto-epistemic framework of a cognitive situation. Another methodological prerequisite of quite general meaning involves the *principle of method effectiveness*, according to which a method should allow for a research task to be solved properly within a reasonable timeframe. The *principle of method limitation* asserts that each particular method is local with regard to its resolving power because of certain 'technical' limitations inherent to it. In our case, this principle means that there cannot probably exist some universal method that would be effectively applied to all manifestations of systematic diversity. Therefore, a special substantiation is requested for transferring any specific classificatory approach from one cognitive situation of systematics to another (say, from phenetics to phylogenetics).

Regarding the concrete methods employed in systematic research, the following principal groups thereof could be recognized at the bottom levels of the pyramid of reasoning (Pavlinov 2018, 2021a). The *sampling* methods are used at the stage of preparing the raw data for systematic research, which compose a research sample as an operational model of the studied fragment of systematic diversity. The *descriptive* methods serve as a means of recording information about a sample in the form that makes systematic research actually feasible. The *comparative* methods include analytical tools for revealing a certain structure of the research sample by comparing its elements (specimens and their characters). The *experimental* methods involve direct or indirect manipulations with these elements (for instance, hybridization of alive specimens, computer simulation). The *mathematical* methods are based on the numerical estimates of the relations of interest and on the computational algorithms. At last, the *classification* methods provide a kind of quintessence of the empirical research activity in systematics as a classifying discipline, as they implement the above-mentioned classification methodology at the operational level.

2.5 THE HIERARCHIES OF TAXONOMIC THEORIES

In order to progress and function productively and properly as a natural science branch, biological systematics should develop its own biologically meaningful theoretical foundation, which is aptly referred to as *taxonomic theory* (TT) (Sneath and Sokal 1973; Ereshefsky 2007a; Pavlinov 2011a, 2018, 2021a).[7] Following the general provisions of what is denoted as '*theory building*' (e.g., Dubin 1978; Gay and Weaver 2011), elaboration of TT constitutes one of the principal goals and means of the conceptualization of this biological discipline. The main vector of its development is generally driven by a more complete comprehension of systematic diversity, and it is directed toward the elaboration of its most representative conceptual models.

The section of systematics that directly deals with TT is called *taxonomy* since the early 19th century, when this term has been coined by A.P. de Candólle (1813, 1818). However, it is not rarely applied to the entire body of systematics (e.g., Hull 1998; Wheeler 2008; Padial et al. 2010; Williams and Wheeler 2025), and it is also used to denote any kind of hierarchical orderliness (Bursten 2020).

> It is needless to say that the conceptual standpoint adopted here is strongly opposed by empiricism, which asserts that systematics 'has no theory, being merely diagnostics and classification' (Wheeler 1929, p. 100). This position is most popular among practical systematicians, who appeal to their intuitive (personal) knowledge and pretend to realize their 'aspiration for truth without engaging any theoretical reasoning' (Stekolnikov 2003, p. 367). However, the pursuing such a non-theoretic aspiration hinders the normal development of systematics, and even throws it back to a stage of folk systematics of Indigenous people.

Surprisingly enough, contemporary (post-positivist) science has no unified understanding of what a scientific theory is or should be (Ilyin 2003; Stepin 2005; Burgin 2015). With this, *scientific pluralism*, which is acknowledged at the very basic level of scientific knowledge, implies that that not-so-long-ago dominated physicalist understanding of theory building as the only 'right' one (e.g., Stoljar 2024), is in fact but one among several possibilities (Popper 1972; Suppes 1978; Ludwig and Ruphy 2024). Such a general cognitive perspective implies the recognition of a special sort of natural science theories that are not intended to hypothesize, explain, or predict particular facts but rather elaborate the theoretical frameworks for the explorations of certain aspects of nature. Such theories are rather common for biology: for instance, T. F. H. Allen considers hierarchy theory, as it is applied to ecological systems, in such a vein (Allen et al. 2009). The same seems to hold true for TT, whose primary and most essential objective is to provide theoretical foundations for the exploration of systematic diversity (Pavlinov 2011a, 2018, 2021a).

Recollecting the three-partitioned structure of cognitive situation, in which systematics operates, and especially the above-mentioned metaphor of 'cognitive triangle', it is reasonable to consider TT as taking the latter's center. Such a metaphoric construal clearly highlights one of the key points in the understanding of the cognitive status of TT: its content is shaped coherently by all three basic components of cognitive situation and, therefore, changes with their intercorrelated historical development.

The basic structure of TT, as it is here understood, is primarily set by its consideration as a specific *quasi-axiomatic system* (Pavlinov 2011a, 2018, 2021a). The latter's elementary units are conceptions and notions to be used in building such a system and all its subsystems. Typically, an axiomatic approach implies certain rigid enough formalizations of both such conceptions and the operations on them following some standard protocol, which is a challenging task to accomplish (Suppes 1992). Therefore, with the problems of definitions of systematic notions in mind, the first step in elaborating TT as a quasi-axiomatics is to figure out a puzzling situation of interpreting cognitive status of its basic propositions. For this, two their major categories are suggested to be distinguished in accordance to how they function in a cognitive situation.

Those referring to the ontic component answer the question '*what*?'—what exactly does systematics investigate? and they can be categorized as either *quasi-axioms* or *presumptions*, depending on their conditionality (Rasnitsyn 1996, 2002; Pavlinov 2011a, 2018, 2021a). Quasi-axioms are accepted *a priori* in an 'absolute' form, which implies that their truth status is neither questioned nor verified within the conducted taxonomic research. Unlike them, presumptions are accepted probabilistically and their truth status may vary (increase or decrease) as a result of specific research. The statements referring to the epistemic component of cognitive situation answer the question '*how*?'—how exactly does systematics investigate systematic diversity? Accordingly, they are regarded as *inference rules*, and these are them that can most pertinently be called the *principles* of taxonomic research.

It is clear from the above-highlighted onto-epistemic dualism that such a partition of basic propositions of taxonomic quasi-axiomatics is not something constantly given; in fact, they are rather fuzzy and depending on the overall configurations of particular conceptual frameworks. This is especially true for the hierarchical pattern of systematic diversity, which may be viewed from both the ontic and epistemic standpoints. Accordingly, the respective statements about hierarchies may be either attributed to the diversity itself by the respective quasi-axioms/presumptions or treated as the specific methodological devices for the analyses of the structure of biodiversity. A similar reservation should be made for the quasi-axioms and presumptions: distinguishing between them is also fuzzy and contingent on the specific onto-epistemic backgrounds of the particular cognitive situations. For instance, phylogenetic definition of systematic diversity is introduced as a mandatory quasi-axiom in most of the evolutionarily interpreted TTs, whereas it is considered an optional presumption in the typological ones. With this, it seems that any judgments about homologies are always presumptive.

Any more or less well-developed theory represents a complexly organized hierarchical construct containing an array of interrelated conceptions and notions of various levels of generality. This hierarchy involves the basic TT structuration, which can be stratified into three basic categories, namely, *foundational*, *first-order*, and *second-order* ones (Tennis 2015). In more detail, the hierarchical structure of TT roughly corresponds to that of the above-discussed conceptual pyramid.

Its top position is taken by a *general* TT (GTT), which corresponds to the foundational level of theoretization in biological systematics. The primary goal of GTT is to develop a set of most general propositions regarding the respective onto-epistemic framework of the general cognitive situation in our discipline. These propositions are specified by lower-level conceptual *partial* TTs (PTs), which correspond to the first-order generalizations about biodiversity. This implies that GTT as such is not intended to address particular classificatory tasks; rather, it serves as a kind of *meta theory* that provides a substantive conceptual basis for the elaboration of lower-level PTs, which are specifically designed just for the solving those tasks. Keeping in mind one of the basic tenets of contemporary conceptualism regarding a certain specificity of the theoretical foundations in particular natural science branches, one should suppose a similar possibility of elaboration of specific versions of GTT for distinct wildlife kingdoms fundamentally differing in natural histories of their organisms. Such an idea seems to epitomize the general trend of biologization of contemporary systematics. By this consideration, we seem to approach quite naturally an idea of *taxonomic pluralism*, according to which it is quite normal for taxonomic theories to be diverse (Pavlinov 2003, 2018, 2020, 2021a). This specific pluralism is admissible to be considered a version of Mario Bunge's *integrated pluralism*, which proclaims both the unity of the world (biodiversity in our case) and the diversity of possible ways of its conceptualization (Bunge 1969).

The first-order PTs take the middle levels in the pyramid of TTs and deal with the conceptualizations of particular aspects and objects systematics deals with. Accordingly, the GTT is first decomposed into partial *taxonomic* (PTTs) and *meronomic* (PMTs) theories, which refer to two basic aspects of systematic diversity of the same name. The PTTs are in turn stratified into taxonomic theories of lower levels of generality following the particular reduction paths within the respective conceptual pyramid. The lowest levels of the latter are occupied by the operational interpretations of the conceptual PTTs, which can be referred to as second-order *operational* PTTs as specific epistemic devices for the elaboration of concrete taxonomic systems. It is important to emphasize that operational PTTs are elaborated as the lower-level partial interpretations of certain higher-level conceptual PTTs that provide the substantive contexts for the former, thus making them biologically relevant. The PMTs conceptualize all aspects of partitions of generalized 'bodies' of organisms, including reconstruction of their archetypes and body plans, elaboration of homology criteria, recognition of particular atributes (features, parts, properties, etc.) for elaboration of classificatory characters, etc.; these theories are not specified here.

According to the three-partitioned structure of cognitive situation, three prime kinds of conceptual PTTs are reasonable to recognize (Pavlinov 2018, 2024; Figure 2.2). One of these encompasses the *ontic* PTTs, which may be either aspect-based or object-based depending on how the respective manifestations of systematic diversity are recognized. The *aspect-based* PTTs address its typological, phylogenetic, biomorphic,

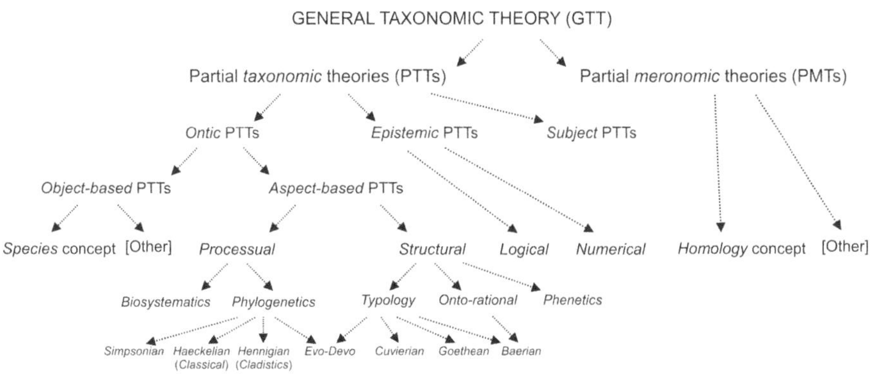

FIGURE 2.2 The overall hierarchical structure of taxonomic theories (TTs) of various levels of generality.

(Based on: Pavlinov 2024, p. S9; significantly modified). See text for explanation.

and other manifestations, whereas the *object-based* PTTs focus on the specific units of taxonomic diversity (such as species). The *epistemic* PTTs deal with the classification methodologies to be applied for the explorations of the ontically defined entities: these are, e.g., logical or numerical PTTs. At last, the *subject* (subject-oriented) PTTs emphasize the subjective component of cognitive situation, as illustrated by what may be called 'phenomenological' PTT with its paying most attention to the personal knowledge (Stekolnikov 2003; Oskolski 2007, 2011; Wilkins 2018).

* * *

Of the ontic PTTs, the aspect-based (aspectual) ones are of special significance, because they serve as the conceptual bases for the development of *research programs* in biological systematics. These programs constitute important elements of the latter's cognitive situation and therefore deserve at least a brief comment here (based on Pavlinov 2018, 2020, 2021a).

The development of the most significant of them has been stipulated by the major shifts in worldviews and/or basic methodologies in the conceptual history of systematics. When viewed in a broad and somewhat informal manner, the initial seeds of such programs have begun to come out together with the emergence of biological systematics as a natural science branch in the 15th–16th centuries. Their quite advanced versions began to take distinctive shapes in the 19th century (typology, numerology, classical phylogenetics, etc.), and they continued to evolve and diversify throughout the 20th century (biosystematics, phenetics, several versions of phylogenetics, rational, biomorphics, etc.). However, it was not until the mid-20th century that the research programs in systematics were explicitly recognized, albeit in a very reduced account of three 'contemporary systematic philosophies' by David Hull (1970).

Such a 'three philosophies' viewpoint became rather popular (Löther 1972; Hull 1988, 2001; Mayr 1982, 1988; Pesenko 1989; Minelli 1993; Schuh 2000; Ereshefsky 2001a, 2008; Rasnitsyn 2002), although it provided an oversimplified picture of the theoretical foundations of biological systematics and resulted in a very distorted image of its conceptual framework (Pavlinov 2018, 2020, 2021a). In fact, as just said, the diversity of research programs reproduces by and large the hierarchy of the aspect-based PTTs, so no less than a dozen of them may be recognized if the entire four-century-long conceptual history of this discipline is taken into consideration, and no less than six of them can be categorized as 'contemporary'. The persistence of many of the ideas that underlie them, and even the rebirth of some once-forgotten ones (such as rankless hierarchy), reflect the pluralistic nature of biological systematics, although classical monism still flourishes in it.

2.6 ON THE OBJECTIVITY AND SUBJECTIVITY OF HIERARCHIES AND RANKS

The most profound inherent character of classical and positivist philosophies of science is that a cognizing subject and a cognized object are unambiguously disconnected within a cognitive situation. This core idea is substantiated by a conviction that such a 'removal' of a subject would provide a supposedly unbiased 'objective' knowledge that reflects what the studied entities really are, and because of this is qualified as scientifically significant. Therefore, one of the most fundamental issues in epistemology of this kind is to minimize and eventually nullify the subjective impact on knowledge in order to make it 'objective'. The epistemic principles of Cartesian rationality serve as a primary cognitive device for the solving this fundamental task (Wahl 1991; Gaydenko 2003).

One of the basic demands of this rationality implies that for a natural phenomenon to be acknowledged as actually existing and 'objectively' cognizable, it must be 'physically' perceivable and measurable. This became amended by an idea, laying the foundation of contemporary physicalism, that only such knowledge about nature deserves to be counted as truly scientific if it can be expressed, roughly speaking, in a mathematical formula. This physicalist tenet, which dominated in natural science up to the mid-20th century, had a prominent metaphysical background: no other than the German philosopher Immanuel Kant (1724–1804) declared in his *Metaphysical Foundations of Natural Science* that 'in any special doctrine of nature there can be only as much proper science as there is *mathematics* therein' (cited after: Kant 2009, p. 235; italics in the original).

From this physicalist perspective, the entire classical systematics with nearly all of its conceptual constructs, including higher-level hierarchy, appeared to be interpreted as a 'stamps collecting' and turned out to be thrown out of the 'normal' natural science thus understood (Johnson 2007). Biosystematics of the early 20th century has been the first to adopt this idea, which was followed by numerical phenetics. Their adherents have unanimously agreed that all supra-species groups do not actually exist in nature altogether because of being unobservable, so any assumptions regarding hierarchy of systematic diversity were subjective and therefore should be abandoned. The nowadays proponents of cladistics, while having inherited the idea of higher-level hierarchy from classical phylogenetics, take a similar negative position with regard to its ranked character. According to this standpoint, the actually existing systematic diversity is hierarchically structured, but being the result of a presumably continuous lineage branching, its hierarchy lacks any fixed discrete levels (e.g., Eldredge and Cracraft 1980; de Queiroz and Gauthier 1992; de Queiroz 1997; Ereshefsky 2001a, 2004; Zachos 2011). Hence, the traditional ranking of taxonomic hierarchy does not correspond to anything actually existing in nature, i.e., it is largely subjective.

However, contemporary conceptualism disagrees with such an unconditional contraposition of the ontic and subjective/epistemic considerations. Instead, a mutual interrelations between all three components of cognitive situation is acknowledged, which results in the recognition of irremovable impact of the cognitive devices of knowing subject onto both the process and results of cognition (Toulmin 1972; Maturana and Varela 1980; Ilyin 2003). From this standpoint, any judgments about certain unobservable natural entities depend on the respective subject-motivated onto-epistemic premises, so it is incorrect to categorize them unequivocally as either purely objective or subjective. This means, in particular, that in the hierarchy theory, "*the relation between the structural and descriptional levels is the central problem*" (Pattee 1973a, p. 136; italics in the original). In this quotation, 'structural' refers to the objective properties of hierarchies, while 'descriptional' refers to the epistemic and subjective impacts on the judgments about them. In this regard, the opinion of Timothy Allen that '[h]ierarchy theory is particularly conscious of the observer in the system' (Allen and Giampietro 2014, p. 37) looks especially remarkable, as it clearly points out that it would be too rash for the discussions of hierarchy issues to disregard the effect of a subject's (observer's) presence in cognitive situation (see also Wilby 1994). A clear comprehension of this indissociable dualism is quite characteristic of the discussions of hierarchy issues in ecology and sociology (e.g., Wilson 1969; Pattee 1973a; Allen and Starr 1982; Ahl and Allen 1996; Floridi 2008; Wu 2013), but it is unfortunately lacking in nearly all relevant theoretical discussions in systematics (e.g., Simpson 1961; Hennig 1966; Platnick 1979; Eldredge 1985; Ereshefsky 2002, 2004; Zakharov 2005; Zachos 2011; Zuev 2015).

The contemporary conceptualism generally implies that, as far as directly unobservable complex natural entities are concerned, there is nothing like 'absolute objectivity' or its opposite 'absolute objectivity' that could be unambiguously ascribed to them. However, such their context-relative interpretation does not imply that all the entities studied by biological systematics, beginning with the similarity relations and finishing with the overall biodiversity structure, are merely some 'phantoms' having nothing in common with the objective biological reality. In fact, all of them relate to a degree to certain manifestations of really existing biotic structure and hence can be thought of as *quasi-objective*, and they represent in this capacity more or less clearly outlined cognized objects as the elements of cognitive situation. The only reservation to be definitely kept in mind is that these research objects appear within the latter as a result of specific interaction of the cognizing subjects with the cognizable entities. In this regard, the above-supposed epiphenomenal status of biodiversity and its hierarchy should be especially emphasized (see also Section 3.2).

* * *

Consideration of the above-discussed ontic reduction cascade within the general context of cognitive situation allows for emphasizing the following crucial points concerning the problem of *objective-subjective dualism* of hierarchical patterns studied by biological systematics. First, the ontic reduction is an integral part of any knowing activity, and it is a result of an intricate interactions of all three components of cognitive situation. Second, it is the cognizing subject that carries out that activity, 'launches' the entire ontic reduction cascade, defines its structure, and transforms by this the research objects into respective cognitive models. Third, it is unattainable to segregate the effects of these components, but it is possible to consider and estimate 'qualitatively' their relative contributions to the resulting knowledge. Fourth, inasmuch as the information content of a cognitive model is always less as compared to that of its natural prototype, they are always separated by a certain ontic gap, which range expands as the reduction increases. Fifths, this entails a progressive increase in the ontic gaps and a respective decrease in information content about systematic diversity in the those models while sequentially constructing a reduction cascade. Therefore, finally, the effect of subject's activity increases as the reduction operation progresses, which entails a constriction of the ratio of 'objective' and a swelling of the ratio of 'subjective' components in the systematic knowledge from top to bottom of the reduction cascade.

From a conceptual perspective, the entire hierarchical structure of reduction cascade implies that any assessment of the 'degrees' of objectivity or subjectivity of a studied object gains its substantiation within an inclusive conception of the highest level of generality, which 'explanatory objectivity' is undoubtedly acknowledged within a certain scientific community (Hempel 1966; Hull 2001). In our case, this holds true, before all, for the evolutionary process and phylogenetic pattern, which are treated as 'objective' and hierarchically arranged (Tëmkin and Eldredge 2015). One may assume the same for the ranked character of the hierarchies studies by systematics: the core question is whether this feature can be consistently inferred within a certain plausible evolutionary scenario (see Section 3.7 for detail).

On the other hand, one has to take into consideration two basic obstacles for treating hierarchies, regardless of their either rankless or ranked character, as 'absolutely' objective that come from the subjective and epistemic components. The first is the human mental 'cognitive matrix' or 'cognitive map', being a part of personal knowledge: it is presumably arranged hierarchically and makes knowing persons seeing hierarchies everywhere in nature (Brown 1991; Winthrop 1991; Atran et al. 2002; Norenzayan and Heine 2005; Pumain 2006; Lyubarsky 2018). The second is of methodological kind and includes various hierarchical classificatory algorithms, beginning from the logical genus-species scheme and ending with the contemporary clustering methods. After their applications, a question of crucial importance arises: what does in resulting hierarchies of classification systems reflect a supposed natural hierarchies and what does result from the applied methods (Lubischev 1972; Hawksworth and Bisby 1988; Cronk 1989; Pavlinov 2007a, 2018)?

Proceeding from these considerations, one can come with a certainty to the following interesting corollaries regarding phylogenetic systematics, taken as an example. It is evident that all phylogenetic conceptions, being specific cognitive models, are undoubtedly reductive, so each of them, just as undoubtedly, contains a mixture of objective and subjective components, albeit to varying degrees. When comparing basic concepts of Haeckelian classical phylogenetics and Hennigian cladistics in this regard, their respective degrees of reduction can be assessed as follows. The former, by embracing numerous properties of manifold biological evolution (both its cladogenetic and anagenetic components, parallelisms of evolutionary pathways, their adaptationist explanations, timing, etc.), is more 'objective' and therefore is more meaningful biologically. To the contrary, cladistics reduces evolution to its cladogenetic component and disregarding all others, so it is more 'subjective' and refers to something biologically negligible, like a 'spherical horse in vacuum' (Pavlinov 2024). The latter conclusion is amplified by both the Hennigian conception of *semaphoront* (Hennig, 1966; Löther 1972; Wiley 1981) and molecular genetic sequences used for inferring cladograms (phylohenetic trees) (Felsenstein, 2004) that provide oversimplified representations of organisms based on their specific onto-epistemic considerations. Such an idealized cladogenetic pattern has no more relevance to the actual evolutionary process than, say, a carefully cleaned skeleton has to the living organism: both are dead artifacts. However, ironically and paradoxically enough, cladistic taxonomic theory, with its abandoning much of the evolutionary content from its ontic background, was vociferously declared as 'truly evolutionary systematics' (de Queiroz 1988, p. 247).

The subjective nature of cladistic idea is specifically emphasized by the fact that its 'founding fathers', when having elaborated its basics, were strongly motivated by an aspiration to have such an evolutionary model that (a) would be free from the 'subjective' speculative assumptions about evolution typical of traditional approaches, and (b) would allow for the developing stable taxonomic systems maximally compatible with such a model (Pavlinov 2005; Rieppel 2016). They found the following solution to the task that quite fitted their pre-set condition: the entire multifaceted evolutionary process should be reduced to cladogenesis, its graphical representation—to cladogram, and the resulting taxonomic classification—to the hierarchy of holophyletic groups. So these were them who have initially invented a kind of 'evolutionary spherical horse in vacuum' and declared its 'objective' character. This declaration would be enjoyed by many systematicians who combine the phylogenetic idea with certain elements of positivist philosophy of science and embody this idea into a set of highly formalized algorithms of cladistic inferences.

The proponents of such a standpoint are sure that the numerical techniques, being minimally dependent on the users' subjective personalities, are 'objective' in this sense. Inspired by this assurance, they claim that application of such techniques guarantees the 'objectivity' of the resulting classifications, and this is allegedly supported by the fact that different methods provide similar results (e.g., Sneath and Sokal 1973; Gower 1988; Beckett and Lidgard 2014; Kranke 2024). However, when appealing to the latter argument, they obviously confuse an ontic objectivity with an epistemic *intersubjectivity*: the former implies maximizing of the correspondence between the prototype natural phenomena and their cognitive models, whereas the latter just aspires to the repeatability of the results of applying standardized analytical procedures to the properly standardized data (Smaling 1992; Reynolds 2016). With this, when considering the above-discussed instrumentalism and intersubjectivity jointly, one faces the problem of correct 'truthful' interpretations of the results obtained with various analytical methods. Because they are derived axiomatically, any similarity or difference in their outcomes may merely reflect similarity or difference in their axiomatic backgrounds, which has nothing to do with an assumed 'objectivity' or 'subjectivity' of those outcomes (Shatalkin 1983; see also on the logical foundations above).

2.7 THE IMPORTANCE OF BEING A SYSTEM

The basic idea of contemporary post-positivist philosophy of science asserts that the universe and its manifestations (aspects, fragments, etc.) are complexly structured and do not permit oversimplified reductionist descriptions and explanations. One of the most significant implications of this general idea is an assumption of *systemic* nature of all compound entities that science deals with, be they material or ideal. Its acceptance requires a certain reorganization of the cognizing human mind in order to replenish it with a specific *system thinking*, being a variant of holistic thinking (Bertalanffy 1972; Checkland 1988; Anderson and Johnson 1997; Ruse 2009; Weinberg 2011; Hester and Adams 2014). This makes the basic units of scientific explorations not consisting of isolated elementary entities of any kind, but their more or less integrated and structured complexes (Mella 2009; Weinberg 2011). This has led to an increasing interest among theoreticians in the properties of such complexly organized entities that make them *the* systems (Laszlo 1972a, 1972b; Pattee 1973a; Salthe 1985, 2012). Several versions of the systems theory, suggested equally as 'general' (Bertalanffy 1969; Laszlo 1972a; Rapoport 1986; Urmantsev 1988; etc.), constitute a conceptual 'core' of systems science, which is sometimes related to 'systems philosophy' (Laszlo 1972a, 1972b; Klir 2001), including systems biology in its broad sense (O'Malley and Dupré 2005; Kesić 2016). Although differing in detail, they agree on some fundamental points of our interest, which are briefly summarized here.

In its most general understanding, a system is such a compound entity that (a) is *structured*, i.e., consists of an array of orderly interrelating (interacting, interconnecting) constituents (parts, elements, etc.) and, due to this, (b) is *integrated* to the degree that it functions and develops as a whole. A system's structure can most generally be interpreted as shaped by specific interrelations between its constituents, so any systemic object is rightly thought of as *relational* entity. A system wholeness means that it is endowed with some immanent quiddities that makes it something 'more' than just a sum of its constituents. It is both differentiated and integrated by specific flows of both/either matter and/or energy and/or information both within system itself among its constituents and between system and its environment (Bertalanffy 1969; Pattee 1973a; Pumain 2006). Its constituents participate in these flows as *interactors* or *duplicators* (*replicators*) (Hull 1980; Heyes and Plotkin 1988; Zylstra 1992; Tëmkin and Eldredge 2015): the former are involved in the matter and energy flows, whereas the latter transmit information (e.g., mental, cultural, conceptual, genetic) both within and out of a system.

With increasing of the number in a system constituents and their interrelations, it acquires some additional emergent features and is specifically referred to as a *complex* one (Simon 1962, 1991; Laszlo 1972a, 1972b; Pattee 1973a; Auger 1989; Bar-Yam 1997; Pumain 2006; Holland 2014). Edgar Morin is sure that 'any system, whatever it might be, is complex by its own nature' and that a non-classical 'paradigm of complexity <…> would impose a *principle of distinction* and a *principle of conjunction*', according to which 'the knowledge of the parts is not enough, the knowledge of the whole as a whole is not enough, if one ignores its parts', so only 'the relation of whole-part mutual implication' can provide an actual comprehension of a system (Morin 2007, p. 10; italics added). However, understanding of the complexity includes a significant portion of subjectivism: 'Complexity may lie in the structure of a system, but it may also lie in the eye of a beholder of that system' (Simon 1976, p. 508). 'Complex systems are perceived by people as complex because their large number of interacting components resists easy description and understanding' (Wu 2013, p. 298). So in general, 'Truly complex systems [are those that] almost by definition evade complete and detailed descriptions' (Mesarovic and Macko 1969, p. 30).

Considered from a certain perspective, three major types of systems were suggested to be distinguished, namely, *small-number*, *medium-number*, and *large-number* ones (Allen and Starr 1982; O'Neill et al. 1989; Weinberg 2011). A small-number system allows for its description by enumerating the properties of its individual components, which is typically exemplified by an idealized biological organism (archetype) described through enumeration of its organs (merons).

A medium-number system contains too many components to be described with regard to their own properties but too few to obtain their reliable averaged characteristics. At last, a large-number system is the one that is allowable to be characterized by average values of the properties of its components. In biological systematics, this categorization is relevant to the taxa of various levels of generality: the lower-level ones (species, etc.) are normally characterized as large-number systems by averages, whereas the higher-level ones (classes, etc.) are characterized as reasonably small-number systems by lists of their diagnostic features. With this, however, all of them can be treated as medium-number systems, which specific characters can be formulated as 'most usual' or 'as a rule' properties of their components.

The systems that interchange with their surrounding environment by matter, energy, and information are called *open systems*. In turn, a special class of open systems are *non-equilibrium* (dissipative) systems; all living and many non-living open systems are of this kind. According to the so-called *Prigogine's theorem*, one of basic for non-equilibrium thermodynamics studying such systems, the latter exist far from a point of thermodynamic equilibrium due to 'exporting' their own entropy into their environments, which makes it possible for them to become more complicated and structured, including hierarchical arrangement (Prigogine 1981; Prigogine and Stengers 1984; Demirel 2002). This general model is demonstrably exemplified by biological evolution: from the perspective outlined here, this notion refers to the historical development of biota as an open non-equilibrium system, which implies its progressive complication at various levels, which entails its structuration and hierarchization (Brooks and Wiley 1986; Gladyshev 2017). In particular, this involves complication of organisms that allows them to participate more effectively in the complexly organized ecosystemic interrelations as both integrators and replicators (Brooks and Wiley 1986; Salthe 2015).

* * *

It is commonly acknowledged that one of the crucial features that makes a specifically structured entity *the* system is its hierarchical structure, or simply *hierarchy*. As said above, any complex system possesses certain inherent all-encompassing features that are extensionally equipotent to it and do not permit their simplified representations, and one of these features is system hierarchy. Therefore, both a complex system and its hierarchical arrangement are considered so intimately interconnected (Bertalanffy 1969) that H. Simon almost synonymizes them, when writing 'a hierarchic system, or hierarchy' (Simon 1962, p. 468). However, as always, this general declaration is not necessarily universal: for instance, the author of the book *The Future of Complexity* does not even mention hierarchy as something deserving special consideration in the context of complexity (Baofu 2007).

> Hierarchically structured complex systems are almost everywhere—in nature, in society, in mind, and in knowledge about all these. In systematics, they are of paramount importance, because it has so happened historically that both the impression of a hierarchically arranged Nature and the use of hierarchical algorithms to describe it and to represent it in the hierarchical classifications have dominated in this natural science branch since its very beginning.

The overall structuredness of a complex hierarchical system is determined by its *near-decomposability* into certain components, which can be *vertical* and *horizontal* (see Section 3.1). The former involves system near-decomposition into *partial hierarchies*, and the latter involves its near-decomposition into different *levels* (layers, etc.) of whatever kind (generality, perfectness, structural, conceptual, etc.). Layered hierarchy is characterized by a specific sequential orderliness of its levels, so that the interrelations of hierarchy constituents are naturally differentiated into *between* and *within* levels. The between-levels interrelations are of primary importance for considering a set of respectively interrelated (coupled) entities constituting a hierarchical system (Simon 1962, 1973, 1991; Bertalanffy 1969; Salthe 1985, 2012; Ahl and Allen 1996; Wu 2013). Therefore, a set of the so-called Chinese boxes (or Russian matryoshka doll), usually considered a paradigmatic instance of nested hierarchy, is not a system thus understood, as these constituent boxes (dolls) do not interact functionally. From this viewpoint, not all hierarchies are systems in strict sense. However, according to another view of hierarchies, they may not necessarily imply vertical functional (causal) relations between their constituents; spatial or eventually other 'non-functional' relations are enough (McShea 2016).

The latter point closely concerns classificatory hierarchies, as far as the taxa/merons allocated to their different levels do not interact functionally. However, they do interact through information flows that integrate them into classification systems. Therefore, such an 'informational' interpretation of their interactions suggests a possibility for the taxonomic hierarchies to be regarded as a peculiar case of the systems hierarchies.

* * *

There are many kinds of systems that depend on the specific 'systemic views' of the world, which address particular features of the above-mentioned flows, both integrating and differentiating them, particular features of their hierarchical structuration, particular exploratory tasks addressing its various features and manifestations, etc. Without delving into this complex and multifaceted issue and taking into account that our consideration involves mainly exploration of biodiversity, two approaches to categorizations of systemic entities and their hierarchies are sufficient to be recognized here as the most significant for our purposes.

One of them implies distinguishing between primarily ontic and epistemic systems biological systematics deals with. The *ontic* systems are assumed to belong to the 'external' worlds existing out of any cognitive situations, and these systems can be conventionally denoted as *physical* in the broadest sense, including those belonging to the biological and social realms, as well as to Platonic, Pythagorean, and other ideal worlds. The *epistemic* systems belong to the 'internal' worlds created by human creative activity, these are basically *mental* systems that include any subjective images of the physical systems. The mental systems deal with the information about physical systems; all *knowledge* systems (including cognitive, see below), which are of our prime interest, are of this sort.

Another categorization is based on the specific interactions within and between systemic objects, which are of two kinds: in some of them, their interactions involve exchanges of matter + energy + information, whereas in others only information exchanges are involved (Miller 1978). Respectively, the systemic objects of the first kind can aptly be denoted as *matter + energy systems*, and those of the second kind as *information systems* (Langefors 1977; Buckland 1991; Grötschel and Lügger 1999; Gregor 2006; D'Atri et al., 2008; Ogiela and Ogiela 2012; Moshkov 2019). Some information systems, considered with regard to their specific way of fixing and transmitting information, are termed *symbolic* or *descriptive* (Simons 1962; Pattee 1973a; Salthe 1985; Ahl and Allen 1996). They are most often discussed within the contexts of computer and management technologies, but classification systems that include only the names of taxa seem to belong here, as well. It is important for our discussion to emphasize that the information systems are assumed to be endowed with most of the characteristic features of open non-equilibrium complex systems (Laszlo 1972a; Maturana and Varela 1980; Ogiela and Ogiela 2012; Oseledchik et al. 2017).

> The general information concept has very diverse meanings, from usually employed broad common-sense to the most rigid mathematically and physically defined ones. It constitutes a focal point of the general information theory, which is as heterogeneous in its contents as the general systems theory (e.g., Brillouin 1962; Ash 1990; Capurro and Hjørland 2003; Ursul 2010). However, there is one important point of the conceptual consensus, which is of our concern, namely, a close relation between the notions of information, knowledge, and certainty. Accordingly, the term 'information' is used here to refer, by and large, to a specific knowledge about some research object. With this, it is assumed that the more information contained in the latter's cognitive model, the higher the model's adequacy to it and the more certain the knowledge about it.

* * *

The information systems and their hierarchies are of prime interest to us. They are not uniform with respect to the function they carry out and the ways of ordering the information flows passing through them. This makes it possible to differentiate these systems by their levels of generality and arrange them in a specific conceptual pyramid. The physical and mental information systems are to be distinguished at their highest level; of those studied by systematics, genealogical hierarchies are of physical kind, whereas all their representations are of mental kind. Among the latter, again two basic kinds can be recognized, which are aptly referred to as *organizational* (*management*, *control*) and *knowledge* systems (Grene 1969, 1987; Mesarovic and Macko 1969; Pattee 1973a; Zylstra 1992; Cash et al. 2003; Bergman 2018, 2022). This dichotomy, as always, is not discrete because of the existence of some information systems that can be categorized as 'something between' these two by their properties and partly overlap with them, examples are provided by *linguistic* and specific *knowledge organization* systems (Hurford 1987; Beghtol 2000; Holland 2014; Hjørland 2016; Kleineberg 2017; Mazzocchi 2018; Szostak 2020). In this regard, it is worth noting that any knowledge systems, including classifications, can be interpreted as partly linguistic and partly knowledge organization ones (Chebanov and Martynenko 1998).

Among various kinds of knowledge systems, the most important for our consideration are *cognitive* systems, which are developed deliberately and almost exclusively for performing just the cognitive functions by encompassing both the procedures and results of the knowing activity. Science in general deals with just such systems, which function as specific cognitive models of the studied objects. In particular, any properly developed cognitive situation may be regarded as a high-level cognitive system, whereas its constituents (reduction cascades, conceptual pyramids, particular conceptions, cognitive models, classifications, etc.) are lower-level systems of such kind.

With regard to the information flows in such systems, in order to comprehend how they are organized, one should bear in mind that any cognitive system is generally arranged by a cognizing subject. Accordingly, these systems differ from organizational ones in that the former's constituents are involved in the information flows not directly but through the subject's mediation. In our case, the information flows in them correspond to the relations between organisms, which can be kinship, similarity, etc., and it is the subject that (a) chooses among particular relations to be used as a foundation for the elaborating a particular cognitive system and (b) analyzes the relations of interest and evaluates the closeness of organisms to be interpreted in the terms of density of respective information flows (see below).

Based on this consideration, it is reasonable to introduce a fundamentally important *principle of irremovable congizer's* (*obsever's*) *mediation* in the arranging and functioning of cognitive systems, including those elaborated by systematics. Respectively, using the above-mentioned Hull's (1980) categorization, the constituents of such a system that participate in the information flows should probably be treated as 'duplicators', whereas the subjects that mediate these flows should be recognized and termed specifically as *mediators*. This implies rather peculiar character of cognitive systems, according to which they can be regarded as a special kind of *panarchic* systems (in the sense of Gunderson and Holling 2001; Gunderson et al. 2022). The latter notion originally referred to the compound functional interactions between ecological and social systems, but its relevance to the cognitive systems looks quite justifiable. Accordingly, a comprehensive explanation of the cognitive systems, including analysis of their causes, development, structure, and function, should take into account both this 'mediation' principle and their panarchic interpretation.

> In this regard, cognitive systems are somewhat similar to the works of art (literary, musical, etc.). Indeed, the latter are meaningful to the degree they are interpreted due to a specific co-creative activity of the interested subjects (readers, listeners, etc.). So in fact, any compound union 'composition + interested subject' is a specific panarchic system (Polanyi 1968, 1998; Mesarovic and Macko 1969)

Another important point that should not be omitted is that the cognitive systems are *evolving* systems; this general idea (Heyes 2003, 2012; Kenrick et al. 2003; Asano and Seifert 2014; Thagard 2023) evidently holds true for those systems that biological systematics deals with. The latter's cognitive situation, along with all of its constituents (theories, conceptions, classifications, etc.), has been developing for centuries following certain regular trends (Pavlinov 2018, 2021a). The 'conceptual environment' that softly direct and shape them includes changing general onto-epistemic contexts, which open new promising and close obsolete perspectives of this conceptual development, just as biological evolution is externally regulated by the opening new and closing old adaptive zones. In this regard, one can reasonably borrow an idea of *evolvability* (in the sense of Kirschner and Gerhart 1998; Minelli 2017; etc.) from biology and consider it an inherent property of any cognitive situation, which generally means its capacity to generate novel conceptions to renew or replace old ones.

* * *

The particular cognitive systems developed by biological systematics are of different kinds depending on how the structure of systematic diversity is interpreted and represented (modeled) by them. In this regard, two essentially different perspectives of that structure are distinguished, *quantitative* and *qualitative* (Whitehead 1925; Broadfield 1946), and the relevant information about the structure in question is aptly generalized by two major types of cognitive systems, *parametric* and *classification* (Hempel 1966; Rozov 1995; Subbotin 2001). In the case of predominately one-level (linear) relations, as assumed by the Scale of Being, the respective cognitive system is parametric, while in the case of multi-level hierarchical relations, like in the System of Nature, it is classification.

Parametric systems accentuate the quantitative regularities in the relations between particular variables (parameters of interest), characterizing the studied diversity, and they generalize these regularities by certain formulas. When such regularities are 'formulized' in one way or another, any deviations from their central trends are deemed insignificant and therefore disregarded. In the case of systematic diversity, the most characteristic example of this approach is the *species–area effect*, which links numerically the species richness within a habitat to the latter's area (MacArthur and Wilson 1967; Ricklefs and Schluter 1993; Wu and Vankat 1995). Parametric systems are elaborated by onto-rational schools of systematics, which look for certain linear or periodic regularities in systematic diversity (Lubischev 1923; Popov 2008; Pavlinov 2011c). If different parameters contribute differently to the ordering of the studied systematic diversity, they are weighted (ranked) according to their significance, with the most heavily 'weighted' ones being denoted as the *key* parameters. If these parameters are correlated in such a way that the key one indirectly determines the values of the others (secondary), such a parametric system is denoted as *correlative* (Lubischev 1923).

Unlike parametric, classification systems[8] in their most general sense focus on the qualitative (categorical) structure of the diversity of things (including organisms), which results from their interrelations (kinship, similarity, spatial and temporal, ecological, etc.) and is revealed by a conjoint analysis of the relevant variables. The 'supertask' of classification systems (or simply classifications) is to encompass the studied diversity in its entirety without reducing it to any 'central' trend.

Following his rather 'formal' consideration of the ways, with which any set of objects can be ordered categorically, Graham Griffiths distinguishes between

> *classifications* and *systematizations*: the former refer to the logical ordering into classes, and latter order objects based on their certain 'systematic relations' (Griffiths 1974). Following this argumentation, Eric Knox declared the term 'cladistic classification' to be oxymoron (Knox 1998), which was suggested to be denoted as *cladification* (Mayr and Bock 2002; Hörandl 2010).

Several kinds of hierarchical classification systems are recognized based on specific consideration aspects of the structural diversity of organisms. Systematics mostly focuses on *taxonomic* systems, which represent groups of organisms arranged into taxa (genera, families, ecomorphological groups, etc.) based on their similarity and/or kinship relations. *Meronomic* systems classify the organismal properties based on their homology and other relations, in typology they are elaborated as the hierarchical archetypal representations of idealized organisms. The taxonomic-meronomic dualism results in the intersections of these two aspects, which yields the integrated (combined) *taxono-meronomic* systems that encompass taxa together with their diagnostic characters. Phytosociology (phytocoenology) represents structure of plant communities by specific *synaxonomic* systems, whose basic elements are *synaxa* (Mirkin 1985, 1989). Stratigraphy and biogeography explore structural aspects of biodiversity based on the temporal and spatial relations of organisms, respectively, and this results in specific cognitive systems of partonomic kind, which are sometimes denoted as not classifications but *zonations* (Meyen 1977, 1989). As already mentioned above, partonomic interpretation of classification systems is characteristic also of a holistic vision of systematic diversity (organicism, typology, partly phylogenetics, etc.).

> It should be noted that as-if 'taxonomic' systems elaborated based on the isolated organismal attributes or parts (even their traces), are in fact largely meronomic, which is especially true for the paleontological and molecular genetic studies. Such a 'meronomic' accentuation is most characteristic of typological approaches, in which archetype (body plan, etc.) hierarchies are first developed, and afterward they merely interpreted in taxonomic terms. A book by the Russian zoologist-typologist Vladimir Beklemishev (1890–1962) vividly illustrates this point: although entitled *Methodology of Systematics* and qualified as a significant contribution to the taxonomic theory (Lyubarsky 1996; Epstein 2002), it almost entirely focuses on the problems of typological (comparative morphological) reconstructions, whereas the basic taxonomic conceptions (taxon, rank, character, etc.) are not discussed at all (Beklemishev 1994; originally written in the 1920s).

In addition to these major variants of classification systems, *categorization* systems should probably be singled out, which differ from parametric and classification ones in the ways of recognition of the groups of objects, including organisms and their features (Rosch 1973, 1975, 1988; Gardner 1985; Jacob 1991, 2004; Kleineberg 2017). According to the just-cited authors, classification systems of any kinds are elaborated based on certain explicitly formulated principles of delineation and hierarchization of the system constituents. Unlike this, categorizations are based on the non-formalized intuitive perceptions of things, which are grouped (and by need hierarchized) according to intuitively grasped *prototypes*; in this capacity, categorization is characteristic of folk systematics.

* * *

One of the most important conceptions employed for the description of the orderly structured diversity is that of *natural* system, which is conventionally opposed to that of *artificial* one. This fundamental dichotomy are demarcated specifically in different cognitive contexts, though in different manner. According to one of the traditional lines of reasoning, the regularly ordered physical systems (in the above-indicated broad sense) are treated as natural and the mental ones as artificial (e.g., Laszlo 1972a, 1972b; Holland 1992, 2014). This point has been explicitly recognized by C. Linné, who separated two major kinds of classification systems by treating them as 'products of Nature' and 'products of Art', respectively (Linnaeus 1751). Another tradition understands natural systems in quite a broad manner, summarized by systemologist Ervin Laszlo as being 'characterized by <…> nonrandom regularity of the coactions of their components' (Laszlo 1972a, p. 30), which implies that the components of artificial systems are presumed to coact 'irregularly'. This definition of naturalness is context-dependent for its indirect assuming the reference to a certain parameter, with respect to which the coactions in question can be interpreted as 'regular' or not. In this regard, such an understanding of natural system seems to liken that of *natural kind* in its philosophical sense suggested by John Mill (e.g., Quine 1994; Koslicki 2008; Hawley and Bird 2011; Slater 2015; Kendig 2016). To emphasize this idea, the Dutch botanist Herman Lam (1892–1977) specifically distinguished between two major kinds of taxa in systematics: *nattaxon* is a concrete natural group of organisms, and *nomtaxon* is an abstract nomenclatorial unit (Lam 1959).

As far as biological systematics is concerned, the dichotomy of natural and artificial classification systems has originated from the old natural philosophical idea of *Systema Naturae* as an overwhelming law that somehow orders Nature (such as divine plan of creation of biblicists). Following this tradition, which C. Linné has expressed in his aphorism 'Nature is the behest of God' (*Natura est jussum Dei*, Linnaeus 1749, p. 546), some of nowadays experts regard classifications as natural, if they are supposed to reflect adequately an assumed 'natural law' (Löther 1972; Zarenkov 1988; Zabrodin 1989; Vityaev and Kostin 1992; Rozov 1995; Mahner and Bunge 1997; Wilkins and Ebach 2014); accordingly, artificial classifications do not correspond to any of such laws. The problem here, as always, is that there can be assumed several 'natural' (as well as 'supranatural') laws that structure systematic diversity differently, so there can be different particular systems, each 'natural' with regard to a particular implied 'law'. Acknowledging this fact entails *taxonomic pluralism* in its conceptual account (Pavlinov, 2003, 2018, 2020, 2021a).

This basically natural-philosophical idea of the naturalness of classification systems was suggested to be operationalized by the reference not to the enigmatic 'laws' of whatever kind but to the explicitly defined specific *criteria of naturalness*, which may imply any assumed regularities (Gilmour 1940; Michener 1957; Sokal and

Sneath 1963; Sneath and Sokal 1973; Rozova 1986). Accordingly, the classifications that are elaborated in line with such criteria are deemed natural, whereas all others are artificial. These criteria are potentially numerous, so there can be more than one operationally defined natural classifications for the same group of organisms, and this serves as one of the major sources of the *taxonomic uncertainty* (see Section 3.5 on it). For instance, one of such criterion, ascending to the ideas of the above-mentioned A.P. de Candólle and the English logician John Mill (1806–1873), defines natural classifications as agreeing in several essential traits (Mill 1882; Lubischev 1923; Meyen and Shreyder 1976; Ahl and Allen 1996). Such classifications are proposed to be termed *general purpose* ones, and opposite to them are *special purpose* classifications (Gilmour 1940; Michener 1957; Sokal and Sneath 1963). One very important kind of the latter are *identification keys*, elaborated as algorithms for taxonomic identification of organisms (Lobanov 1972; Payne and Preece 1980; Tilling 1984; Sviridov 1994); they are sometimes not regarded as the classifications proper (Bather 1927; Simpson 1961).

* * *

The philosophers and systemologists have mostly focused so far on the physical (material, social, etc.) systems, so the general theme of the systemic treatment of cognitive systems remained poorly investigated. Therefore, many important questions concerning their structure and functions were not only unanswered but even not aroused. Two of them are discussed here, as they seem to be quite relevant to the basic topic of our consideration.

One of these general questions concerns the interrelations between cognitive situations, treated as open non-equilibrium information systems, and their 'environments'. The latter were once suggested to be defined by cognizing subjects, who are 'signaled' by certain 'realm[s] of physical-events' to be perceived and 'conated' (Laszlo 1972a, p. 120). Using formalisms employed here, two major kinds of such 'environments' should be recognized. Some of these are shaped by the conceptual contexts of a cognitive activity composed of higher-level onto-epistemic ideas of various kinds. It is not especially clear what kind of 'conceptual disorder' can be exported from a developing cognitive situation into those ideas. Others are 'material environments', which in our case can be imagined as including systematic diversity of organisms, structured by the latter's interrelations (kinship, similarity, etc.). With them, the issue of disorder exportation seems clearer.

Indeed, the ultimate end of the development of any cognitive situation in systematics involves the elaboration and enhancement of classification systems as cognitive models of that diversity. Such systems should be deemed self-developing (self-stabilizing, self-structuring, etc.) due to the 'importation' of information about the studied diversity from nature. This results in an increase in both information capacity and orderliness of the systems in question, thus making them (presumably) more adequate representations of organismal diversity. With this, to become more ordered, the developing cognitive situations should 'export' their initial interior disorder to their 'material environments'. Identifying the latter with the natural systematic diversity, one of the possible answers may be as follows. According to the traditional practice of taxonomic research, any improvement of classifications implies most usually the removal of the respective organisms from their natural habitat. These organisms make up a studied sample, which serves as an operational source of new information to be 'imported' from nature into classifications to enhance their orderliness. Hence, one can imagine that it is the removal of organisms from nature that represents a specific form of disorder, which is 'exported' to the 'material environments' (natural diversity) by the developing cognitive situation in systematics.

Another intriguing issue involves the cognitive system *boundaries*, or *surfaces*, or *closures* (Simon 1962; Platt 1969; Wilson 1969; Laszlo 1972a, 1972b; Maturana and Varela 1980; Allen and Starr 1982; Ahl and Allen 1996). Putting apart such general issues of this concern as the socio-historical boundaries of any such systems (e.g., Bryant 2000; Mai 2004, 2009; Rozova 2014) and using formalisms employed here, one can say that the subject-mediated information flows, corresponding to the above-indicated inter-organismal relations, shape all interplays within the systems proper and between them and their 'environments'. As said above, the density of these flows corresponds to the closeness of the studied relations. Accordingly, it seems quite sound to speculate that the boundaries of a cognitive (including classification) system are defined by a certain threshold value of the respective information flows density. Those exceeding the specified threshold are *internal* and shape the within-system flows network that integrates the organisms (taxa) in the respective classifications; in particular, they are responsible for the *level boundaries* (closures). Those below this threshold are *external* and correspond to the relations between system constituents (taxa) and the groups of organisms outside of classifications, they outline the *system boundaries* (closures). Such boundaries, considered as 'cognitive surfaces', can be thought of as *semi-permeable* (in the sense of Laszlo 1972a; Hurford 1987) for their being endowed with specific 'cognitive filters', allowing only those information flows to pass through them, which correspond to the relations fixed by an explorer (say, a similarity by selectively weighted characters). Of importance is that the similar semi-permeable boundaries with filtering function should be supposed for the hierarchical levels within a classification system that make them *quasi-autonomous* due to the weakening of the information flows across the inner boundaries. It is to be added also that as far as both the internal and external boundaries of cognitive systems are defined in terms of information flows, they are probabilistic and therefore fuzzy.

One more important issue arises from the consideration of classification systems as open non-equilibrium ones, which concerns their adequacy as cognitive models of certain fragments of systematic diversity. An intriguing point here is that, according to the relevant thermodynamic theory, the orderliness of such systems can grow due to their intrinsic (endogenous) patterning alone. In our case, this is because the content of a classification system, including its hierarchical structure, results from the interactions of all three components of cognitive situation. Keeping in mind a significant joint contribution of epistemic and subjective components to the resulting classifications, the latter's higher orderliness can be achieved without direct alignment with an assumed structure of systematic diversity. In particular, a classification can be 'improved' due to the application of several classification methods according to the methodological *principle of convergence* (e.g., Young 1971; see Section 3.6), which is a usual approach applied in numerical taxonomy (Sneath and Sokal 1973; Felsenstein 2004).

Such odds imply that, when evaluating and comparing the representation capacity of a particular classification system, a thorough examination of the contributions of each of three components of cognitive situation is necessary in order to comprehend the actual causes of its hierarchical structure. This consideration is consistent with the opinion of those conceptualists who give special attention to the interaction of ontic and epistemic components when studying any complex systems (Pattee 1973a; Auger 1989; Gasparski 1994), with an important addition that the subjective component should also be taken into account (Allen and Starr 1982; Ahl and Allen 1996; Pavlinov 2018, 2021a).

Let us consider in this regard, how the overall structure of an evolutionarily interpreted taxono-meronomic system, based on the same general ontic idea, can be variably shaped. For this, let us first assume that the evolutionary process, being a cause of the whole diversity of organisms, involves two basic trends, cladogenetic (divergent, branching) and anagenetic (parallel and convergent, unidirectional). The resulting systematic diversity is structured by the kinship (genealogical) relations and two kinds of similarity relations, inherited (within each of the phyletic lineages) and quasi-independent (parallel or convergent, between different lineages). Let us now agree that of two above-mentioned major 'arrows' of the information flows within a classification system, the vertical one corresponds to the kinship and inherited similarity, whereas the horizontal one corresponds to the quasi-independent (acquired) similarity. The resulting combinatorics of these flows, which shapes the overall hierarchical network structure of the taxono-meronomic classifications, reflects the ratio of anagenetic and cladogenetic trends in the studied phylogeny, and this multiple combinatorics results in the variety of evolutionary interpreted classifications. The ontic evolutionary models, underlying them, can be either mostly anagenetic (Cope's version) or mostly cladogenetic (Hennig's version) or combining both (Haeckel's version). Considered epistemically, they result from the application of particular classificatory algorithms (hierarchical, ordination, or combined). Finally, considered subjectively, they depend on the choices of either of those models by the researchers, which are determined by the personal preferences. Obviously, in such a situation, in order to properly investigate the diversity of the resulting classifications and to understand its causes, it is necessary to carefully analyze the contribution of each of these three components to the results of taxonomic research.

2.8 A FEW WORDS ABOUT STRUCTURALISM

Hierarchies are inherent attributes of complexly structured systems, and to the degree that the latter are relational entities, the same holds for their hierarchies. They represent specifically structured layered relational entities composed of two basic kinds of relations, within and between layers. So they can probably be sufficiently described and explained based on the consideration of these two kinds of relations between system constituents.

A possibility and desirability of such a specific 'relational' account of systemic objects is profoundly substantiated and expressed by the philosophy of *structuralism*

(Piaget 1970; Bokulich and Bokulich 2011). Its basic presumption was rather informally expressed by the French mathematician and philosopher of science Henri Poincaré (1854–1912), who said in his *Science and Hypothesis* that 'the aim of science is not things themselves <...> but the *relations* between things; outside those relations there is no reality knowable'. In our case, this implies that the proper conceptualization of the hierarchies in information systems as relational entities can be achieved mainly in the relational terms, without considering intrinsic properties of their constituents (organisms and their features).

Generally speaking, the structuralist view of the universe, including the Earth biota and its manifestations and aspects, suggests that there are specific compositional laws that structure systemic objects in certain patterns, and the latter, given some level of the objects' complexity, involve their hierarchical orderliness. This orderliness is non-random because of the existence of certain constraints in the organization of systems that make some of their states more probable than others. Accordingly, structuralism accentuates more on *internal* causation of the orderliness rather than on *external* one (Denton 1986, 2016); such a specific structuration was supposed to be regarded as resulted from a *systemic causation* (Tchaikovsky 1990). This ontic version of structuralism is supplemented by an epistemic one, according to which it is the *structuralist thinking* style ('way of seeing') that serves as a powerful heuristics aimed at and allowing for discovering specific structural orderliness in the universe (Foucault 1970; Piaget 1971; Greene 1984; Hughes and Lambert 1984; Reck and Schiemer 2025).

With regard to systematic diversity, its structuralist consideration was first explicitly developed by typologists and partly organicists during the late 18th and the early to mid-19th centuries, and their line of reasoning was significantly refined by the next generations of typologically minded biologists (Naef 1931; Troll 1951; Riedl 1978; Meyen 1978; Stevens 1984; Vasilieva 1989, 2001; Amundson 1998; Beklemishev 1994; Lyubarsky 1996; Webster and Goodwin 1996; Zakharov 2005; Lewens 2009; Riegner 2013). The general typological model of living nature implies the following basic assumptions: (a) there exists a stable overall hierarchically structured system of general body plans (bauplans, archetypes, etc.) of living beings, (b) its orderliness is determined by certain compositional relations of various levels of generality, (c) their totality is thought of as a specific universal 'relational' law of living nature, (d) particular organisms are variants of the respective body plans, (e) so that their entire diversity is ordered in a specific hierarchical fashion by those compositional relations. Organicists appeared to go the furthest with this structuralist model by identifying the whole living nature with a super-organism, which implied the former's hierarchical structuration in a similar manner to that of the latter, so the taxa in an organicistically inferred total classification system are related to each other just as the body compartments in a generalized organism. Regarding the law-like status of typological patterns, this general idea was quite expressly articulated by the German natural philosopher and poet Johann Wolfgang von Goethe (1749–1832) in his *Versuch die Metamorphosen der Pflanzen*, for whom particular plants related to the general archetype 'just like particular instances relate to a law; they are contained in it but they do not contain it or give it' (cited after: von Goethe, 1957, p. 76).

This basically stationary account is organically combined with a dynamic one, which provides one of the most profound conceptual models of the universe. As highlighted above, this implies consideration of the latter as a global self-developing non-equilibrium (dissipative) system, with its development inherently involving its hierarchical structuration (Prigogine and Stengers 1984; Prigogine 1997). The most fundamental structuralist addition to such a dynamic consideration is that it suggests non-random transitions of self-organized non-equilibrium systems from one states to others, with all such states representing non-accidental system configurations due to the existence of certain relational constraints.

An application of this general idea to biological evolution implies the latter's ordered character, in which some transformations and resulting organismal constructions are more probable than others. Such a standpoint is formalized as a *process structuralism* and shapes a kind of *evolutionary typology*, which links phylogenetic process with structural regularities in a specific manner (e.g., Agassiz 1859; Naef 1931; Griffiths 1996; Amundson 1998; Gayon 1998; Vasilieva 1998). With this, the historical development of the structural (typological) aspect of the diversity of the living is not merely a successive realization (implementation) of the predetermined body plans, which would make this version of typology superficially similar to Platonic eidology (see Section 1.3 on it). In fact, such a development supposedly proceeded as the sequential emergence of such plans, which developed as the adds-on to those that had been evolved before, in a similar manner to how the developing organisms acquire their definitive structures based on the preceding embryonic stages. This fundamental idea has ben first formulated by the Baltic German naturalist Karl von Baer (1792–1876) (von Baer, 1828; see also Abzhanov 2013), it was incorporated in Haeckelian phylogenetics by a metaphor of developing genealogical individual (Haeckel 1866, 1895; see also Rinard 1981; Porges et al. 2019), and now it is largely supported by *evo–devo* concept (Minelli 2003; Shishkin 2006; Laubichler and Maienschein 2008; Lewens 2009; Hall 2012; Riegner 2013; Brigandt 2017; Jaeger 2019). Among other things directly pertinent to the systematic diversity issues, the evolutionary typology model emphasizes the key role of *evolutionary* (*phylogenetic*) *inertia*, which is responsible for what in other terms is called *phylocreod* or *canalized evolutionary potential* (Waddington 1962; Meyen 1984; Amundson 1998; Wagner and Stadler 2003; Pavlinov 2005; Burt 2008; Shanahan 2011). This inertia makes parallelisms an essential component of evolutionary processes and, as far as systematics is concerned, substantiates thereby the natural status of paraphyletic groups (Brundin 1972; Saether 1983; Cronquist 1987; Brummitt 1996; Hörandl 2006, 2010; Stuessy and Hörandl 2014; Willner et al. 2014; Carter et al. 2015; Pavlinov 2015, 2018, 2021a). So it is this general content-rich evolutionary model that actually provides a 'truly evolutionary systematics'.

NOTES

1 It should be reminded that the general notion of classification frequently refers to both the process and the result of application of the classifying procedures (Voronin 1985; Parrochia and Neuville 2013; Parrochia 2016). The second understanding is adopted in this book throughout, whereas the procedure is referred to as classifying.

2 In mathematics, the representation function carried out by any kind of cognitive models is conventionally termed as *mapping* (Jech 2003; Simmons 2011).
3 Keeping in mind that this text is addressed to systematicians, this general understanding of the concept/conception should not be mixed with the notion of 'taxonomic concept' as applied in a technical sense to particular taxa, which is typical for bioinformatics (e.g., Geoffroy and Berendsohn 2003; Berendsohn and Geoffroy 2007; Franz et al. 2008; Franz and Peet 2009; Franz and Thau 2010).
4 This diversity is routinely termed 'taxonomic'. It is termed here 'systematic' (taxonomic *sensu lato*) to distinguish it from the taxonomic (*sensu stricto*) diversity as opposed to meronomic diversity, both being considered manifestations of the same systematics diversity.
5 It was initially denoted by me as 'partonomic-taxonomic'.
6 These two terms were suggested to replace the former *monotypic* and *polytypic*, respectively, which have another meaning in contemporary systematics (on this terminology, see Beckner 1959; Williams and Dale 1965; Sneath and Sokal, 1973; Pavlinov 2021). Monothetic and polythetic definitions of taxa hold also for inductive logic.
7 It should not be confused with the 'classification theory' of mathematical logic (Baldwin 1988), 'theory of taxonomy' of information science (Muntermann et al. 2015; D'Amico et al. 2016), and other similar highly formalized generalizations.
8 See also (Hjørland 2017) on them; not the same as *classifier systems* in the sense of (Urbanowicz and Browne 2017), who refer to a specific function of cognitive systems in IT projects.

3 The Hierarchies as They Are

Hierarchy theory is as much as anything a theory of observation.
Timothy F. Allen (2001)

The general notion of hierarchy is of polysemous 'nomadic' kind (in the sense of Minelli 2022) by referring to quite different entities. Hierarchies can be general and partial, inclusive and non-inclusive, physical and mental, conceptual and empirical, and taxonomic and partonomic; they consist of the levels of generality or integration, organization, or consideration (interpretation), etc. With this, there are certain basic features in common to all manifestations of hierarchies regardless of their kinds—such features that are indicative of the recognition of *the* hierarchies among all possible variants of orderliness.

Uncover and summation of both general features and variety of hierarchies are the responsibility of what should be called *hierarchy theory*, which in turn is a part of *systems theory*. The latter's basics were briefly outlined in the previous chapter (see Section 2.7); regarding the former, the success of its development has not been especially significant so far, so we have instead an array of particular 'microtheories', as Mario Bunge (1969) called them. Of importance is that according to the systemist-'hierarchist' Timothy Allen, 'hierarchy theory is a special sort of theory that may not meet criteria for what many would have theory be. <…> It does not make predictions per se, but rather explicitly extracts the functional structure of the system from the data, rather than relying on an arbitrary designation of components' (Allen et al. 2009, p. 2939). In this regard, the (still absent) general hierarchy theory seems to be quite similar to the general taxonomic theory, as it was understood by the author (Pavlinov 2011a, 2018, 2021a; see Section 2.5).

As shown in the 'historical' chapter of this book, biological systematics has begun with the mastering hierarchical classificatory methodology of scholastics. However, it so happened that now it turned out to be falling behind some other branches of science (ecology, the humanities, partly mathematics and physics, etc.), which are more advanced in their studies of general properties and features of hierarchies. Therefore, in order to gain a thorough understanding of the hierarchies that our discipline addresses, it is essential to compose a comprehensive and detailed profile of 'the hierarchies as they are' based on the relevant ideas and conceptions developed in those disciplines that are more advanced in this area.

This chapter attempts to answer the most fundamental questions about hierarchies: what are they in general and what are their specific manifestations, with focusing, quite naturally, on those studied by biological systematics. It begins with the summarizing basic features of hierarchies, which is followed by a special consideration

DOI: 10.1201/9781003597094-3

of their onto-epistemic dualism. Their general patterning is next discussed, including topological features of hierarchies, their leveled architecture, and overall fuzzy character. The next follows a review of the general principles of hierarchization of classification systems that biological systematics elaborates. The chapter concludes with the author's view of the currently debated controversy regarding ranked *vs.* rankless taxonomic hierarchies.

3.1 BASIC FEATURES OF HIERARCHIES

One of the most important characteristics of the overall system hierarchy affecting all or most of its other features is its *near-decomposability*. The respective conception initially addressed the recognition of distinct sequentially arranged levels within a hierarchy (Simon 1962, 1973, 1991; Mesarovic and Macko 1969; Wilson 1969; Mesarovic et al. 1970; Allen and Starr 1982; Floridi 2008; Wu 2013), but in fact near-decomposability has two mutually 'orthogonal' aspects. One of these is '*horizontal*' near-decomposability of the general hierarchy into *partial hierarchies*, and another is their '*vertical*' near-decomposability into *hierarchy levels* (layers, scales, steps, strata, etc.). In this notion, the '*decomposability*' implies that both partial hierarchies and their levels are individuated as particular research objects, whereas the '*near-*' implies that they remain components of the same overall hierarchy. It is the distinctive 'vertical' feature of systems structuration, as in a layer cake, that entails their overall hierarchical arrangement, so their *layered stratification* or *laveling* (in the sense of Floridi and Sanders 2004; Floridi 2008)[1] can aptly be synonymized with their *hierarchization*.

* * *

The basic features of hierarchical structures of systemic entities are defined by *hierarchization principles*, which can be general or particular. The *general* principles are relevant to all and any hierarchies by defining their universal properties, including their topology. One of such principles sets hierarchies *relatively* or *absolutely*, another regulates their *leveling*, one more defines them as *inclusive* or *non-inclusive*, a specific general principle makes hierarchies *directed* or not, etc. The *particular* principles are implemented by *hierarchization parameters*: they substantively define (delineate) partial hierarchies by indicating specific factors that structure systemic objects, namely, matter and/or energy and/or information flows between system constituents. In systematics, the partial hierarchies can aptly be regarded as the attributes of partial aspects of systematic diversity, such as meronomic and taxonomic, and typological and phylogenetic, so the hierarchization parameters play an important role as specific exploratory 'devices' for the recognition of these aspects.

Referring to systematics, the parameters in question are of two major kinds, interior or exterior, and they are operationalized by particular *hierarchization rulers* to be applied for the hierarchical arrangements of classifications. The *interior* parameters are mandatory and include the features (attributes) and relations of biodiversity constituents, which participate directly or indirectly in the just-mentioned flows, whereas the latter refer to the requisite 'frame' conditions for the functioning of these constituents. For the purposes of our consideration, two major groups of interior parameters should be recognized. The *hierarchization relations* are the specific relations (similarity, kinship, ecological, spatial, etc.) between organisms and groups thereof; they specify and structure both partial hierarchies studied by systematics and taxonomic systems representing them. The *hierarchization of characters* refers to the organismal features (merons) and inter-feature relations (such as homology); their operational representations serve as an empirical basis for the elaboration of particular classifications. As is seen, the parameters of the first group concern the taxonomic aspect of systematic diversity, while the parameters of the second group concern its meronomic aspect. The *exterior* parameters and respective rulers cannot be attributed to the systematic diversity proper; they are optional and include geological timing scale and/or adaptive zonation (eventually some others), and they serve to arrange hierarchies in certain 'absolute' fashions.

The hierarchization characters are commonly referred to as *classificatory characters*: they participate in the elaboration of taxonomic systems as specific *comparison units* (Zarenkov 1976; Pavlinov 2018) The characters are near-decomposable into their own constituents, which are typically denoted as their *modalities* (states, elements, etc.). The characters are responsible for an overall structuration of classification systems, whereas their modalities delineate particular taxa in the latter.[2] The characters are discrete in the sense that the same modalities cannot belong to several characters, otherwise they are considered partly tautologous (Sneath and Sokal 1973; Sereno 2007). In this regard, they are logically independent, although substantively they can be (and often actually are) interdependent to an extent. Their modalities are logically interdependent in the sense that within a particular character, they all are delimitated and arranged in relation to each other.

All procedures concerning hierarchical arrangements of cognitive systems (including classifications) are based on the *hierarchization scenarios*, which summarize and specify particular hierarchization principles and parameters/rulers to be applied. This notion is introduced here as analogous to the evolutionary scenario, which has been proven to be effective for both exploration of phylogenetic processes and substantiation of phylogenetic taxonomic systems (Eldredge and Cracraft 1980; Wiley 1981; Pavlinov 2005). The scenarios are elaborated within particular cognitive situations based on the provisions of the latter's onto-epistemic backgrounds, which makes them both context- and subject-dependent.

One of the most significant instruments of hierarchization scenario involves certain interrelations between hierarchization relations and hierarchization characters, which can be denoted as their *concordance*. The latter has two aspects that differ in meaning and application: one involves interrelations within each of the arrays of hierarchization characters and relations, and another refers to the correspondence between these two distinct arrays. In biological systematics, this issue is of great relevance in both its aspects, addressing all tasks regarding the inferences of cognitive models for systematic diversity. The first aspect implies, roughly speaking, a certain correlation (a) between partial hierarchies and taxonomic classifications (categorizations) reflecting them, and (b) between particular hierarchization characters in their classificatory meaning. The second aspect implies a certain concordance between partial hierarchies and hierarchization characters, which is referred to as *taxonomic-meronomic correspondence* and plays the key role in the hierarchization of classification systems (see Section 3.6).

* * *

In the case of biodiversity, its aspect-based structuration begins with the identification of systematic and some other (ecological, geographical, etc.) partial hierarchies, with the former being further near-decomposable into taxonomic and meronomic hierarchies, and the taxonomic one into typological, phylogenetic, and other hierarchies.[3] All partial hierarchies, into which an overall system hierarchy is near-decomposed, are *equipotent* and *internested*, i.e., they equally encompass all constituents of the same complex system, and such consideration is important in two respects. On the one hand, it allows for referring to each of them as a 'system hierarchy' without specifying its partial character (e.g., ecological or phylogenetic or else). On the other hand, their equipotency serves as a prerequisite for the possibility for an overall system hierarchy to be inferred, with certain reservations, through the simultaneous analysis of the partial ones.

The multiplicity of both general and especially partial hierarchization principles makes any complex system potentially and actually *polyhierarchical*, which is sometimes denoted as *heterarchy* (Crumley 1987, 1995, 2005; Bondarenko et al. 2002; Dekker and Kuchař 2021).[4] Its alternative presumes the only possible hierarchization of a system, it is denoted as *monohierarchy* or *homoarchy* and is characteristic of the social systems of totalitarian kind (the same references). An overall multiplicity of partial hierarchies, into which the overall one (such as that of biodiversity) is near-decomposed, constitutes a kind of multi-dimensional *hierarchy hyperspace*, which dimensions correspond to the specific partial hierarchies (Ravasz and Barabási 2003). This formalism is useful in that it makes all such hierarchies, in principle, commensurable with respect to at least some of their essential features.

* * *

As already has been said, the general structure of whatever hierarchy is defined by two basic kinds of its inherent relations, which makes hierarchy a *relational* entity; hierarchy 'inherits' this fundamental property from the relational nature of any systemic object, of which it is an attribute. Regarding hierarchy proper, these relations are either within or between their levels, they are sometimes called 'horizontal' and 'vertical', respectively (this terminological pair was applied above in another sense). They constitute in total a complex *hierarchical network* (Ravasz and Barabási 2003; Trusina et al. 2004), which is typically treated as 'making' the very hierarchy of whatever kind (Simon 1962; Mesarovic and Macko 1969; Wilson 1969; Mesarovic et al. 1970; Allen and Starr 1982; O'Neill et al. 1986; Floridi 2008; Wu 2013).

The overall orderliness of partial hierarchies can be set *relatively* or *absolutely*. In the first case, the levels within each hierarchy are ordered strictly interdependently: the position of any one of them depends on the positions of some others. In the second case, they are ordered more or less independently of each other, as each is defined with the reference to a exterior hierarchization parameter. With this, it is assumed that the latter should be applied uniformly to all hierarchy levels, which makes them *commeasurable*. Such an 'absolutely' established hierarchy is supposed to be admissible for the level-independent analyses and is suggested to be denoted as *stratified* (Mesarovic and Macko 1969). However, it was supposed that the highest and lowest levels of any given hierarchy are set always 'absolutely', whereas its intermediate levels are set always relatively (Starobogatov 1989).

The hierarchies can be arranged *deductively* or *inductively*: the former are constructed from upper to lower levels, while the latter proceed in the opposite direction. These options are ontically defined by the directions of major matter/energy/information flows that structure the respective systems. They are defined epistemically by the logical system of the same name and operationally by hierarchization algorithms applied, such as those developed in numerical taxonomy (e.g., Sneath and Sokal 1973). In some formalized systems, the deductively inferred arrangement is denoted as hierarchy proper, whereas the inductively inferred one as *anti-hierarchy* (Sneppen and Minnhagen 2004; Corominas-Murtra et al. 2013).

* * *

It remains to consider very briefly some formal relations between sequentially arranged hierarchy levels/steps, which are borrowed by systemists from mathematical set theory (e.g., Hrbacek and Jech 1999; Jech 2003). These relations between the levels A, B, and C are *transitive*: if $A \supset B$ and $B \supset C$, then $A \supset C$; *non-symmetric* (sometimes called *antisymmetric*): if $A \supset$ B and $A \neq B$, then $B \not\subset A$; and *reflexive*: $A \Leftrightarrow A$, $B \Leftrightarrow B$, etc. All these relations hold true for the system constituents allocated to the respective levels. Transitivity and non-symmetry are the most significant for our consideration, as they contribute to the nonequivalence of the respective levels/steps (see Section 3.4).

3.2 MORE ON ONTO-EPISTEMIC DUALISM OF HIERARCHIES

Following the classical 'discrete' line of reasoning, Donna Wilson recognized two major kinds of hierarchies, which were given ontically as 'hierarchies in nature' (from elemental particles to the universe) and epistemically as 'hierarchies in mind' (cognitive or any other mental models); Wilson added to them also 'hierarchies in artifact' (from computers to human organizations) (Wilson 1969; Wu 2013). Such a near-discrete distinction is characteristic of many research on hierarchies; for instance, Timothy Allen distinguished between two major kinds of leveled structure of ecological hierarchy, namely, *levels of observation* and *levels of organization* (Allen and Starr 1982; Ahl and Allen 1996; Wu 2013).[5]

However, a conceptual standpoint, while acknowledging this dichotomy, rejects its rigid, discrete character based on the most fundamental idea of onto-epistemic relativism (Quine 1969; Dupré, 1993). Therefore, prior to discussion of the onto-epistemic status of hierarchies in the context of three-partition structure of cognitive situation, it is essential to highlight the following crucial points. On the one hand, hierarchies are actually characteristic of each of ontic, epistemic, and subjective components: they are attributes of both the biodiversity itself (its own hierarchical structure), its various cognitive representations (reduction cascade, conceptual pyramid, classification systems, etc.), epistemic means (methodologies and methods), and subjective mental properties (worldviews, thinking styles, skills, etc.). On the other hand, the hierarchies themselves are distinct research objects, which are represented in the cognitive situation of systematics by their cognitive models, with the latter being specific outcomes of the subjectively mediated intricate interactions of its ontic and epistemic components. This implies that considered from a conceptual perspective, the hierarchies, when encompassed into a cognitive situation, become endowed with a certain *onto-epistemic dualism*: they are assumed as something really existing in nature and, with this, individuated as specific research objects by a knowing subject by means of applying certain epistemic instruments. Accordingly, when discussing onto-epistemic status of hierarchies with all their specific features, one has to take into account that they are not 'given' as such but are context- and subject-dependent, and therefore, their status may vary from one cognitive situation to another. For instance, the nominalist and realist perspectives differ drastically in their treatments of the nature and structure of systematic diversity and its hierarchical orderliness.

This just-outlined standpoint suggests that any considerations of hierarchies, including their own structure, depend on the basic premises that constitute the onto-epistemic frameworks of the respective cognitive situations. This is explained in general by an active cognitive position of knowing subjects, who select specific hierarchy models to be imposed onto studied systemic objects to describe their complex structure (Simon 1962; Mesarovic and Macko 1969; Pattee 1973a; Allen and Starr 1982; Ahl and Allen 1996; Wu 2013, 2014). For instance, Mario Bunge, before suggesting his formalisms concerning leveled hierarchies, emphasized that he referred primarily to their evolutionary understanding, thus demarcating his specific ontic standpoint from those presumed by the stationary worldviews of Antique natural philosophy or classical physics (Bunge 1969). It is noteworthy that T. Allen goes with his emphasizing the active cognitive role of 'hard-line hierarchists' so far as to declare hierarchy theory to be a special kind of 'theory of observation' (Ahl and Allen 1996).

Such a cognitive perspective addresses the most essential issue of the objectivity *vs.* subjectivity of the hierarchies, which has no so that simple dichotomous solution as suggested by the approaches based on classical binary logic (see Section 2.6 on this). This key point was repeatedly highlighted by nearly all authors who discussed cognitive aspects of the hierarchies inherent of physical, biological, social, and other systems (e.g., Wilson 1969; Pattee 1973a; Allen and Starr 1982; O'Neill et al. 1986; Ahl and Allen 1996; Bunge 1969; Floridi 2008; Wu 2013; Child 2019). This conclusion holds evidently true, in particular, for the hierarchical levels: it is not so much easy, if possible at all, to separate unambiguously the structural levels *per se* and the levels of observations in those complex systems whose hierarchical structure is more conjecturable (as in biota) rather than directly observable (as in biological organisms) (Mesarovic and Macko 1969; Pattee 1973a; Allen and Starr 1982; Ahl and Allen 1996). Such a specific onto-epistemic dualism implies that the consideration of hierarchies of directly unobservable complexly structured natural entities, including biota, should be, in theory, given conjointly from both ontic and epistemic perspectives.

* * *

However, in the concrete explorations the hierarchies are most usually approached from either of these two standpoints, depending on the subjective preferences with respect to particular cognitive attitudes. Realists tend to think of hierarchies as objective natural phenomena, whereas nominalists regard them as cognitive artifacts, and conceptualists take a standpoint 'somewhere between' (though frequently categorized as nominalists). Therefore, according to this latter point of view, one of the key issues in the discussions of hierarchies becomes *objectivation* of natural (non-mental) hierarchies within the context set by their combined onto-epistemic consideration.

From an ontic perspective, all such hierarchies and their constituents and levels can be assumably treated, with certain reservations, as 'natural' and by this 'objective' to the degree that they are delineated with the reference to certain causes not directly related to the human cognitive activities (Simon 1962, 1973, 1991; Mesarovic and Macko 1969; Pattee 1973a; Patten 1975; Salthe 1985, 2012; O'Neill et al. 1986; Ehresmann and Vanbremeersch 1987; Ahl and Allen 1996; Pumain 2006; Wu 2013). Such a starting standpoint necessitates searching for the causes of partial hierarchies. Considering the subject area of biological systematics in its full natural-philosophical and historical capacities, two major categories of causation are feasible to be briefly discussed in this regard, supra-natural and natural.

The Antique natural philosophical systems explained the overall natural hierarchy of things (including organisms) by the action of generative power of either Platonic *One* or Aristotelian *Unmoved Mover* (Balaguer 2025; Bodnar 2025). These versions of causation was refined in a specific way by biblical theologians, who replaced both the *One* and *Unmoved Mover* with the single divine plane of creation; such a viewpoint was most expressly formulated by the above-mentioned Linnaeus' maxim '*Natura est jussum Dei*'. Of a similar kind was the world picture of many typologists of the 19th and 20th centuries (Cuvier, Owen, Troll, Vasilieva, etc.), who considered meronomic hierarchies of general body plans (archetypes, bauplans) really existing in the ideal world and causing taxonomic hierarchies of organisms in the material world. The classical biblical causation is specified by contemporary *baraminology* (e.g., Marsh 1944; Schafer 2003): God created *baramins* (Aristotelian forms) of various levels of generality, which implies the reality of hierarchies of their 'physical' embodiments, including units of biodiversity.

The materialistic standpoint fundamentally rejects such idealistic causation; though, its advocates seem to forget that their own standpoint, when viewed from a certain 'abstract' angle of conceptualism, is framed in a manner essentially similar to the one they oppose, i.e., by a specific world picture they believe to be true. Anyway, the materialistic perspective was specifically actualized in the 19th century and later by an evolutionary world picture, which was developing in two basic versions. The *macro*evolutionary view of the historical development of biota (Haeckel, Hennig, and all their followers) suggests that the real (objective) phylogenetic process of sequential splitting of lineages generates the real (objective) omni-leveled hierarchically arranged structure of systematic diversity now referred to as *phylogenetic pattern*. Unlike this, the *micro*evolutionary view (Darwin and his strict followers) supposes an objective status of only lowest-level (infraspecies) hierarchy. The so-called synthetic theory of evolution (STE) extended the second viewpoint to species by suggesting particular biological mechanisms of species identity (e.g., Dobzhansky

1935; Simpson 1961; Mayr 1982, 1988). The cladistic version of phylogenetics acknowledges the objective nature of the hierarchy proper but disputes the objectivity of its fixed levels (ranks), referring to a supposedly continuous character of the branching pattern of phylogeny (see Section 3.7).

* * *

One of the most significant onto-epistemic features of multifaceted hierarchies of whatever kind is their *systemicity* which has at least two general meanings. On the one hand, this implies, first and foremost, that any hierarchy does not exist by itself, but it is always a 'hierarchy of something' and cannot be alienated from the latter. This 'something' in question refers to any systemic object (process, etc.) in its broadest sense (natural, social, cognitive, etc.), whose own structure admits its treatment as a complex system. On the other hand, this feature refers to the systemic properties of the hierarchy proper, which makes itself a specific systemic entity. This implies, as said above, that hierarchies are orderly structured by specific interrelations between their constituents and levels. With this, inasmuch as hierarchies are assumed to be extensionally equipotent to the systemic objects they are attributed to, their orderly structuredness corresponds to that of the entire objects, so both the latter and their hierarchies are deemed completely concordant in this regard. Accordingly, the systems' constituents are also their hierarchies' constituents.

Keeping in mind the distinction between small-, medium-, and large-number systems (see Section 2.7), it is reasonable to recognize in a similar manner *local-scale*, *mesoscale*, and *global-scale* system hierarchies (Jonas 1994). Their scaling rulers can themselves be of different scale, if the entire cognitive situation of systematics is meant. Indeed, its overall hierarchy is global-scale, hierarchy of taxonomic theories developed within it is mesoscale, and hierarchy of particular classification systems is local-scale. Proceeding from the idea of a two-fold basic near-decomposition of biodiversity, either object-based or aspect-based, it seems more than reasonable to acknowledge aspect-based manifestations of systematic diversity as mesoscale partial hierarchies. Of them, the most significant are taxonomic and meronomic hierarchies, and further, particular manifestations of the former, such as phylogenetic, typological, and other hierarchies. On the other hand, considering only classification systems of living beings, their overall hierarchy (from *Imperium* downward) is global-scale, the hierarchies of macrotaxa (say, from *Classis* downward) are mesoscale, and the hierarchies of lower-level taxa (say, from *Genus* downward) are local-scale.

The research objects of biological systematics, into which systematic diversity is near-decomposable, differ in their integrity. This implies that they can be arranged in a specific linear hierarchy according to an imaginary 'integrity scale'. For the needs of our discussion and using the notional system adopted in this book, such a scale can be roughly coarsened to three major steps or levels. The most-cohesive whole-part hierarchical relations of a holistically treated entities imply their *partonomic* partition; the least-cohesive class-subclass (set-subset) entail their *loose* partition; the middle-cohesive relations within classification systems imply the latter's *systemic* partition that takes position between these two.

* * *

The above-discussed presumably epiphenomenal status of biodiversity is relevant to the hierarchy as its systemic attribute. In this case, such its treatment is implied indirectly by a special emphasis placed not on the hierarchy itself, but on its levels. Indeed, philosophers and naturalists, be they Platonic thinkers with their successive steps of emanation of the *One* or Pseudo-Dionysial celestial levels, contemporary naturalists Herbert Simon with his near-decomposable levels (Simon 1962), Timothy Allen with his distinction between organization and observation levels (Allen and Starr 1982; Ahl and Allen 1996), or Arthur Koestler with his 'holon' as an integrating principle for organization levels (Koestler 1968; Pichler 2000; Mella 2006, 2009)—all of them were and are actually about the levels first and the whole of hierarchies second. In other words, these are the certain levels of whatever kinds (organization, consideration, inclusiveness, perfectness, etc.) within the orderliness of any complex systemic entities that are first recognized, whereas hierarchies appear secondarily as their specific consolidations into the overall leveled orderliness (Ahl and Allen 1996; Allen 2001, 2008; Wu 2013).

Considered in such a vein, hierarchy of whatever substantive meaning can be thought of as a *derived macroparameter* (= macroscopic variable, in the sense of thermodynamics) of a complex system, which is defined by the interactions between the latter's levels, whereas the system constituents allocated to these levels can reasonably be considered as the system's inherent *microparameters*. An analogy with the 'colorless' universe was provided above to explain this idea (see Section 2.6); rainbow and atmospheric temperature can serve as more analogies. In such an account, the hierarchies themselves do not seem to have any emergent features of their own that would be independent of the characteristic features of the system constituents, which are ascribed certain levels of generality (inclusive hierarchies) or 'progression' (linear hierarchies). Accordingly, it seems reasonable to regard just those systems constituents as real phenomena (in a broad sense) and to treat the hierarchies as systems *epiphenomena*. With regard to the hierarchy levels, their ontic status within the dichotomy of phenomena/epiphenomena is not as certain as that ascribed to the systems constituents and hierarchies. One cannot exclude that such a dichotomy, as usual, is too simplistic to accurately describe the complex reality systematics deals with, and that the hierarchy levels take a position 'something between' these two extremes and could be categorized as *quasi-phenomena*.

At any rate, such a theorizing implies that when discussing the hierarchical structure of systematic diversity, we must (a) first clarify whether any levels can soundly be supposed or not as particular systemic quasi-phenomena, (b) if they are, explain them by reference to their supposed natural causes, and (c) then think of hierarchy as a certain epiphenomenal principle for integrating such a layered orderliness in a specific way. For instance, Carl Linné concentrated on the explanation of the levels of systematic diversity in a Platonic manner by referring to the successive implementation of divine plan of creation; evolutionists refer to the distinct organizational levels of biota, at which specific evolutionary mechanisms can act; typologists refer to the structural levels of archetypes or body plans (meronomic orderliness) that provide for the levels of generality of taxonomic diversity; positivists reject any metaphysic consideration of supra-organismal levels and, consequently, the leveled hierarchy altogether.

And yet, however, it is just the overall hierarchy itself that is usually taken as the most characteristic inherent attribute of any complex systems. This is probably due to the fact that the hierarchies, as already said, are assumed to be extensionally equipotent to the systems they are attributed to, and therefore they can justifiably be interpreted as the latter's integrated (synthetic) characteristics, unlike particular levels that are thought of as their local (analytical) characteristics.

This consideration seems to be relevant to another issue regarding onto-epistemic dualism of the hierarchies and their constituents. Specifically, it concerns the separation of the hierarchies *per se* and the propositions about them conjectured in concrete explorations. According to contemporary scientific epistemology, these propositions should be treated as hypotheses that are formulated and tested according to specific rules (Popper 1972; Stepin 2005). Regardless of the latter's contents, which vary dependent on some specific epistemic premises, the first and the most important question addresses an ontic issue, namely, *what* such hypotheses are about. Given the interpretation of hierarchy as an epiphenomenon and provided that any hypothesizing refers to the multiple and not individual objects, the relevant hypotheses should address not the hierarchies themselves but rather the ordered sequences of hierarchy levels recognized within a complex system (see below on ternary relations).

3.3 THE HIERARCHIES TOPOLOGY

In contemporary science, representation of hierarchies by tree-like diagrammatic graphs is especially popular, so their basic features, including those referred to as *hierarchy topology*, are typically characterized by using terminology borrowed from graph theory (e.g., Bollobás 1998; Semple and Steel 2003; Bondy and Murty 2008). It is to be noted that such terminology concerns primarily (in our terms) *connection* trees, whereas *categorization* trees remain mostly without consideration (see Section 4.2 on their distinction). A minimal set of the formalisms most relevant to the main theme of our book is as follows.

A tree-like figure is generally denoted as *tree graph G* or just a *tree*. It is formally defined as an ordered pair [$V(G)$, $E(G)$], where $V(G)$ are *vertices* (*nodes*) and $E(G)$ are *edges* connecting them, with edges being ascribed *incidence function* ψ_G, which defines how the edges connect the vertices. Such a graph is most usually represents a connection tree drawn in a simple *line format*. A collection of trees ordered according to the same hierarchization principle is denoted as *forest*; in systematics, such 'forests' usually result from the application of numerical methods to the same data sets.

The vertices connected directly by one edge are *adjacent* (neighbor, contiguous); an edge connecting them is referred to as a *hierarchization step*; the edges that sequentially connect several vertices constitute a *tree path* or a *tree branch*, which is sometimes denoted as a *chain*. Within a particular edge, its starting vertex is termed *original* or '*parent*' and its ending vertex is *derivative* or '*child*'. An edge that connects two vertices is *outcoming* with regard to the original vertex and *incoming* with regard to the derivative vertex. A vertex with no incoming and at least one outcoming edges is called tree *root*; the vertices with one or several both in- and outcoming edges are *internal*, those at branch tips with at least one incoming and no outcoming edges are *terminal* or *leaves*. The number of both incoming and outcoming edges attached to a vertex defines its *degree*, which is applied also to the hierarchy pathing. In systematics, the tree vertices are usually called *nodes*, which are identified with taxa/merons. The tree edges are identified with the latter's interrelations, which are usually interpreted as the affinities of any kind: similarity, kinship, spatial proximity, etc. The incidence function refers, among other things, to the directions of these interrelations, and they define the branch/tree *directedness*.

* * *

As just said, the hierarchy topology seems to be the most general characteristic of the hierarchy structure of any systemic entity and, of course, a tree representing it. A hierarchy is composed of the interrelations between basic hierarchy elements, which are its levels, constituents allocated to them, and their interrelations; in a tree, these are vertices and edges. It is to be noted that tree representation differs topologically from hierarchy *per se* in that the former consists only of vertices and edges, whereas hierarchy levels are not explicitly specified in it. Inasmuch as systematics deals basically with the information systems, it is possible to equate their hierarchy topology with what is customarily denoted as the *information architecture* in computer science (e.g., Morville and Rosenfeld 2007). Anyway, this characteristic is formed by several partly interconnected hierarchy features discussed below.

One of very important formal topological features of a tree representing hierarchy of any kind is the so-called *treeness*, which reflects its condition of 'being a tree', i.e., simply speaking, being similar to a natural tree or its shoot. This feature is characterized by the relations between the vertices and edges: for a graph to be regarded structurally as a tree, it should be both *connected* and *acyclic*. The first provision implies that the vertices in a graph are interconnected by the respective edges, and the second one implies that there are no vertices connected to themselves by loops. In systematics, of special importance are *partially connected acyclic* trees, in which (a) not all vertices are interconnected in the 'all-to-all' mode and (b) there are no isolated (unconnected) vertices. This integrated feature can be evaluated quantitatively as a ratio of 'vertical' and 'horizontal' relations within a graph (Carbonell-Nicolau 2025). The *treeness tests* are applied in numerical taxonomy for 'measuring' a tree-like topology of inclusive hierarchies (Sneath and Sokal 1973; Cavalli-Sforza and Piazza 1975; Astolfi and Zonta-Sgaramella 1984; Gordon 1987; Fisler et al. 2020); several specific measures are elaborated for the evaluation of organizational hierarchies (e.g., Luo et al. 2009; Corominas-Murtra et al. 2011; Mones et al. 2012; Saaty and Vargas 2022). Different trees of the same 'forest' can be compared numerically, and several trees can be generalized by the so-called *consensus tree* or *super-tree* (Gordon 1980, 1987; Leclerc 1985; Sanderson et al. 1998; Bininda-Emonds et al. 2002; Bininda-Emonds 2004).

The near-decomposability of the overall system hierarchy 'horizontally' into partial hierarchies and 'vertically' into hierarchy levels provides the major contribution to the hierarchy topology by determining the basic character of its overall patterning. Partial hierarchies can *overlap* and therefore are *not discrete*, which means that the same system constituents (such as organisms) can belong to several partial hierarchies recognized for the same system based on different hierarchization parameters, e.g., typological and phylogenetic. Another well-known instance is provided by the multiple taxonomic systems elaborated for the same group of organisms due to usage of different characters or application of different methods. Unlike this, the hierarchy levels within the same path *do not overlap* and are *discrete* in this sense; or rather *near-discrete*, if their fuzzy probabilistic interpretation is taken into account (see Section 3.5 on this). Their discreteness means that the same entities (taxa and merons in systematics) cannot be allocated to two adjacent levels (always $A \neq B$). However, according to some authors, such levels can be considered *overlapping* in the case of extensionally defined monotypic taxa (Gregg 1950, 1954). The entities allocated to the same hierarchy level are ordered by an *equivalence relation* and are normally assumed to be discrete (non-overlapping) (Gregg 1954; Jardine et al. 1967; Jardine and Sibson 1971). However, the latter strict demand is introduced axiomatically for the classical logical systems, and it may be weakened by allowing for such entities partially overlap (Sklar 1964; Jardine and Sibson 1968, 1971), which is also supported by their fuzzy interpretation.

The nature of interrelations between hierarchy levels differentiates hierarchies topologically into two major categories, namely, *non-inclusive* (exclusive, linear, flow, and component) and *inclusive* (nested, encaptic, containment, and aggregative); their terminology is not stable (e.g., Simon 1962; Pattee 1973a; Allen and Starr 1982; Mayr 1982; Aho and Ullman 1995; Valentine and May 1996; Salthe 2006; Wu 2013; Marume and Chikasha 2016). In the case of non-inclusive hierarchy, its non-overlapping levels are linearly (sequentially) ordered in a *chain*, and each of system constituents is allocated to only one of these levels. Such specific kind of hierarchy was suggested to be denoted as *scalar*, which is derived from the 'scale' referring to a ladder with several steps (Salthe 2006; Marume and Chikasha 2016). The inclusive hierarchy implies a sequential nestedness of its levels within the same hierarchy path. In such a hierarchy, the levels are differentiated by their *levels of generality* and are arranged in such a manner that higher-level ones include (encapsulate) lower-level ones; the same holds for the system constituents that are allocated to the particular levels.

> These two major hierarchy categories are most demonstrably illustrated by the standard military organization: its units of various levels of generality or scales (from army and corps to battalion and squad) form an inclusive hierarchy, whereas their respective commanders (from marshal and general to captain and sergeant) are arranged into a non-inclusive hierarchy. In biological systematics, the inclusive hierarchy is instantiated by the standard taxonomic systems traditionally termed 'Linnaean', whereas differently weighted classificatory characters constitute linear non-inclusive hierarchy.

The hierarchy *directedness* is one of its key characteristics, with the incidence function ψ_G, referring to either presence or absence of specific *directions* imposed on the edges (tree branches). Accordingly, a tree can be either directed or undirected depending on whether or not directedness is specified for it.

The *undirected* (*unrooted*) tree corresponds to the non-hierarchical structures, in numerical taxonomy it is usually called *minimum spanning* tree or *network* (e.g.,

Sneath and Sokal 1973; Wiley 1981; Kitching et al. 1998; Felsenstein 2004; Huson and Scornavacca 2011; Wiley and Lieberman 2011; Francis et al. 2018). In contemporary numerical phyletics, with its emphasize on the epistemic *parsimony principle* (Farris 1982, 1983; Wiley et al. 1991; Kitching et al. 1998; Brower and Schuh 2021), the inference of undirected trees (networks) constitutes an important intermediate step of phylogenetic reconstructions, and their transformation into the directed ones implies fixation of the roots according to certain rules (e.g., Wiley 1981; Pavlinov 1990a; Page and Holmes 1998; Francis et al. 2018; Penn et al. 2023; see below).

The *directed* (*rooted*) or *oriented* tree is the one in which directions are determined for at least some of its edges. In such a tree, a *root* is typically fixed, and its position defines directedness of the entire tree. It is generally defined by an *affinity scale*, which is based on any of the above-mentioned hierarchization rulers. Such a scale poses a kind of 'affinity arrow' onto an undirected tree by determining its basal hierarchy level, which is identified with the tree root; it corresponds to a minimal affinity and arranges all other levels sequentially according to the increasing affinities. Interpreted substantively, the basal level (root) corresponds to a minimal similarity in phenetics and a minimal kinship in phylogenetics. The directed trees in systematics are most usually constructed as *monorooted*, which in Haeckelian or Hennigian phylogenetics is substantiated by the *principle of monophyly*; however, they are *polyrooted* in Copean phylogenetics based on the *principle of polyphyly*.

The hierarchy *pathing* refers to the totality of particular *hierarchization paths* (branches) in the tree. These paths are either simple linear chains, or they *split* and occasionally *merge* at some vertices. Their splits make the sequences *branching* and yield their multiplication from higher to lower levels of hierarchy, which in the case of deductive hierarchy is instantiated by the above-considered conceptual pyramids. If a path bifurcates when splits, it is called *dichotomous*, and if it is divided into several branches, it is *polytomous*; a tree consisting only of dichotomies is also denoted as *dichotomous* or *binary*, which is regarded an ideal in classical logic. The paths that initially outcome of the same vertex and connect several other vertices are called *parallel*. If two or more tree paths merge at a vertex, this provides a special kind of *cycle*, and it is called *polyhierarchy* in the technical sense of computer science (Babikov et al. 2004; Hedden 2010; Morville et al. 2015; Laubheimer 2018).

In systematics, the nature of pathing defines the overall inclusive hierarchical topology of taxonomic systems. Non-splitting paths entail the hierarchies of *monotypic* taxa, whereas splitting paths result in the hierarchies of *polytypic* taxa: each taxon includes only one lower-level subtaxon in the first case and several subtaxa in the second case. Cladistics considers dichotomous (binary) tree the best representation of evolutionary history, whereas polytomies are treated as 'unresolved dichotomies' implying *phylogenetic uncertainties* and therefore undesirable (Hennig 1966; Nelson and Platnick 1981; Wiley 1981; Lee 1998; Huelsenbeck et al. 2000; Rangel et al. 2015; Willis and Bell 2018). However, an elaboration of a fully dichotomous tree can result in its over-fragmentation with many poorly justifiable dichotomies that may provide a biased representation of the real hierarchy (see below). A polyhierarchy can be evolutionary substantive by referring to the possible reticulate evolution and hybrid origin of taxa.

A multi-modal classificatory character represents a linear or branching sequence of its modalities. It can be identified with a particular hierarchy path, which can be ordered as either inclusive or non-inclusive hierarchy or non-hierarchy; in the first two cases, this specific orderliness is referred to as character *polarity* (Salthe 1985). The meronomic systems that are ordered as inclusive hierarchies are instantiated by the archetypes or body plans according to the typological conceptions (Lyubarsky 1996, 2018). A variant of non-inclusive meronomic hierarchies represents differential *weighting* of classificatory characters or their modalities.

The hierarchy *fragmentation* refers to the degree of detailing of hierarchical structure depending on the number of levels/steps (and corresponding vertices) individuated within a tree path; this number is sometimes denoted as *path length* or *hierarchy score* (Pavlinov 2005; Nonaka et al. 2021). Degree of fragmentation of a partial hierarchy depends on the exploratory tasks, on the tree inferring algorithms, and on the character of raw data, to which these algorithms are applied. In general, the higher fragmentation degree of a tree, the more detailed description of the studied portion of systematic diversity it provides. On the other hand, as just said, an over-fragmented tree paths may provide a biased representation of the structure of the prototype natural hierarchy, so it would be reasonable to 'coarse' them, for instance, by applying certain probability criteria of the reliability of particular steps in the inferred paths.

A tree, in which the branch lengths are drawn proportional to their numerical estimates is denoted as *metric*; otherwise, a tree is *ultrametric* (Sneath and Sokal 1973). In phylogenetics, the latter reflects only branching sequences, which is exemplified by cladogram, whereas a metric tree reflects both these sequences and 'amount of evolution', which is typical both for the classical and 'molecular' phylogenetic trees (Felsenstein 2004; Pavlinov 2005; Podani 2013).

The *hierarchy depth* (*height*) or *hierarchy range* is defined by the distance between its initial and final levels/steps measured by the number of intermediate steps. For instance, in systematics, a hierarchical classification that covers the range from genus to phylum is 'deeper' ('taller') than the one covering the range from genus to order. The possible deepest hierarchy referring to the totality of natural systems is that of universe, with its range covering the levels from subatomic to metagalactic units. From a systematics perspective, the highest and lowest levels in this global hierarchy are redundant by having no evident direct impact on the structure of biologically sound organismal diversity, so this hierarchy is *truncated* from above and below to retain the range from molecules (as a specific data source) to biota with its biodiversity. An extra upper level of hierarchy depth is characteristic of the so-called *panarchic* structures (in the sense of Gunderson and Holling 2001; Gunderson et al. 2022); in the case of cognitive situation, this implies a supplementing of the basic hierarchy of its onto-epistemic background with a subject functioning as its higher-level interpreter (see Section 2.1).

A particular characteristic of hierarchy depth is *tree length*, which is resulted from summarizing the lengths of all its edges (branches). It has a special meaning in numerical phyletics, where the overall tree length serves as one of several *tree optimization parameters*. With the reference to the above-mentioned parsimony principle, it is transformed into a specific *optimization criterion*, according to which, of all possible trees inferred for a given dataset, the shortest one is counted as the most optimal (Farris 1983, 2008; Kluge 1984; Kitching et al. 1998; Gascuel 2000; Pavlinov 2005, 2021a; Wiley and Lieberman 2011).

The hierarchy depths, ranges, and scores are not intercorrelated: for instance, a very deep ('tall') overall hierarchy of Tournefort–Linnaeus and a 'shallow' infraspecies hierarchy of contemporary systematic botany include a similar number of steps (see Sections 1.5).

For the sake of having a possibility to operate with a unified thesaurus, it is reasonable to fix terminologically minimal levels of tree pathing, depth, and fragmentation. In the first case, a minimal (zeroth) pathing value provides a *one-level* hierarchy; in the case of depth/fragmentation, one can speak of *flat* hierarchy (= *flatarchy*), which implies significance of the 'horizontal' thinking accompanying the 'vertical' one (Marston et al. 2005; Jonas 2006); it is rather popular in sociology and sociogeography (e.g., Simon 1962; Herbst 1976; Ingram 2024; Chakraborty and Mohr 2005). A specific variant of such a 'delayered' hierarchy is also known as the above-mentioned *heterarchy*, now referring to a non-encountered hierarchization (Crumley 1987, 1995, 2005; Dekker and Kuchař 2021). The above-mentioned unordered (including minimum spanning) trees in systematics may be considered as instances of heterarchy (flatarchy). The development of actually multi-level hierarchical constructions begins with the establishing a *two-level* hierarchy. The latter seems to be a widespread hierarchical phenomenon, if at least some *block network systems* (in the sense of Ravasz and Barabási 2003) are permitted to be considered in such a manner. The coarse-grained Ladder of Being, which links are not specimens or species, but rather the higher-level groups of organisms (such as birds and beasts), may probably be treated as a special kind of such two-level systems.

A multi-level hierarchy can be either *monotone* or *non-monotone*.[6] Two kinds of the monotone/non-monotone opposition should be distinguished, either between or within paths, depending on the specific structure of pathing.

On the one hand, the parallel paths can include the same or different number of vertices: the first condition implies monotone and the second one non-monotone hierarchy. This aspect of pathing is of great significance for determining the overall structure of hierarchies: any strictly dichotomous monotone tree is *symmetric*, whereas the trees with various variants of polytomies and non-monotonicity are *asymmetric* (imbalanced). In systematics, this concerns an important issue of the equivalence/nonequivalence of ranks in different fragments of the overall taxonomic hierarchy (see Section 3.7).

On the other hand, different hierarchy levels within particular paths can be mutually equivalent or not, which implies, respectively, either monotone or non-monotone hierarchy of the second kind. In the first case, all levels and their constituents are supposed to have the same status, so their hierarchy is monotone in this respect. If some levels are more and others less salient with regard to the applied criteria, the respective hierarchy is non-monotone. This kind of non-monotonicity is illustrated by the hierarchy level triads, in which one level is *focal* and two others are just its 'near-bys'(Patten 1975; Bunge 1979; Salthe 1985, 2012; see below for detail). In systematics, the rankless taxonomic hierarchy presumed by cladistic theory can be treated as monotone, whereas the recognition of principal and auxiliary categories within the classical ranked hierarchy represents its non-monotone variant.

In scholastic systematics the generic rank was ascribed a salient status, whereas in evolutionary school such status became ascribed to the species rank, so in both these schools taxonomic hierarchy appeared to be non-monotone in this regard.

3.4 LEVELISM

The *levelism* generally refers to the specific 'vertical' stepwise structuration of any complex system, which implies its near-decomposability into sequentially arranged levels (Floridi and Sanders 2004; Floridi 2008). It is this specific feature of such a system that makes it hierarchically ordered: to put is figuratively, if there is a sequential orderliness of the levels in its structure, then there is a hierarchy; if there is not, then not (see also above on the epi- and quasi-phenomena).

With the key notion of level being universally regarded as a basic, even defining attribute of any hierarchy, it remains rather ambiguous, as its particular definitions depend on particular concepts of hierarchy. This makes both notions, hierarchy and level, interdependent and tautological to a degree: for instance, it was suggested to take 'hierarchy <...> generally to mean a *set of ordered levels*', though the latter remain with no explicit definition (Anonymous 1969, p. *viii*; italics in the original). A cursorily analysis of the relevant literature indicates that there seems to be no definitions of levels that would be both general enough and substantively meaningful (non-formal). The situation is aggravated by the fact that the very notions of level seems to be largely metaphoric (Stevens 1984; Bryant 2000). The potential difficulties that underlie this issue are illustrated by M. Bunge's suggestion of several complementary formal definitions of the level, which were supplemented with a 'less pompous' one as 'an assembly of things of a definite kind, e.g., a collection of systems characterized by a definite set of properties and laws' (Bunge 1969, p. 20).

The leveling of a complex system and its hierarchy implies a certain *nonequivalence* relations between its constituents. In general, these relations are determined by respective hierarchization parameters that divide the levels and their constituents into '*lower*' and '*higher*' depending on their position relative to the initial and final hierarchy steps (Simon 1962, 1973, 1991; Bertalanffy 1969; Salthe 1985, 2012; Ahl and Allen 1996; Wu 2013). For instance, biological organisms can be nonequivalent with respect to organismal complexity (uni- or multicellular or multi-tissue), their body size (from microbe to whale or oak), ecological relations (preys or predators), social position (dominant or subdominant), etc. The same holds true for those constituents which are arranged according to their levels of organization (cells, tissues, organs in organismal biology) or generality (microbiomes, biotopes, landscape ecosystems in ecology). Such nonequivalences can be treated predominately ontically or epistemically: in the first case, they are attributed to the systems and their constituents themselves, whereas in the second case they are imposed by the cognizing subjects onto otherwise equivalent entities. In systematics this is most characteristically exemplified by differential character weighting, which can be based either on functional significance of the respective structures or diagnostic significance of the characters themselves.

The nonequivalence relation is of special importance in the case of inclusive hierarchies, where it implies recognition of distinct *levels of generality* (*inclusiveness*).

This implies that within each of linearly ordered hierarchical sequences (paths), higher-level groups include the lower-levels ones. The above-mentioned sets of Chinese boxes or Russian matryoska dolls clearly exemplify this topological transitive relation. In traditional systematics, it is imperative for the nonequivalence relation to be established uniformly across the entire taxonomic hierarchy in a classification regardless of its pathing structure; this important feature has been inherited by systematics from the hierarchy of scholastic genus-species scheme. In addition, a certain nonequivalence is characteristic of various levels of generality delineated within the non-monotone hierarchies.

* * *

The very possibility for the ontically defined hierarchy levels to be individuated as distinct research objects is provided for by the fact that the system constituents at each of them are assumed to be endowed with specific emergent properties (Jones 1998; Shatalkin 2012; Bornkessel-Schlesewsky et al., 2015). These properties (a) make each level with its constituents unique within a hierarchy and (b) result in the within-level couplings of constituents being stronger than between-level ones (Simon 1962, 1973, 1991; Ahl and Allen 1996; Wu 2013). The following inherent properties of hierarchical systems are highlighted in this regard, which are sometimes denoted as 'hierarchy principles' (Allen 2001; Wu 2013):

- Higher levels operate more slowly (at a lower frequency) than lower levels.
- Higher levels have lower integrity as compared to lower levels.
- In the case of inclusive (nested) hierarchies, higher levels contain (consist of) lower levels.
- Higher levels exert certain constraints on lower levels.
- In the case of cognitive systems, higher levels exert interpretative contexts for lower levels.

Most of the substantive properties implied by levelism involve matter, energy, and information flows between system constituents allocated to specific levels (Simon 1962, 1973, 1991; Bertalanffy 1969; Pattee 1973b; Patten 1975; Salthe 1985, 2012, 2015; Ahl and Allen 1996; Pumain 2006; Wu 2013). A hierarchy considered from such a specific 'flow' perspective is sometimes denoted as *flow hierarchy* (Zafeiris and Vicsek 2018; Suchecki and Hołyst 2022). These are such flows that provide for an *interlevel integrative explanation* of the 'vertical' structure of hierarchically arranged systems, including their reciprocal (downward + upward) causation (Brigandt 2010; Santos 2015, 2024). On the other hand, particular flows differentiate the overall hierarchy into partial ones, which is exemplified before all by the recognition of ecological and genealogical hierarchies (Eldredge and Cracraft 1980; Eldredge 1985; Sluys 1991; Tëmkin and Eldredge 2015; Caponi 2016).[7]

One more basic aspect of the global biotic hierarchical structure was recognized by Niles Eldredge, which took position 'something between' ecological and genealogical hierarchies. It referred to the hierarchies of such specific

> composite 'eco-genealogical units' as lichens and was called '*Bretskyan*' after paleoecologist Peter Bretsky (Eldredge 1985). This idea was not noticed at first by biologists, but recent publication draws an attention to it (Spiridonov and Eldredge 2024). In this connection, it should be noted that 'Bretskyan' hierarchy seems to be a particular case of *phylocaenogenetic* hierarchy, recognized and studied by the Russian paleoecologist Vladimir Zherikhin (1945–2001) (Zherikhin 1986; Zherikhin and Rautian 1997).

The contents of these integrative flows depend on the nature of the respective systems, whether they are physical or mental, natural or artificial, biological or social, etc. In particular, it is assumed that the hierarchy levels of both systematic diversity and its cognitive models (classifications, etc.) are interconnected specifically by the information flows.[8] The directions and intensity of those integrating/differentiating flows within a leveled system are determined by the predominant causal interactions within and between levels. With regard to their directions, they can be either upward or downward, according to which the entire hierarchical structuration can be either inductive or deductive. The above discussed hierarchies of reduction cascades and conceptual pyramids, as specific cognitive systems, are largely shaped deductively by downward information flows performing causation function. Contrary to this, positivist philosophy of science argues that all causations in any systems are directed upward, hence all hierarchies generated by them are inductive (anti-hierarchies, see above).

A particular mechanism for ensuring between-level integrity of complex systems has been proposed to be recognized and referred to as *holon*, and its inclusive hierarchy was called *holarchy* (Koestler 1968; Pichler 2000; Mella 2006, 2009). The latter is not a specific 'thing' with its own emergent quiddity, but rather 'a conceptual, nonobservable entity that functions as a connective or intermediary element among hierarchical levels of [a system]. The holon is not *the* structure but *of* the structure' (Mella 2009, p. 10; italics in the original), so the holons are better to be thought of as certain 'abstractions more than material objects' (Allen 2008, p. 1855). However, it seems quite reasonable to think of holons more realistically: in this regard, in typological systematics they may be identified with the integrity of the overall archetype, and in evolutionary systematics they may be interpreted most aptly in terms of kinship relations.

Within a multi-level hierarchy, the overall between-level relations/connections tend to regressively weaken along the hierarchy paths due to an assumed decrease in the intensity of integrating flows with increase of the number of steps in them (Simon 1962, 1973, 1991; Pattee 1973b; Patten 1975; Salthe 1985, 2012; Ahl and Allen 1996; Pumain 2006; Wu 2013). The general cause for such a specific dynamics is that each transition between any two adjacent levels/steps involves a certain amount of uncertainty ('noise'), which accumulates along the hierarchy paths. From a formal perspective, this effect entails an *incomplete transitivity* of between-level relations, which can be referred to as a specific instance of non-linearity of hierarchies.

With regard to classification systems developed by systematics, a presumed weakening of the information flows between their levels can aptly be generalized by the *short-range* (*proximity*) *principle* borrowed from physics (Pavlinov 1998, 2018).

Its effect is highly pertinent to the discussion of some points crucial for biological systematics. On the one hand, this means that both stability and dynamics of taxa at sufficiently distant hierarchy levels are not strongly interdependent, so that, say, changes in species composition of particular genera do not typically affect taxonomic treatment of the order or class they belong to. Another issue it concerns is the level nonequivalence in the complexly structured taxonomic systems, especially if their hierarchies are rankless (see Section 3.7).

* * *

The above-mentioned onto-epistemic dualism of the general notion of hierarchy holds evidently true for its levels that can, with some reservations, be treated either ontically or epistemically or (most correctly) both. A focus on one of these entails distinguishing between ontically treated *autonomous* and epistemically treated *non-autonomous* levels: the former 'are responsible for producing their own rules', whereas the latter 'are designed to have hierarchical behavior by an external authority' (Pattee 1969, p. 162; also Pattee 1970). In other terminology, these two categories of hierarchy levels are referred to as either *organization* ('physical') or *consideration*, respectively (Ahl and Allen 1996; Allen 2001; Floridi 2008), or *scales of real processes* or *scales of analyses* (Smith 1987; Jonas 1994). As is seen, such an accentuation assumes different causations for the delineating hierarchy levels, which is either inherent (intrinsic) or imposed (extrinsic). In systematics, the levels of taxonomic hierarchy are regarded either as organization or consideration depending on a particular onto-epistemic framework (based on nominalism or realism, respectively); the same dichotomy holds for taxonomic ranks. However, conceptualism consider these two accounts conjointly.

This specific dualism concerns the interrelation between the notions of hierarchy *level* and *scale*, which are applied nearly synonymously, albeit with a slight preference for the latter, in the studies of hierarchical structures in cosmology, geography, ecology, sociology, linguistics, etc. In them, scales typically refer to the particular levels of complex systems, whether spatio-temporal, ecological, social, linguistic, or others, without explicitly specifying their onto-epistemic status (e.g., Jonas 1994; Wu 1999; Brenner 2001; Potochnik and McGill 2012; Henle et al. 2014; Padt et al. 2014; Enfield 2023). However, cognitive 'near-decomposition' of their onto-epistemic dualism makes it desirable a substantive and terminological distinction between these two notions, which may be achieved as follows. It seems reasonable to treat the notion of *hierarchy leveling* ontically, presuming that there may (or should) exist some objective hierarchically leveled structures in nature, society, etc. Unlike this, the notion of *hierarchy scaling* may (or should) imply a pure epistemic interpretation and refer primarily to the scales of consideration of complex systems, without explicitly assuming any natural 'prototype' for them (Ahl and Allen 1996; Jones 1998; Manson 2008). To emphasize the epistemic status of such scales, it might be useful to refer to them as *reference* or *implicational scaling* (these notions are commonly used in linguistics; see Decamp 1971; Bornkessel-Schlesewsky et al., 2015; Gildea and Zúñiga 2016).

However, such a rigor separation of ontic levels and epistemic scales, as they are outlined here, should not be taken too literally (absolutely): as a matter of fact, the overall network of relations within an onto-epistemic framework is fairly tangled, each of its ontic and epistemic components is complexly structured, and boundaries between them are smooth and fuzzy. On the one hand, the ontically 'given' research objects may be treated as either phenomena or epiphenomena (see Section 2.1 on this). On the other hand, the epistemic treatment seems to be equally applied to the cognitive (epistemic by definition) hierarchies (Bornkessel-Schlesewsky et al., 2015), and, in addition, epistemic scales may function either as rigid abstractions or flexible metaphors (Jonas 1994). Because of such a complex structuration, an epistemic approach may uncover certain otherwise non-evident inherent features in the hierarchical structure of ontically treated complex systems (Easterling and Kok 2002). Besides, it should be noted that there can be (and really are) other demarcation lines based on different principles; one of them is an original suggestion by the French sociologist Henri Lefebvre (1901–1991) to distinguish between 'global levels' and 'worldwide scales' in the hierarchical organization of the global (worldwide) human society (Lefebvre 2003).

* * *

One of the most significant features of levelism, viewed from the biological systematics perspective, is the distinction between the conceptions of hierarchical *level* and *category*, which is not explicitly recognized in the vast majority of other disciplines. This distinction was introduced following that between the notions of *taxon* and *taxonomic category* (Beckner 1959; Simpson 1961; Van Valen 1964; Blackwelder 1967; Mayr 1969, 1978; Jardine and Sibson 1971; Stuessy 2009).[9] Accordingly, a *level* is a specific attribute of the hierarchy proper, which refers to the latter's 'vertical' stepwise structuring, whereas a *taxonomic category* is a specific attribute of a set of objects (organisms, taxa) being hierarchized, and it refers to an assembly thereof allocated to a particular level in the taxonomic hierarchy (Jardine 1969; Mayr 1978; Rasnitsyn 1996; Pavlinov 2018, 2021a).

Such a hierarchy categorization in biological systematics of traditional ('Linnaean') kind is especially accentuated by a strict fixing of positions of each of the hierarchy levels within an overall hierarchy. Such fixed levels are conventionally referred to as *hierarchy ranks*, and they are individuated by assigning specific terms to them. Their general construction constitutes the *ranking system*, which conceptualization should be elaborated by *ranking theory* as a part of a broader hierarchy theory, which in turn, as far as biological systematics is concerned, is supposed to be a part of a general taxonomic theory (GTT; reviewed in Section 2.5). These theories are currently lacking, so only a few empirical rules for elaboration of ranking system are occasionally mentioned as 'given by tradition': e.g., a positive correlation between the size and complexity of a taxon and the number of ranks in its own classification (see Section 3.7).

Recognition of ranks transforms general hierarchization scenario into a specific *ranking scenario*, and the fixed sequence of ranks forms a *ranking ruler* as a particular version of hierarchization ruler. Respectively, the hierarchies ordered according to the ranking scenarios are referred to as *ranked*, which is applied both to linear and inclusive hierarchies. The ranked hierarchy is always directed, and it can be set either relatively or absolutely (see above). The taxa allocated to the same category/rank are denoted as *equiranked*, they form specific equivalence sets.

The general conceptual framework of any ranking scenario is formed by a set of *ranking principles*, the most important of which can be summarized as follows (Dubois 2005; Pavlinov 2015). The *principle of rank sequencing* implies that all ranks defined by a given ranking ruler are ordered in a certain sequence, with each rank taking in the latter a certain fixed position. The *principle of consistent designation* implies that each rank should be assigned a unique term, which is the same in all classifications implementing the same ranking ruler. The *principles of ranking completeness* (*continuity*, *exhaustiveness*) and *uniformity* (*unicity*) imply that all hierarchization paths in the same hierarchy must consist of the same sequence of ranks with no gaps in it, at least as far as the principal (mandatory) ranks are concerned. The *principle of rank homogeneity* implies that a rank denoted by the same term takes the same position in all hierarchies ordered according to the respective ranking ruler; this makes the ranks recognized in all classifications implementing the same ruler equivalent and comparable in this regard. The *principle of ranking of classificatory units* implies that all taxa and merons recognized in a hierarchically arranged classification system should be allocated to certain ranks in accordance to their levels of generality. The *principle of inclusiveness* implies that within the same hierarchization path, the higher-level classificatory units include the lower-level ones according to the above-introduced condition of *transitivity*. The *principle of rank equality* (*equivalence*) implies that all classificatory units that correspond to the tree branches diverging at the same tree node, are allocated to the same rank; in cladistics, such units are denoted as *sister groups*, so this principle reads as *rank equality of sister groups*.

Most of these principles have been implicitly assumed as a part of the specific refinement of the scholastic logical genus-species scheme by the earliest systematicians. Therefore, they seemed to be initially loaded with an epistemic meaning, which made them universally applicable. Their specific ontic treatments (typological, phylogenetic, etc.) were being shaped subsequently with the development of particular taxonomic schools.

Neither particular subordination of ranks, nor their number, nor their designations (with some exclusions) have any specific meaning. This fact reflects the lack of any well-elaborated substantive ranking theory and is responsible for a portion of ranking uncertainty (see below). In the history of systematics, all these ranking parameters initially varied, but they turned out to be standardized over time by a tradition and settled officially by Codes of nomenclature (see Section 6.3).

The implementation of a particular ranking ruler supposed by the respective Code (rank sequencing, terminology, etc.) is presumed to be the same across all taxonomic systems that are based on its provisions. Such its consistent application is of special significance, as it makes all classifications ordered uniformly and comparably in this regard. However, some of the above principles (e.g., ranking completeness) are not mandatory to apply, so that neither an omission of some codified nor an addition of some non-codified ranks is prohibited (there are exceptions stipulated by some Codes, see Pavlinov 2015, 2022a). Besides, some of those standard principles can be mitigated, if their unconditional applications entail certain inconsistencies: an example is *degenerate* (non-strict) hierarchy that questions the literal application of the principle of rank equality in cladistics (Wiley 1979, 1981, 2009; Schoch 1986; Gauthier et al., 1988; Pavlinov 1990a, 2005; Stuessy and König 2008).

More information about taxonomic ranking is summarized in Section 3.7.

* * *

One of peculiar applications of the nested hierarchy scaling, considered from an epistemic perspective, involves *fractal* systems, or simply *fractals*, for which their hierarchical fragmentation (a) depends entirely on the arbitrarily (epistemically) set hierarchization rulers, (b) can be unlimitedly fractional, and (c) generates inclusive (nested) hierarchies of *self-similar* subsystems (Mandelbrot 1982; Briggs 1992; Edgar 2007). This inherent property of any fractal means that the structuration pattern of hierarchically arranged subsystems, resulting from the fractal scaling of an inclusive system, is repeated at all its reference scales, which is known in physics and ecology as *scale invariance* (e.g., Levin 1992; Meakin 1998; Gelashvili et al. 2013; Clark et al. 2021). One of its most illustrative examples is a well-known tree-like graph (Figure 3.1); it is of special importance for our consideration, because tree graphs serve as typical representations of the hierarchical structure of classifications (see Section 4.2).

The general concept of fractal has a certain implication in systematics by referring to the self-similarity of structural units recognized at different hierarchical levels of

FIGURE 3.1 A tree-like graph illustrating self-similarity as one of the most characteristic features of fractals. The open Internet source: Wikimedia Commons (https://upload.wikimedia.org/wikipedia/commons/3/30/FractalTree.gif; modified).

biodiversity. Because of this, both systematic diversity itself and hierarchical classifications as its cognitive models are fractals (Gelashvili et al. 2010; Zander 2025). Such property of classification systems is most clearly manifested in the so-called *rank distributions* (Mayr 1969; Burlando 1990; Minelli et al. 1991; Gelashvili et al. 2010; Pavlinov 2018; Sigwart 2022; see below on it).

* * *

When considering the structure of multi-level hierarchies, the recognition of certain minimal meaningful components in their structure is of importance. This approach primarily focuses on the *ternary relations* within a system hierarchy that form its basic 'vertical' *triadic structure*, which minimal components are *triads* (triplets). It is assumed that such a structure is both necessary and sufficient to specify between-level relations that are most characteristic of the given hierarchy. This implies that at least three adjacent levels within a hierarchy should be meaningfully analyzed, whereas any single level or even a pair thereof cannot be considered a meaningful component in the hierarchy analyses. In particular, the above-mentioned hypotheses about hierarchies should address such triads as their minimally sound components. With this, both the composition of triadic components and their position within particular hierarchies are defined more or less arbitrarily depending on specific cognitive or other tasks, and distinct triads can overlap within the same hierarchy.

There are two versions of such a triadic structuration of hierarchies deserving being highlighted here as the most relevant to the systematic issues. One of them suggests that the triadic structure of a classification system is formalized by the so-called *three-taxon statement*; this implies a possibility for the entire set of taxa in a classification to be partitioned into taxon/meron triads/triplets of the kind (*A*(*B*,*C*)), each constituting an elementary two-level classification (Zimmermann 1931; Hennig 1966; Nelson and Platnick 1981; Wiley 1981; Pavlinov 2018; Rineau and Prin 2024; Valentin and Prin 2024). Another version of triadic structure implies a possibility for any multilevel system to be decomposed into level triads, each comprising three contiguous levels (Patten 1975; Bunge 1979; Salthe 1985, 2012; Wu 2013). Each level triad includes a *focal Level-0*, or the 'level of interest', and *Level+* and *Level-*, which are the higher and lower (respectively) levels adjacent to the *Level-0*.

* * *

One of the most remarkable features of levelism of inclusive hierarchies is a specific regular numerical distribution of the system (hierarchy) constituents at each hierarchy level. It is known among applied mathematicians as the *power*, *Zipf*, or *Zipf–Mandelbrot law* (e.g., Schroeder 1991; Li 2002; Newman 2005), but to biologists, it is better known as the *rank* or *hollow-curve*, or *skewed distribution*. It is also referred to after the persons who first investigated its properties in particular branches of human cognitive and practical activities (Willis, Zipf, Pareto, etc.). It is sometimes treated as a variant of an extensive family of *probability distributions*, but this interpretation is not correct, as far as the distributions generalized by this law do not imply standard statistical characteristics, such as dispersion and confidence interval

(Kudrin 2002; Podlazov 2009), so it was suggested to be recognized as a distinct *H-distribution* (Kudrin 2002). A system's observance of the Zipf law was suggested to be denoted as its '*zipfity*' (Pavlinov et al. 1995; Pavlinov 2018).

According to this law, at each hierarchy level, the size (volume) of system constituents (i.e., the number of subunits in them) and the number of constituents with a certain size are related by inverse ratio. This means that the more voluminous constituents are less numerous than the low voluminous ones. When the constituents are plotted according to the absolute value of their size (volume), the resulting distribution looks like a hollow-curve one; when these values are given on a logarithmic scale, the hollow curve becomes a straight line (Figure 3.2). Of special importance is that in the case of multi-level inclusive hierarchies, this distribution is reproduced at each of their levels; as said above, such a distribution signals that the respective hierarchical system is a fractal and its subsystems are self-similar (Schroeder 1991; Turbin and Pratsevityi 1992; Gelashvili et al. 2010, 2013).

It should be emphasized that the 'ranks' defining abscissa axis of the hollow-curve distribution have nothing in common with the ranks of taxonomic hierarchy. Actually, the former 'ranks' correspond to the specific volume estimates for system constituents, and they are fixed more or less arbitrarily to form a specific ranking scale, along which the constituents are distributed.

The rank (hollow-curve) distribution has a very wide application in ecology (Willis 1922; Fisher et al. 1943; Frontier 1985; Gelashvili et al. 2010, 2013; Puzachenko 2016), linguistics (Zipf 1949; Mandelbrot 1966; Orlov 1976; Dahui et al. 2005; Tunnicliffe and Hunter 2022), informatics (Pecarič and Pecarič 2019), sociology and economy, etc. (the list of references is far from being complete). In biological systematics, this distribution was revealed in many groups at various levels of taxonomic hierarchy (e.g., Clayton 1974; Sneath and Sokal 1973; Chislenko 1977; Kafanov and Sukhanov 1981; Dial and Matzluff 1989; Minelli et al. 1991; Minelli 1993; Pozdnyakov 2005; Pavlinov 2018; Sigwart 2022; Zander 2025). Their constituents in question can be: in ecology—species with their organisms, in linguistics—words with letters in them, in economics—settlements with their populations, in systematics—taxa with their subtaxa, etc.

The ubiquity of 'zipfity' of complex hierarchical systems of very different kinds, both physical (including ecological, and social) and mental (cognitive, linguistic, etc.), encourages scientists to discuss its nature and possible causes. The numerous analyses of real distributions have led to a suggestion that such a property may be peculiar to the systems endowed with a certain inherent harmony (Schreyder and Sharov 1982; Dunaev 1984; Schreyder 1996; Pozdnyakov 2005; Podlazov 2009). In particular, these can be open developing systems that hold a certain integrity and stability (Pavlinov et al. 1995; Pozdnyakov 2005; Podlazov 2009; Puzachenko 2016). With regard to the groups of organisms, their 'zipfity' is supposed to indicate that they have reached a high level of biological success (Yule 1924; Mayr 1969; Filippov 1984; Zarenkov 1988; Minelli et al. 1991; Pozdnyakov 2005, 2014). Accordingly, it was suggested to consider 'zipfity' a specific indicator of the naturalness of taxonomic systems (Zarenkov 1988; Pozdnyakov 2005). However, it was shown that the pretty long randomly generated texts were also endowed with the property of 'zipfity'

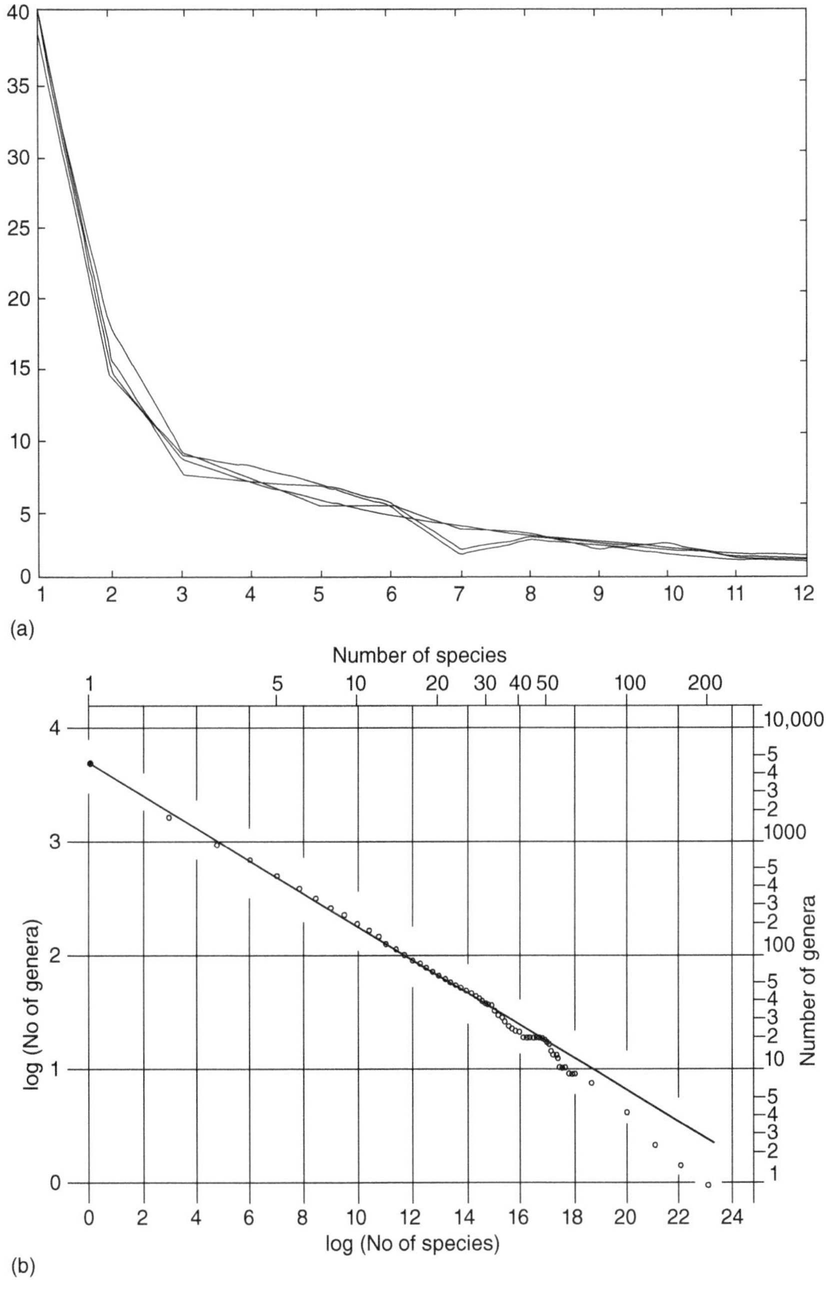

FIGURE 3.2 Rank distributions of number of the flower plant genera of various size (volume) (a) for several families on the absolute values and (b) in total on a logarithmic scale.

(After Willis 1922, pp. 181, 241).

(Schroeder, 1991), which leaves the questions of the latter's natural regularity and causality still open.

3.5 FUZZY HIERARCHIES

The classical natural philosophy considers both the studied universe and the knowledge about it deterministically and therefore 'discretely', which is underpinned by the scholastic binary logic with its rigid 'either-or' dichotomy. This means that everything is recognized as either real or nominal, and knowledge about it is either true or false. However, the contemporary scientific world picture includes the recognition of probabilistic nature of both the universe itself and its cognitive models (Prigogine and Stengers 1984; Stepin 2005; Dompere 2009a), which implies their inherently and irremovably fuzzy character. The emergence of such a 'new rationalism', based on this most general probabilistic idea (Gaydenko 2003; Stepin 2005; Dompere 2009a), gave rise to the probabilistic interpretation of the descriptive language of science (Nalimov 1979) and fuzzy logic (Zadeh 1992; Kosko 1993; Dompere 2009b) as its basic cognitive instruments.

A general onto-epistemic view of the fuzzy character of hierarchies should likely proceed from the idea of the 'order out of chaos' by the Russian-Belgian chemist Ilya Prigogine (1917–2003) that characterizes the dynamics of complex non-equilibrium systems in a specific fashion (Prigogine and Stengers 1984; Prigogine 1997). This metaphor can be applied to the evolution of the orderliness of both biological matter and knowledge about it. In their beginning was a chaotic lowest-level diversity of primordial organisms and vague ideas, and the outcomes are multifaceted hierarchies of the overall biotic diversity and its rational cognitive models. In terms of a (non-existent) hierarchy theory, the initial point of the structuration in question may probably be denoted as *chaotic* or *half-chaotic hierarchy* discussed in physics (Rössler 1983; Baier and Klein 1991), and this specific initial state of hierarchization may be identified with the aforementioned flatarchy.

This fundamental viewpoint is exceedingly relevant to contemporary biological systematics by presuming that all manifestations of biodiversity, including its hierarchical structure, and the knowledge about them are probabilistic and therefore inherently fuzzy. Such character of the basic conceptions and notions is quite old, with its roots dating back to Antique natural philosophy. The latter's fuzziness began with the fundamental onto-epistemic dualism of the categories of genus and species and was supplemented with its two other kinds, more particular and yet very important, namely, hierarchical genus-species and aspectual taxonomic-meronomic dualisms (see Sections 1.3 and 2.2 on them).

The conceptual history of biological systematics enabled it to 'overcome' all these kinds of dualism by the 19th century by near-decomposing each of them into respective near-independent components. However, the onto-epistemic dualism was reborn in the second half of the 20th century due to non-classical (basically conceptual) philosophy of science, which was founded on the probabilistic world picture. Regarding contemporary systematics, this implies the recognition of a fuzzy, non-rigid character of the conceptions and notions, with which biodiversity and all its manifestations and constituents, including hierarchy, are defined and explored.

All this makes the entire cognitive situation of systematics full of fuzziness, beginning with the development of its basic onto-epistemic framework and finishing with the particular classification systems.

The fundamentally fuzzy character of the basic notions of this concern begins with the fact that 'horizontal' near-decomposition of biodiversity cannot be accomplished unambiguously in order to discriminate systematic diversity quite definitely from its other aspects (Pavlinov 2018, 2021a). Indeed, biosystematic and ecomorphological approaches, as well as an idea of omnispective taxonomic systems based on 'all and any' classificatory characters, including ecological (Blackwelder 1967; Rollins 1965; Stuessy 2009)—all this makes global systematic and ecological hierarchical patterns overlapping. Elaboration of classifications based on distinct anatomical or molecular structures, such as pollen grains or mtDNA, makes them not so much taxonomic but largely meronomic. The classifications of lichenes, which are usually regarded as taxonomic, are to a degree syntaxonomic, as is evident from the composite characters of these organisms (see 'Bretskyan hierarchy' above). The sought 'exactness' of the delineation of taxonomic hierarchy can only be reached through its maximal reduction to a certain most strictly defined aspect, which is most characteristic of cladistics. But in such a case, due to the effect of the inverse relation between the rigor and meaningfulness of notions, a too narrowly delineated hierarchy may appear to be reminiscent of 'a spherical horse in vacuum' because of its artificiality (see Section 2.6).

Considering conceptual hierarchies in such a vein, two aspects of their fuzziness should be distinguished, *vertical* and *horizontal*: both imply the impossibility of strict delineation of the notions/concepts with close meanings. In the first case, this means that within the same path of reduction cascade, it is hardly possible to discriminate unambiguously any two conceptions of contiguous levels of generality by their fuzzy definitions. In the second case, the conceptions individuated at the same level of generality in the 'parallel' reduction paths are meant. Such fuzzy effects are clearly evident in the case of definitions of the notions, with which taxonomic hierarchies are described. The 'vertical' aspect is instantiated by the so-called species question, which is caused by a fuzzy character of 'around-species' categories (Mikhailov 2003; Pavlinov 2023), and the 'horizontal' one—by a similar character of the holo- and paraphyletic groups in cladistics (Pavlinov 2021a). In all such cases, these effects get stronger as the respective notions that delineate research objects become more fragmented, which is very characteristic of contemporary systematics.

Such a fuzzy character of the cognitive situation of biological systematics makes attractive the notion of *uncertainty* to be applied in addition to fuzziness. It was originally developed in science by nuclear physics (Heisenberg 2007; Sen 2014; Hilgevoord and Uffink 2016) and subsequently became recognized as a semantic core of the epistemic *uncertainty principle* of a fairly general meaning (Kuraev and Lazarev 1988; Armand 2008).

> The major point of our interest here is a specific interpretation of the relation between the supposed real 'behavior' of the elementary particles and their representations by the respective mathematical models. These particles, which are the waves at the same time, are assumed to be the elements of objective

> physical world, but they are 'smudged' over that unimaginably strange 'world' in their particle/wave dualism and their locations, all appearing to be uncertain. These elements become concretized (actualized) as certain quasi-discrete research objects due to specific cognitive operations (experiments) and representation models in the form of concrete decisions of the Schrödinger equation (Hilgevoord and Uffink 2016).

The following reasonable question goes immediately: what all of this uncertain microphysical stuff might bear on the issue of the hierarchies studied by systematics? To answer it in some approximation, just imagine that there are partial hierarchies and hierarchy levels assumably existing in an unobvious 'smudged' manner within the overall structure of biodiversity. According to the basic conceptualism provisions, in order to make such specific 'Schrödinger's entities' obviously individuated (delineated) as research objects, a certain cognitive activity of the knowing subjects should be applied to biodiversity. These entities are first actualized theoretically by certain conceptions elaborated as specific cognitive models of the respective manifestations of the overall diversity. Thereafter they are concretized empirically through the hierarchical arrangements of the concrete classification systems as cognitive models of the particular fragments of the studied diversity, assuming that these models should correspond to both the initial (set by the objective assumptions), boundary (set by the epistemic instruments), and ultimate (set by the subjectively motivated cognitive activity) conditions for solving a particular cognitive task. As a result, given a certain ranked taxonomic system, the objective component contributes to it by providing a certain range of possible variants of its hierarchical arrangements, whereas the subjective one contributes to it by making a particular choice among possibilities. Eventually we have to admit that a certain degree of fuzziness always occurs in such systems, which renders futile and non-productive the endless debates about either subjective or objective nature of their hierarchical patterning, as well as about the precise position of particular taxa in the ranked (or not) hierarchy.

* * *

The overall fuzzy character of classifications and their hierarchies, considered in biological systematics at an empirical level, is referred to as *taxonomic uncertainty* (Sneath and Sokal 1973; Zarenkov 1988; Pavlinov 2018, 2021a). The notions of *phylogenetic* and *cladistic uncertainty* are developed to refer to the fuzzy character of phylogenetic hypotheses, which are operationalized by the probability trees (Pavlinov 1987, 1990a, 2005; Lee 1998; Huelsenbeck et al. 2000; Rangel et al. 2015; Willis and Bell 2018). In both cases, two aspects of uncertainties should be distinguished, global and local.

The *global* aspect of the uncertainty in question involves the differences between particular classification systems, which result from two major sources, ontic and epistemic. The first of them is the diversity of aspect-based partial taxonomic theories that serve as specific conceptual frameworks for the elaboration of concrete classifications; this implies that different ontic models entail different classifications for the same groups of organisms. Such an uncertainty of ontic kind has a profound biological meaning: typological, classical phylogenetic, cladistic, biomorphic, and other classifications emphasize specific aspects of the complexly structured systematic diversity. In a similar manner, Anatole Shatalkin (2012) suggested to recognize *anagenetic* and *cladogenetic* hierarchies for the evolutionarily interpreted taxonomic systems. Another kind of hierarchical uncertainty in phylogenetics is provided by different interpretations of cladogram nodes as representing either the ancestral species or lineage splittings (Hull 1988; Rieppel 2006a, 2010a).

A significant portion of the global taxonomic uncertainty is taken by a *hierarchization uncertainty* (Zarenkov 1988; Pavlinov 2015, 2018), which means impossibility for any taxonomic system to adequately represent the hierarchical pattern of any particular aspect or fragment of systematic diversity. This uncertainty is multifaceted, beginning from the lack of a well-developed hierarchy theory and finishing with the absence of reliable ranking criteria for taxonomic hierarchy.

The ontic component of global taxonomic uncertainty is indirectly regulated by the so-called *principle of NP-completeness* (mentioned in Section 2.1) in a specific manner. Its effect is due to the fact that a multifaceted fragment of systematic diversity can be represented by several cognitive models, each grasping its distinct aspect (the *principle of model multiplicity*). In systematics, these models are classification systems that appear equally 'natural' according to some not strictly defined naturalness criteria, which makes them equally plausible and admissible. This point was considered by G. Simpson, albeit without such formalisms, who suggested that evolutionary classifications should be just *consistent* with an inferred phylogenetic pattern they are designed to represent (Simpson 1961). Reduction of the complexly structured multifaceted phylogenetic pattern to the cladistic relations was proposed by cladistics in order to minimize the potential uncertainty of its representation by taxonomic systems (Rieppel 2016; Pavlinov 2018, 2021a).

The epistemic source of global taxonomic uncertainty includes the diversity of classificatory methods applied (Sneath and Sokal 1973). The latter are developed within particular axiomatic systems, each method being precise 'locally' with respect to the underlying formalisms only, and with this, different methods cannot be summarized by some more general approach that would be as precise as the 'local' ones (Williams and Dale 1965; Moss and Hendrickson 1973; Shatalkin 1983). This inevitably results in the irreducible diversity of classifications based on different methods. Their multiplicity can be 'technically' removed by the elaboration of unified classification systems based on the idea of *taxonomic congruence* (Mickevich 1978; Rohlf et al. 1983; Crisci 1984), but this approach, if treated probabilistically, does not necessarily lead to the reduction in the initial uncertainty. The so-called *organizational classifications* pretending to combine in some way the results of phylogenetic and phenetic hierarchizations (McKelvey 1982) seem probably to suffer from the same flaw.

The *local* component of taxonomic uncertainty refers to the fuzzy character of the particular classification systems, which can result from different causes (Pavlinov 2018, 2021a). On the one hand, it should not be forgotten that each such system is a probabilistic hypothesis about hierarchy of a particular fragment of systematic diversity, and its uncertainty correlates negatively with its plausibility. The latter largely depends on the original research sample that serves as the sole information source for taxonomic classifications, so the level of taxonomic uncertainty is initially determined by the information incompleteness in a sample. On the other hand, it results from the incomplete taxon-character correspondence, which hinders an unambiguous delineation of taxa. This component also includes a multi-aspectual *ranking uncertainty*, which is actively discussed in contemporary systematics (see Section 3.7). Regarding the below-considered iterative classificatory procedure, one may say that it begins with the higher and finalizes with the lower levels of fuzziness (uncertainty) contained in the classification systems. This procedure may involve a stronger meronomic hierarchization due to the successive character weighting; and yet, the resulting taxonomic hierarchy cannot become less fuzzy than it is presumed by the source meronomic hierarchy.

The fuzzy character of evolutionary classification systems also depends on some onto-epistemic premises and limitations that underlie them. They entail the *kinship-similarity uncertainty*, meaning an impossibility for kinship to be unambiguously inferred from similarity. This specific uncertainty manifests itself both at formal and substantive levels. In the first case, it is caused by a mismatch of some formal properties of these two kinds of relations: kinship is transitive, whereas similarity is not (Dress and von Haeseler 1990). In the second case, it is resulted from the well-known 'formula': not every similarity evidences of kinship.

The latter 'formula' is especially emphasized in cladistics, which accentuates the cladogenetic aspect of phylogeny and minimizes its anagenetic aspect. Because of this, parallel evolution of classificatory characters is considered as a kind of 'phylogenetic noise' that hinders precise identification of the branching patterns. This 'noise' makes it difficult in many cases to unambiguously distinguish between holophyletic and paraphyletic groups, which results in their rather complicated terminology (Wiley 1979, 1981; Haszprunar 1986; Ax 1987; Gauthier et al. 1988; Meier and Richter 1992; Archibald 2014). Consequently, cladistic classifications always contain some level of *phylogenetic uncertainty* caused by an irremovable effect of the evolutionary parallelisms (Lee 1998; Pavlinov 2005, 2018; Rangel et al. 2015). It is supplemented by a specific *cladistic uncertainty* that also depends on the feeble definitions of holophyletic groups and increases from top to bottom of a cladogram following the decline in the ratio of reliable synapomorphies in overall similarity in the same direction (Pavlinov 1987, 1990a, 2005, 2018).

* * *

Such a novel fuzzy consideration of hierarchies studied by systematics may seem 'exotic' and thus challenging, but it may be made less so by its operationalization. This is achieved through a more familiar probabilistic interpretation of the hierarchical structure of both the systematic diversity itself and the classification systems representing it.[10] Given the tree-like style representation of such systems (see Section 4.2), it is convenient to consider probabilistic interpretations of their hierarchies in terms of *probability trees*. Their consideration as *decision trees* demonstrably illustrates the probability estimates of the true status of the judgments about hierarchical arrangements of taxa in the systems in question. Such a view of the hierarchy fuzziness looks quite productive because the probability trees are universal constructs, with non-probability ones being just their 'coarsened' variants with the probabilities rounded to 0 or 1.

One of these interpretations involves indication of the probabilities ascribed to the tree edges that connect 'parent' vertices (nodes) to their multiple 'child' ones. Given a processual account of a system as a developing entity, such a tree provides a model of its possible transitions between its states. In the case of biological evolutions, its probabilistic interpretation results in phylogenetic trees for particular groups of organisms with the probability estimates ascribed to the actual and potential phylogenetic lineages, and these probability trees outline a kind of '*evolutionary cones*' for these groups (Pavlinov 1992a). Another important kind of probabilistic interpretation of tree involves indication of the probabilities ascribed to its vertices. These probabilities indicate the reliability of the delimitation of groups that are rooted at the respective vertices, which, in the case of phylogenetic classifications, means the reliability for the groups in question to be treated as monophyletic. Accordingly, in the case of the above-mentioned super-tree, its probabilities indicate the degree of concordance of the original classifications.

3.6 HIERARCHIZATION OF CLASSIFICATION SYSTEMS

The fundamental tasks of biological systematics include (a) revealing and exploring the hierarchical structure of systematic diversity and (b) developing classification systems as its specific cognitive models. Both the revealing of natural hierarchical structures and their representation by the hierarchically ordered classifications are uniformly denoted here as their *hierarchization*. The latter is aimed at a 'vertical' ordering of the network relations between system constituents according to a certain *affinity scale* (similarity or kinship) implied by a particular hierarchization scenario. This includes recognition of hierarchy levels and ordering them in such a manner that their basal level would correspond to the minimal affinity and other levels are sequentially arranged according to the increasing affinities.

Such a specific exploration endeavor begins with a proper conceptualization of the basic notion of hierarchization, which is not an easy task. Before all, it should be clearly realized that any conceptual judgments concerning hierarchies are not 'absolute' but context-dependent, with these contexts being shaped by the background onto-epistemic frameworks. As far as the latter are multiple, the ways of conceptualizations of the classification hierarchies are also numerous. Indeed, if the concepts of space and time, when considered from a philosophical perspective, admit different context-dependent interpretations (e.g., Reichenbach 1958; van Fraassen 1970; Friedman 1983; Levich 2009), then why should the situation with hierarchy and hierarchization be different?

Accordingly, the entire issue of hierarchies conceptualization, in order to make it more explicit and transparent, should begin with outlining the relevant onto-epistemic backgrounds; this important part of the issue in question was discussed in general terms in Chapter 2. At this step, two basic approaches to the hierarchization are distinguished, namely, *transformational* and *stationary*: the former refers to the processes generating hierarchies, whereas the latter implies their structure as such (Rieppel 1988; Wood 1994). The next step includes elaboration of the hierarchization principles, which are general and particular (see Section 3.1). The general principles consider the overall structure of systematic diversity, including the ratio of its hierarchical and non-hierarchical components and general properties of its topology (such as directedness and inclusiveness). As is seen, this fundamental part of the endeavor in question addresses ontic component of cognitive situation, with its particular background propositions having status of either quasi-axioms (about evolution, aspectual character of diversity, its 'horizontal' and 'vertical' near-decomposability, etc.) or presumptions (about particular causes and properties of systematic diversity, etc.). Epistemic consideration begins with the formulation of *criteria of naturalness* of classification systems based on the previously elaborated ontic models of the partial hierarchies to be represented by those systems. The particular hierarchization principles provide more concrete propositions concerning what (ontology) and how (epistemology) is to be investigated in order to achieve the sought classification systems.

Methodological part of hierarchization of classification systems is presupposed by an awareness that the only empirical basis for this endeavor is provided by a research sample of specimens with their attributes. Such a sample is regarded as a specific *operational model* of the studied fragment of systematic diversity, the specimens are treated as the bearers of classificatory characters, and the latter constitute the basic information source for the inferences of interrelations between elements of the studied diversity. This makes the meronomic hierarchies, in which these characters are ordered, primary relative to the taxonomic ones. Accordingly, the most significant and challenging methodological issue involves involves (a) identification of a specific aspect of meronomic diversity that is most relevant with respect to a given naturalness criterion, (b) an adequate representation of this aspect by an array of classificatory characters, (c) retrieval of the requested information embedded within these characters, and (d) its 'translation' into that within the taxonomic systems elaborated on its basis.

A specific epistemic device for dealing with this issue is provided by the *principle of taxonomic-meronomic correspondence*[11] that serves as a kind of 'conceptual bridge' between these two aspects of systematic diversity, and it is formalized at an operational level as a *taxon-character correspondence* (Pavlinov 2005, 2011a, 2018, 2021a). Substantively speaking, this correspondence mutually links the judgments about the structures of taxonomic and meronomic aspects of systematic diversity, including their hierarchical orderliness. More formally, it establishes a logical relations between extensional (composition and leveling) and intensional (characters) definitions of taxa in a classification. One of the outlets of this correspondence includes a specific *prohibition principle*, which in its taxonomic implication assumes 'prohibitions' of certain biologically meaningless combinations of classificatory characters, with such 'prohibitions' causing the gaps in the potentially continuous morphospaces and, hence, outlining the boundaries between equiranked taxa in classifications (Stark 1966; Zavarzin 1969, 1974; Zarenkov 1976, 1988).

The application of the principle of taxonomic-meronomic correspondence faces the well-known problem of *non-congruence* of characters, which has already been aware by Aristotle (Sloan 1972) and recently proposed to be recognized as a distinct taxonomic principle (Crowson 1970). It means that different uncorrelated characters yield different groupings of the same organisms into taxa, and accentuating this fact was fixed as *taxon-character non-correspondence* (Meyen 1984; Lyubarsky 1991, 1996). However, such a negative accentuation, if taken 'absolutely', yields the logical impossibility of biologically meaningful taxonomic systems (Beklemishev 1994; Pozdnyakov 2007), so it seems more reasonable to speak of *non-strict taxon-character correspondence* (Pavlinov 2018, 2021a), which implies the polythetic intensional definition of taxa.

An ontic foundation of this correspondence is provided by such a general model that presumes regulation of both these aspects of systematics diversity and their hierarchies by (nearly) the same array of causes of whatever kind (formal, material, initial, etc.). Such their uniform causality, according to the *principle of common cause* (Sober 1988; Hofer-Szabó et al. 2013), is supposed to make these hierarchies *concordant* to more or less degree, which implies that their structures are similar, of course not completely. This ontic presumption serves as a requisite substantiation for the epistemic argumentation scheme of inferring taxonomic hierarchies from meronomic ones, provided that the latter is adequately represented by the characters in research samples. For instance, in phylogenetics, such a substantiation is specified by a *presumption of similarity of clado- and semogeneses* informally developed by W. Zimmermann (1934) and fully adopted in cladistics (Pavlinov 1990a, 2005).[12]

According to the provisions of the principle of taxonomic-meronomic correspondence, hierarchization of a taxonomic system, given near-decomposition of taxonomic and meronomic aspects of systematic diversity, proceeds following specific iterative procedure of *successive approximation*, which is rather popular as a numerical technique in data mining (Vilenkin 1964). It is implicitly involved in all taxonomic explorations aimed at the elaboration of classification systems, in which taxonomic and meronomic considerations of systematic diversity alternate within a particular hierarchization algorithm (Meyen 1984; Lyubarsky 1996; Pavlinov 2018). More formally, the latter can be represented as a sequential alteration of the precedences of 'taxon ⇒ character' and 'character ⇒ taxon'.

The procedure in question can be briefly summarized as follows. First, a certain fragment of taxonomic diversity (a group of organisms) is outlined as the research object (a taxonomic aspect), which is followed by the recognition of specific classificatory characters (a meronomic aspect). Their analysis involves selection, weighting, polarization, etc., of the available characters to make the structure of their sample as most consistent with the criteria of naturalness of the sought classification system as possible. The resulting meronomic system of characters serves as a basis for a preliminary intensional definition of taxa for the studied group at the next iteration step, which yields a clarification of their extensional parameters (composition, ranks, etc.). The next step involves again the intensional considerations of that group (refinement of characters), which is followed by its further extensional considerations (refinement of taxa). The entire iterative procedure proceeds as a sequential optimization of the elaborated classification system in both its taxonomic and meronomic aspects. It is presumed that at each iteration step, its renewed hierarchical arrangement becomes (a) less uncertain and (b) more consistent with the implied naturalness criteria. According to the methodological *principle of convergence* (e.g., Young 1971), this procedure proceeds until a certain stability of the classification system is achieved.

This signals that the highest possible accuracy in the latter's overall structure is attained for the given data set and the classification algorithm applied.

An important point of such a procedure is that it involves all taxa and characters within a classification being elaborated, which makes the entire iterative procedure *dynamical*. Therefore, any change in the treatment of either taxon and/or character at a certain level of generality may entail the respective changes in some other taxa and/or characters at the same or several contiguous levels. This is one of the effects of the taxon-character correspondence, when applied to a complexly arranged multi-taxon and multi-character classification system.

Taxonomic hierarchization can be either *direct* or *indirect* depending on the initial taxon-character precedence, which is especially characteristic of numerical phyletics (Weston 1988; Pavlinov 1990a, 2005, 2018). In the first case, which is most usual for systematic research, classificatory characters are primarily analyzed, including their weighting and polarization based on certain criteria, and taxa are then hierarchized according to the results of characters analysis. In the second case, minimally ordered (equally weighted and non-polarized) classificatory characters are initially included in the analysis, which results in a non-encountered taxonomic hierarchization represented by an unordered (undirected) tree. Such its pre-order is transformed into an ordered tree with a fully encountered taxonomic hierarchy by fixation of a specific node on the unordered tree with the reference to the so-called *outgroup*, and this node is interpreted as the tree root indicating an initial cladistic event in the history of the studied group (Wiley 1981; Farris 1982, 1983; Pavlinov 1990a,b, 2005, 2018). At the next step, polarization of characters becomes possible in the context of the resulting taxonomic hierarchy, which results in a posterior meronomic hierarchization.

> The transformation of an unordered tree (network) in an ordered (rooted) one differs in phenetics and cladistics (Figure 3.3). In the former, the highest level of ordered tree (phenogram) is defined by a similarity ruler to correspond to a minimal similarity regardless of the outgroup position. In the latter, the highest level of ordered tree (cladogram) is defined by a kinship ruler according to the outgroup position, which indicates the tree root corresponding to a minimal kinship between the studied group members. Therefore, a suggestion to use phenogram as an estimate (in a statistical sense) of phylogeny (Colless 1970; McNeill 1982) is untenable. However, such an 'estimation' is very common in molecular phylogenetic analyses based on the neighbor-joining algorithm (Nei and Kumar 2000; Felsenstein 2004).

* * *

As follows from the above consideration, the principle of taxonomic-meronomic (taxon-character) correspondence, as implemented by the direct algorithm, involves primarily hierarchization of classificatory characters. With regard to the latter, this principle assumes that the orderliness of modalities in each character corresponds to the orderliness of at least some levels in the taxonomic hierarchy. Further, it is assumed that the orderliness of different mutually compatible characters correspond to different fragments of the taxonomic hierarchy. On this basis, the most

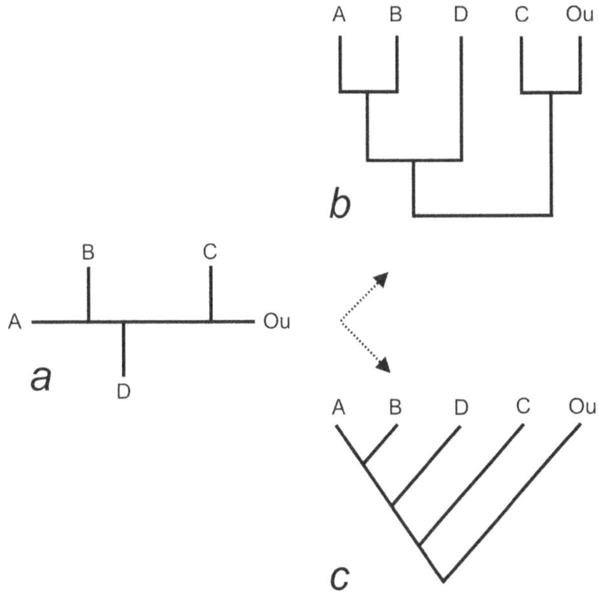

FIGURE 3.3 Transformation of unordered tree (*a*) into phenogram (*b*) and cladogram (*c*) without or with regard to outgroup position. Abbreviations: *A–D*—members of the studied group, *Ou*—outgroup.

important assumption is made for a certain group of organisms, a sample of several partially compatible and properly ordered characters provides a possibility for a more complete disclosure of the entire orderliness of the hierarchy of the sought taxonomic system.

This general provision, and with it the implementation of the taxon-character correspondence in total, is seriously challenged by the multiplicity of near-decomposabilities of both taxonomic and meronomic aspects of systematic diversity. This is manifested in the multiplicity of hierarchization relations and characters, respectively. The former refer to the specific interrelations between classified organisms, which can be similarity, kinship, ecological, etc., and the one of them, most corresponding to the certain criteria of naturalness, should be selected as a guiding hierarchization parameter prior to any other considerations. Such particular principle serves as a background for the selection and ordering of classificatory characters to make their totality a good enough approximation of the required meronomic hierarchy. The general provisions of their selection and orderliness are summarized by the *principle of characters nonequivalence*, which implies estimation and categorization of classificatory characters with regard to their importance for (contribution to) the inferring hierarchies of the sought taxonomic systems.

The respective ordering (selection, estimation, etc.) of classificatory characters is traditionally denoted in biological systematics as their *weighting* already mentioned above. It is carried out through the implementation of the *weighting principles* and applying particular *weighting criteria* presumed by them; given their impact on the resulting classifications, they may be considered as the principles and criteria of *characters optimization*. As said above, they are elaborated based on the respective naturalness criteria for the classification systems and result in specific *weighting functions*. The latter form particular *weighting scales* and ascribe specific *weights* (weight estimates) to the classificatory characters. The entire weighting approach is loaded both conceptually (by basic onto-epistemic considerations) and pragmatically (by exploratory tasks), so all these principles, criteria, functions, etc., are context-dependent and therefore relative.

Generally speaking, character weighting is among the most significant prerequisites for the taxonomic-meronomic correspondence to be properly realized. Both weighting principles and criteria are quite numerous, and their proper arrangement shapes another important conceptual hierarchy that underlies any taxonomic research. They play an essential role in the operationalization of epistemic component of cognitive situation in systematics, and in such a vein they constitute a considerable portion of the inference rules of any of well-developed aspect-based partial taxonomic theories (PTTs). Providing that the latter are built as specific quasi-axiomatic systems and keeping in mind the *constructivity principle* (see Section 2.3), all such weighting principles and criteria should be explicitly formulated, even if they look self-evident. The below review summarizes their discussion by Pavlinov (2018).

The principle of probably the most general meaning, implied by the background worldviews, prescribes either *elementaristic* or *holistic* approach to character weighting. The elementaristic approach considers meronomic hierarchization at the level of particular characters and specific weighting criteria for them, it is most common in systematics (e.g., Cain and Harrison 1958; Simpson 1961; Sokal and Sneath 1963; Hennig 1966; Blackwelder 1967; Farris 1969; Mayr 1969; Lines and Mertens 1970; Hansell and Ewing 1973; Wheeler 1986; Chippindale and Wiens 1994; Vogt 2002). The holistic approach argues that it is the overall generalized organism (archetype, ground plan, etc.) that should be subject to the meronomic hierarchization, whereas operations with particular characters are refused as reductionist. This approach is group-specific to the extent that such are archetypes (ground plans), and it is mostly characteristic of typology (Beklemishev 1994; Lyubarsky 1996; Kasavin 2024), including its phylogenetically interpreted version (Yudin 1974; Gutmann 1977).

According to another, more operational principle, the character weighting may be either *differential* or *equivalent*. The former implies that particular classificatory characters are ascribed different weights, which is most typical for all classical branches of biological systematics. Differential weighting can be treated as ontic, if it refers to the hierarchization parameters proper; or as epistemic, if it considers diagnostic significance of characters; or mixed onto-epistemic. Aristotelian distinguishing between *essentiae* and *accidentiae* has been the historical predecessor of all ontically based differential weighting criteria that refer to the organismal essences. Its earliest version in systematics was implemented by the recognition of *natural* (significant) and *artificial* (non-significant) characters by Linnaeus (1751), who applied mostly (following Tournefort) a 'diagnostic' (epistemic) character weighting. One of the approaches to distinguishing between homologous and analogous characters also instantiates their weighting of mixed onto-epistemic kind (Meyen and Shreyder 1976; Lyubarsky 1996, 2018; Pavlinov 2012, 2018). The alternative equivalent weighting was quite expressly stated by the Russian zoologist Eugeny Smirnov (1898–1979), who believed that 'a dogmatic preference of some characters over others should be recognized as unscientific' (Smirnov 1923, p. 387). According to this belief, all classificatory characters should be included in taxonomic analyses with equal weights, and this approach is usually identified with character *non-weighting* in phenetic school of systematics (Cain and Harrison. 1958; Sokal and Sneath 1963; Colless 1967; Sneath and Sokal 1973). However, the latter standpoint is erroneous, as in fact a certain weighting is still applied in it in a quantitative form.

Particular classificatory characters or ensembles thereof can be weighted either *qualitatively* (substantively) or *quantitatively* (formally). The former approach presumes their differential weighting based on a careful substantive consideration of each of them. The substantive analysis is originally rooted in the essentialist worldview, and it was being practiced by ideologists of scholastic and some post-scholastic systematics schools inclined to typological thinking (see also below). The quantitative weighting, with implying equivalent estimation of particular traits, emphasize their positive summative effect. Pheniticians absolutize the latter by the concept of *overall similarity*, according to which, roughly speaking, their specific *numerical* weighting criterion reads 'the more characters the better' (Cain and Harrison 1958; Sokal and Sneath 1963; Sneath and Sokal 1973), whereas some cladists appeal in this regard to the *principle of total evidence* (Eernisse and Kluge 1993; Kluge 1998a; see Rieppel 2005, 2009 for discussion). A 'moderate' version of quantitative equivalent weighting is formalized by the *principle of characters compatibility* (*congruence*) according to which maximized should not be an overall set of characters but rather that of mutually compatible ones; it was first applied by M. Adanson (1763) and now is advocated in cladistics (Estabrook 1972; Le Quesne 1982; Patterson 1982; Sharkey 1989; Kitching et al. 1998).

Differential weighting may be either *prior* or *posterior*: the characters are weighted before or after elaboration of a taxonomic system. If tracing this dichotomy back to the origins of systematics, it is possible to see its precursor in the one defined by C. Linné, respectively, as either 'a character gives a group' or 'a group gives a character' (Linnaeus 1767). The prior approach is characteristic of all 'classical' schools of systematics, as well as a large part of cladistics (e.g., Hennig 1966; Neff 1986). Numerical phyletics employs an intermediate variant of differential weighting, which is conducted during taxonomic analysis, it is known as the above-mentioned iterative *successive weighting* (Farris 1969; Pavlinov 1990a, 2005; Carpenter 1994) as a part of the *indirect* hierarchization. Such an iterative approach, which is realized also by the most recent Bayesian algorithm of phylogenetic inferences (Felsenstein 2004; Nylander et al. 2004; Bolstad 2007; Chen et al. 2014), makes the weighting procedure dynamical and the weights ascribed to the characters verifiable; however, it is sometimes accused for being 'circular' in the logical sense (Sneath and Sokal 1973).

The classification approaches that employ prior qualitative differential weighting of classificatory characters pay great attention to the biologically sound substantiation of weighting criteria. They are highly specific to the particular approaches and therefore can be considered as the latter's 'diagnostic features'. The most significant of these criteria can be summarized as follows.

In some systematic schools, especially early ones, the *criterion of vital significance* is of prime importance. This implies that the higher a presumed 'vital value' of the organismal features, the more significant are the respective classificatory characters. This criterion has two basic meanings, *functional* and *structural*. In the first case, most close to the essentialism of Aristotelian kind (some scholastic and early post-scholastic schools, biomorphics, etc.), vitality is identified with the functional significance, which, however, is defined differently by different authors (e.g., Jussieu 1789, 1824; de Candólle 1813, 1818; Lindley 1836; Strickland 1841; Yudin 1974; Aleyev 1986). In the second case characteristic of typology, it is considered as a significance with regard to the construing of organismal body planes (e.g., Cuvier 1801; Geoffroy de Saint-Hilaire 1830; Naef 1919, 1931; Kälin 1945; Remane 1956; Lyubarsky 1996).

In many of post-scholastic systematic schools, one of the most important is the *homology criterion*, according to which the more reliable homology of the organismal features (attributes, structures, etc.), the more significant are classificatory characters representing them (Bock 1981, 1989; Nelson 1994; Pavlinov 2005, 2011, 2018). The above-mentioned W. MacLeay was probably the first to recognize this criterion by ascribing the highest rate to the characters that correspond to the 'close affinity' of organisms (MacLeay 1819). However, since homologies can be established in different ways based on different onto-epistemic premises or within different cognitive tasks (see Section 5.2), implementation of this criterion is not uniform: the same features can be considered homologues (significant) in some cases and analogues (insignificant) in others. Methodology of biomorphic approach in systematics considers both homologous and analogous features equally significant (Aleyev 1986; Pavlinov 2010, 2018).

* * *

The typological idea (actually, several ideas, see Pavlinov 2018, 2021a) makes the *principle of typological unity* a basis for the elaboration of taxonomic systems. It is assumed that nearly all information regarding this unity is maximally embodied in the typological hierarchies of body plans, so the latter should be just reproduced by those of taxonomic systems. Accordingly, the taxon-character correspondence takes shape of the *principle of characters subordination* (*ranking*), according to which particular hierarchies of body plans are first inferred, the classificatory characters are then hierarchically arranged respectively, and the taxa are at last delineated and ranked strictly with the reference to the hierarchized characters. This approach has been first formulated by A.-L. de Jussieu (1789) and J. Cuvier (1801), and it was nearly aphoristically expressed by Leonard Jenyns (1833, p. 386, italics in the original) as follows: 'a correct and natural classification must be that of a due *subordination of groups*, founded on a *subordination of characters*'. It became a cornerstone of the classificatory approach developed by Russian typologists, who proposed the *principles of the same level* or *rank coordination* (Skarlato and Starobogatov 1974; Starobogatov 1989; Vasilieva 1989, 1992, 1999, 2007; Shatalkin 1995; Lyubarsky 1996, 2018).

A similar approach was elaborated by the so-called pattern cladistics by the conception of *levels of generality of defining characters*, with the hierarchies of these levels being identified with the ontogenetic sequential structuration based on the ideas of Baerian developmental typology (Nelson 1978; Platnick 1979; Nelson and Platnick 1981; Patterson 1983). One of the versions of onto-rational systematics (see Pavlinov 2011a, 2018, 2021a for review) implies meronomic reconstruction of the ontogenetic patterning, it is represented by a tree-like diagram called *Baerian tree*, which is transformed into a taxonomic system, just as the Haeckelian or Hennigian phylogenetic tree (Ho 1988, 1992).

Evolutionary concept makes the *principle of evolutionary* (*phylogenetic*) *unity* a basis for the elaboration of taxonomic systems. According to the *phylo-taxonomic correspondence*, it is assumed that the hierarchy of taxonomic systems should be based on that of evolutionary (phylogenetic) relations. With the replacement of a very broadly understood evolution with a more concrete phylogeny, the *principle of phylogenetic significance* of characters is adopted as the leading weighting criterion: the more reliably they indicate phylogenetic (genealogical) relations and, hence, the more they contribute to the inferring phylogenetic hypotheses, the more are their classificatory weights. This criterion is specified by the evaluation of *evolutionary conservativeness*, *directedness*, *irreversibility*, and occasionally *adaptive significance* of characters; criterion of *fossil availability* is implied by E. Haeckel's understanding of phylogeny as a 'paleontological history' (Haeckel 1866). Inasmuch as the hierarchization in question is most expressly represented ('modeled') by the branching patterns of the evolutionary (phylogenetic) trees, the hierarchies of taxonomic systems should somehow reflect that patterns and do not contradict them; according to G. Simpson (1961), they should be 'consistent' (see above). The *principle of taxon sequencing* (*seriation*) implies that the sequences of equiranked subtaxa within an inclusive taxon should reflect their levels (degrees) of advancement (evolutionary, typological, etc.).

The phylogeny-based hierarchization scenario differ markedly with regard to the ratio of clado- and anagenetic components of evolution to be embraced by taxonomic systems. The parallel (anagenetic) evolution is taken into account along with the divergent (cladogenetic) one in classical (Haeckelian) phylogenetics and Simpsonian evolutionary taxonomy, in which *criterion of evolutionary parallelisms* is rather significant (Simpson 1961; Bock 1973, 1977; Szalay and Bock 1991). This idea was first expressed by C. Darwin in his *Origin of Species*, who emphasized that similar evolutionary trends may evidence of kinship (Darwin 1859). Unlike this, cladistic version of phylogenetics emphasizes cladogenetic (genealogical) component of evolution and disregards its anagenetic component (Hennig 1966; Wiley 1981; Ridley 1986; Wägele 2005; etc.). Accordingly, the classificatory characters are weighted in order to maximize the cladogenetic information contained in the studied meronomic hierarchies, which is denoted as a *phylogenetic signal*.

The latter approach ascribes a maximal classificatory weight to the characters with the minimal adaptive functionality, because they are supposed to be least susceptible to the parallel and convergent evolution and therefore most reliably uncover the inherited similarity. This important proposition, first outlined also by C. Darwin and therefore denoted as *Darwin's principle* (Mayr 1969), yields the recognition of prime importance of genetic characters and led ultimately to the nowadays domination of molecular phylogenetic approach in systematics (Soltis et al. 1992, 1998;

Page and Holmes 1998; Ayala 1999; Felsenstein 2004), which is suggested to be denoted accordingly as *genosystematics* (Antonov 2006).

In order to maximize isomorphism between the hierarchies of cladistic relations in phylogenies and holophyletic groups in classifications, cladistics developed a very specific weighting criterion that involves not characters themselves (they may be equally weighted) but their modalities. The latter are divided into significant *apomorphies* and non-significant *plesiomorphies*, which results in the *polarization* of classificatory characters and their interpretation as *transformation series* (Mickevich 1982; Pavlinov 1990a, 2005; Pogue and Mickevich 1990; Grant and Kluge 2004; Harris and Mishler 2009). In the traditional terminology, apomorphy and plesiomorphy refer to the final and initial character states, respectively; the characters themselves are often thus denoted. The corresponding 'unit similarities' based on this specific character hierarchization are divided into significant *synapomorphies* and non-significant *symplesiomorphies*; this approach was suggested to be denoted as *similarity weighting* (Pavlinov 1990a, 2005, 2018). The respective weighting *synapomorphy principle* implies that only synapomorphies are taken into consideration in cladistic analysis, and this results first in a hierarchically arranged 'special similarity tree' called *synapomorphogram* (Rohlf and Sokal 1981) and then in a genealogical tree (cladogram), which is directly transformed into a taxonomic system with the same hierarchical pattern.

Evolutionary systematics and phenetics apply certain estimations of systematic diversity as the additional measures for taxonomic hierarchization, which implies various combinations of both between- and within-group differentiation. In the first case, the degree of separateness of the groups is taken into consideration, which is identified with the specificity of their adaptations (Mayr 1942, 1969; Simpson 1953, 1961; Van Valen 1971), 'amount of evolution' (Swofford et al. 1996; Felsenstein 2004) or numerically estimated phenetic distances (Sokal and Sneath 1963; Sneath and Sokal 1973), which in cladistics are specified as *patristic distances*, with resulting classifications being denoted as *patrocladistic* (Stuessy and König 2008).

* * *

The previous considerations address the *interior* hierarchization parameters, which are identified with the affinity relations and classificatory characters. With regard to the *exterior* parameters, as they are defined in Section 3.1, two major kinds thereof are recognized in the evolutionarily interpreted systematics. One of them implies the reference to certain timing scales as specific taxonomic hierarchization rulers (Hennig 1950, 1965, 1966; Sibley and Ahlquist 1990; Avise and Johns 1999; Avise and Mitchell 2007; Fisher 2008; Avise and Liu 2011; Jønsson et al. 2016; Kraichak et al. 2017), which can be formed as *direct* or *indirect*. A direct timing of phylogenetic events is inferred from the paleontological data, which is usual for classical approaches in systematics. An indirect timing results from the comparative analyses of extant organisms: for this, morphological data are interpreted under an assumption of *morphoclines–chronoclines isomorphism* (Maslin 1952; Beklemishev 1994),[13] whereas molecular genetic timing is based on the *molecular clock* hypothesis (Page and Holmes 1998; Ayala 1999; Arbogast et al. 2002; Ho 2014; Duchêne et al. 2024;

etc.). In evolutionary taxonomy, specificity and width (scale) of adaptive zones occupied by respective groups are suggested to use as an additional exterior criteria for ranking taxa (Simpson 1961; Van Valen 1971). These exterior parameters are considered as a means for the setting taxonomic ranks in an 'absolute' fashion and thus making them comparable (see the next section). The main criticism of the second approach 'is that no satisfactory general measure of evolutionary differentiation has yet been found' (Griffiths, 1973, p. 338; also Sneath and Sokal 1973).

3.7 TO RANK OR NOT TO RANK: THE CONTEMPORARY DEBATES

It seems quite appropriate to begin this section with reminding that the emergence of scientific systematics in the 16th century has been stimulated by its mastering of the rankless genus-species classificatory scheme. However, its subsequent development involved the elaboration of ranked hierarchy, which could be considered a very important component of specific rationalization of this natural science discipline (Lyubarsky 2018). The ranked hierarchy served as a universal standard for the elaboration of taxonomic systems until the mid-20th century, when it became strongly criticized by the proponents of rankless taxonomic hierarchy. They declared ranked hierarchy obsolete, untenable, and even 'non-scientific' (e.g., Sokal and Sneath 1963; Sneath and Sokal 1973; de Queiroz 1997; Minelli 2023b; Ereshefsky 2004; Zachos 2011); to the contrary, the adherents of ranked hierarchy were more willing to reject the taxonomic theories denying it than to abandon such a hierarchy (Mayr 1969; Bock 1977). As a result, the controversy of ranked *vs.* rankless classifications arose, leading to a significant 'schism' in the most recent systematics.

The ranking systems are sometimes supposed to be a specific characteristic of biological disciplines that study and classify the structure of biotic diversity (systematics, biogeography, biocaenology, etc.), and that they are hardly comparable directly to the hierarchization systems developed outside biology (Beckner 1959; Lyubarsky 2018). However, it should be reminded that C. Linné and some of his early followers extended the ranking idea to the geological matter (Linnaeus 1767; Scopoli 1777; Fischer 1811), and the ranked hierarchy of 'Linnaean' kind (with almost the same or similar terminology) is the standard in contemporary soil and mineral classifications (Bokiy 2000; Shishov et al. 2004; Buol et al. 2011; Samofalova 2013; Schertl et al. 2019). Besides, ranked classifications are elaborated in stratigraphy with applying another terminology both for ranks (categories) and classificatory units (Sokolov 1980; Salvador 2013). In the first half of the 20th century, there were attempts to apply it to the archaeological classifications (see Klein 1991 for a review), but no system of unified fixed ranks was worked out there.

* * *

The conceptual approach adopted as an 'ideological kernel' for this book necessitates a prior setting of an appropriate onto-epistemic framework for the discussion of the entire issue of ranked *vs.* rankless hierarchy of both systematic diversity and classification systems representing it. This framework should involve the elaboration of a certain *ranking theory* (mentioned above), which is not currently possible because of the absence of a developed hierarchy theory, so it just remains to outline here some of its essential prerequisites.

First, the basic onto-epistemic framework in question is expected to be constructed in line with the general organizational principles of conceptual constructs of this kind. It is arranged hierarchically as a conceptual pyramid, which highest level is occupied by certain onto-epistemic premises most pertinent to the ranking issue. This means that all *pros* and *cons* should be discussed not in an 'abstract' ('absolute') sense, but rather they should be assessed as credible or not based on their substantiation within the specific onto-epistemic backgrounds (Hempel 1966; Hull 2001).

In this connection, one philosophical idea suggested by conceptualism is pertinent to be emphasized here: this is a fundamental *onto-epistemic dualism* of any conceptions addressing the directly unobservable natural entities, one of which is biodiversity. Because of this dualism, it is not an easy, if ever solvable, task to demarcate with a reasonable certainty the ontic and epistemic components contained in the judgments regarding the hierarchical pattern of the diversity in question (Allen and Starr 1982; Ahl and Allen 1996; Wu 2013). Accordingly, such a dualism should be unquestionably taken into account in the discussions of a controversy between objective *vs.* subjective nature of either ranked or rankless arrangement of hierarchies.

> Going deeper in this dualisms and considering it from a certain cognitive perspective, it is possible to assume that the global biotic structure is a real natural phenomenon, whereas the hierarchically ordered diversity is its aspect-based *epi*phenomenon (see Sections 2.1 and 3.2). Bringing such a cognitive consideration to an extreme, one may say that all ideas regarding either ranked or rankless character of taxonomic hierarchy are *qualia* (in the sense of Jackson 1982), in which the objective and subjective sources of the respective judgments are inseparable.

The next lower levels of the pyramid in question are occupied by particular onto-epistemic models that serve to justify either ranked or rankless hierarchy of both systematic diversity and classification systems as its specific representations. Given the materialistic worldview, the ontic models (substantive theories) focus on the possible natural causes for the ranked/rankless structure of systematic diversity, which can be arranged into two partly overlapping basic categories, *evolutionary* and *structural*. On the other hand, if the epistemic part of onto-epistemic dualism is accentuated, it provides its own arguments for each of these alternatives, e.g., of pragmatic kind. At last, the far lower levels of the pyramid in question involve elaboration of *ranking scenarios*, being particular interpretations of hierarchization scenarios, with the basic *ranking principles* are elaborated, for which implementation the respective *ranking rulers* and *ranking criteria* are formulated at the lowest levels of this particular conceptual pyramid.

Such a very roughly outlined general onto-epistemic framework suggests that a balanced consideration should be given to *pros* and *cons* for either ranked or rankless arrangement of hierarchies. To approach this task properly, given the conceptual pyramid in question, it is important to focus before all on the ranking scenarios, which consistency should be assessed within the contexts implied by the relevant onto-epistemic backgrounds.

* * *

The above-mentioned major ways of taxonomic hierarchization, *absolute* or *relative*, are evidently relevant to the ranking, including the allocation of taxa to specific ranks. An absolute ranking involves the use of *exterior* ranking rulers, which provide a possibility for any classification systems to be ordered in a certain unified fashion. Such a possibility has its own merits. First, each of such rulers allows for any particular taxon to be allocated to any particular rank independently of any other taxa. Second, they make distinct fragments of the global classification system comparable with respect to their ranked structure. Third, this comparability holds both for symmetric and asymmetric (imbalanced) hierarchical arrangements of classification systems. Two such 'absolute' rulers of ontic kind were mentioned above, geological timing and adaptive zonation scales, whereas a ranking system based on the numerically estimated similarity levels exemplifies the epistemic 'absolute' ruling.

The linear ranking rulers that shape the empirical foundation of any ranking systems can be treated as one of several variants of the *ordering scales* (in the sense of Orlov 2006); another well-known variant is the *ordinal numbers* (*ordinals*) in set theory (see Sierpinski 1965; Hrbacek and Jech 1999 on them). Such an analogy makes it possible to link the origin of the ranking idea in systematics with the emergence of number systems and numerical series in mathematics, so the ranking can be likened to the measuring, in the sense that the ranking rulers obliquely 'measure' the hierarchical structure of the classified diversity in a specific manner (Klix 1985; Gusev 1995; Lyubarsky 2018).

Ranking rulers and resulting ranked hierarchies can be *monotonic* or not: the first condition implies that all ranks are 'vertically' *equivalent*, the second one implies their *nonequivalence* with regard to some of their features. The second condition is most characteristic of traditional systematics, in which *primary* (principal) and *secondary* (auxiliary) ranks are distinguished. Such their categorization originated historically and got strengthened with the fragmentation of ranked hierarchy as a specific 'technical' device for making the latter more manageable, which indicates its basically epistemic (pragmatic) meaning (Stevens 1994; Pavlinov 2018). However, rank nonequivalence is sometimes interpreted ontically, according to which primary ranks are endowed with a 'natural' status and auxiliary are not (Usov 1867; Mednikov 1974; Vasilieva 1992, 2007; Efimova 2005; Shatalkin 2012). Such their differentiation ascends to C. Linné, who distinguished between (a) genera and species as 'the works of Nature' and (b) orders and classes as 'the works of Nature and Art' (Linnaeus 1751). A similar fundamental dichotomy is supported by some recent taxonomic theories, in which the primary categories are endowed with a certain reference function as *basic-level* ones, being salient for some ontic and/or epistemic reasons (Rosch 1973, 1977, 1988; Brown 1991; Berlin 1992; Lassaline et al., 1992; Markman and Wisniewski 1997; Atran 1999a; Medin and Atran 2004; Lyubarsky 2018). For instance, evolutionary theory may ascribe the salient status to species treated as a basic unit of evolution, whereas paleontologists treat families as basic in their analyses of the geological history of global organismal diversity (e.g., Raup and Sepkoski 1984; Sepkoski 1996; Hallam and Wignall 1997; Markov and Neimark 1998; Contoli and Pignatti 2011; Turner 2011).

One of the focal propositions concerning ranking is the *principle of ranking uniformity*, which holds both for the particular ranking principles and rulers applied to a given taxonomic system or a set of interrelated systems. This principle serves as a kind of heuristics for the ranking of classifications in two important respects. Conceptually, it makes both the ranks, rank categories, and equiranked taxa 'horizontally' *equivalent*, at least potentially, which holds for their both ontic and epistemic consideration. Operationally, this principle substantiates a possibility for the ranking rulers to be shaped by the uniformly expressed numerical distances, so the taxa are ranked according to values of the latter (Sokal and Sneath 1963; Sneath and Sokal 1973; Abbot et al. 1985). This approach is especially popular in the contemporary studies of species diversity based on the *genetic species concept*, which makes species delineation strictly operational and universal (e.g., Mallet 1995; Pleijel and Rouse 2000; Blaxter et al. 2005; Baker and Bradley 2006; Staley 2006; Nebel et al. 2011; Lebedev et al. 2024).

The assumed ontic equivalence suggests that at least primary taxonomic categories can be individuated based on certain 'natural' properties shared by the equiranked taxa allocated to each of them, so they are admissible to be regarded as *natural kinds* (in the sense of Quine 1969; Mahner and Bunge 1997; Spencer 2016). Such a treatment makes equiranked taxa the members of the same natural kind, and the groups of organisms represented by them can be considered equivalent and comparable as the elements of biotic structure with respect to some of their biologically significant features (Simpson 1961; Ghiselin 1969; Van Valen 1973; Starobogatov 1989; Lyubarsky 1991, 2018; Shatalkin 1995, 2012; Dominguez and Wheeler 1997; Vasilieva 1998, 1999, 2007; Rasnitsyn 2002; Giribet et al. 2016). In particular, such equiranked groups are treated as equivalent in the studies of the structure of biodiversity by biogeography, paleontology, wildlife conservation, etc. (Simpson 1961; Raup and Sepkoski 1984; Pik et al., 1999; Rasnitsyn 2002; Kafanov 2005; Bertrand et al. 2006; Contoli and Pignatti 2011). However, it is this equivalence that is most strongly criticized by the opponents of ranked hierarchy (see below).

In contrast, an epistemic consideration of ranking uniformity does not seem to imply necessarily any substantive 'horizontal' equivalence of ranks and equiranked taxa. This reservation holds, before all, for the ranking rulers established based on numerical distances. Indeed, although the respective figures for the hierarchically arranged phenograms look the same, they actually are not comparable directly if based on the essentially different characters and/or different similarity/distance measures (Moss and Hendrickson 1973; Shatalkin 1983). Therefore, the above-mentioned enthusiasm regarding unified delineation of species and some analogous 'surrogate' units based on the numerical approaches (Mallet 1995; Mayden and Wood 1995; Riddle and Hafner 1999; Pleijel and Rouse 2000; Blaxter et al. 2005; Baker and Bradley 2006; Staley 2006; Nebel et al. 2011; Casacci et al. 2014; Lebedev et al. 2024) does not seem to be much justified biologically.

* * *

An appropriate starting point for the ontic-based discussion of the dilemma of ranked *vs*. rankless taxonomic hierarchy is provided by the representation of global biota as a self-developing non-equilibrium system (in the sense of Prigogine 1981; see Section 2.7). Such its consideration (a) highlights inherent structuring accompanying the historical development of biota and (b) encompasses both evolutionary and structuralist approaches to the analyses of its hierarchies. This standpoint evidently emphasizes the evolutionary development of biota as the major driving force for its hierarchical structuration. Accordingly, any discussion of this dilemma should address first of all the underlying evolutionary theories/models and respective hierarchization scenarios implied by them. In this regard, of primary importance becomes the question of either *monotone* (continuous) or *punctuated* (non-continuous) character of macroevolution, which presupposes the recognition of either rankless or ranked hierarchy of systematic diversity, respectively. This dichotomy can be represented as either *microreductionist* or *macroholistic* view of the global evolutionary process (Gontier 2015b).

* * *

Of these two evolutionary models, the microreductionist one suggests the simplest version of the general hierarchization scenario. Its natural philosophical background is shaped by the Leibnizian aphorism *Natura non facit saltum* (nature does not make jumps), and its biologically sound background is based on the 'Darwinian' evolution, which goes by 'very short and slow steps' (Darwin 1859, p. 471). When applied to macroevolution, such scenario reduces the entire historical development to the sequential speciation events. It is quite evident that when considering the evolutionary process from such a microreductionist standpoint, any large-scale macroevolutionary events that can entail quasi-discrete gaps in phyletic lineages will hardly be discovered (Grinin et al. 2009; Turner 2011).[14]

Such evolutionary process can be interpreted as an accumulation of the sequences of cladistic events without distinctive gaps. The respective hierarchization scenario implies that the resulting cladogenetic pattern is shaped by the inclusive hierarchy of holophyletic groups generated by those events. Accordingly, the latter's sequence defines a universal hierarchization ruler as a monotonous sequence of equidistantly distributed hierarchization steps. Such a cladogenetic pattern is modeled by a cladogram, whose tree paths are characterized by (a) evenly distributed nodes (branching points) and (b) no-meaning branch lengths. This character may be locally impaired by making taxonomic hierarchy *degenerate*, which was suggested to be termed as *pseudoranked* (Dubois et al., 2019, 2022). Degenerate hierarchy involves the possibilities of (a) allocation of taxa of the same level of generality to formally different ranks and *vice versa*, (b) introduction of out-of-rank category of *plesion*, and/or (c) preserving some non-holophyletic *metataxa* (Wiley 1979, 1981, 2009; Schoch 1986; Gauthier et al. 1988; Craske and Jefferies 1989; de Queiroz and Donoghue 1990; Pavlinov 1990a, 2005; McKenna and Bell 1997).

The 'moderate' cladists see no obstacle to the transformation of such a 'monotonous' tree into a cladistically consistent ranked classification, with the reservation of a conditional character of taxonomic ranks (Hennig 1966; Brummitt 1997, 2002;

Benton 2000; Nixon and Carpenter 2000; Stuessy 2000; Carpenter 2003; Keller et al. 2003; Monsch 2006; Dyke and Sigwart 2007; Rieppel 2006c, 2008; Dubois 2007; Kraus 2008; etc.). However, the devoted supporters of 'ultra-cladistic' methodology take such character of cladograms as a deciding argument against any taxonomic ranking (Eldredge and Cracraft 1980; de Queiroz 1988, 1997; Ereshefsky 1994, 2002, 2004; Minelli 1995; Zachos 2011; etc.). It is of interest that some opponents of rankless classifications do not treat them as taxonomic (Mayr and Bock 2002; Hörandl 2010; Lyubarsky 2018).

Biosystematicians consider within-species hierarchy in a similar microreductionist manner, which is substantiated with the reference to the continuous degrees of microevolutionary differentiation of populations (Semenov-Tyan-Shansky 1910; Turesson 1922; Thorp 1940; Camp 1951; Valentine and Löve 1958; Lines and Mertens 1970). This very fragmented 'microhierarchy' can be either ranked or rankless. The elaborators of synthetic theory of evolution (STE) recognize such a complex sequential within-species differentiation but consider the species category ontically ('objectively') salient, referring to certain biological mechanisms responsible for the species integrity (Dobzhansky 1935; Simpson 1951a, 1961; Zavadsky 1968; Mayr 1982, 1987; etc.). However, 'anti-rankists' disregard the arguments from STE and expand the rankless idea to the 'around species' taxonomic hierarchy, as well (Minelli 1995; Kober 2008; Zachos 2016; Mishler 2021, 2022), which is indirectly supported by the contemporary phylogeographic approach (Avise et al. 1987; Avise 2000; Knowles and Maddison 2002; Dawson et al. 2006; Edwards et al. 2021).

A certain complication of such a microreductionist hierarchization scenario is provided by the application of the *timing scale*, which serves to measure a kind of 'evolutionary age' of the groups of organisms. According to the above-discussed major versions of hierarchization parameters, such a timing scale shapes a specific universal 'absolute' exterior hierarchization ruler, which can be either direct or indirect. The former is inferred from the paleontological data on fossils, whereas the latter is based largely on the comparative molecular data on recent/subrecent organisms; these may be combined. The respective hierarchization/ranking ruler implies that the 'older' a group, the higher the generality level (rank) of a taxon representing it in the respective classification (Hennig 1950, 1965, 1966; Griffiths 1976; Sibley and Ahlquist 1990; Christoffersen 1995; Avise and Johns 1999; Avise and Mitchell 2007; Fisher 2008; Avise and Liu 2011; Kraichak et al. 2017). This hierarchization scenario is most usually applied to the substantiation of ranked hierarchy, although its rankless interpretation is also possible.

In the ranked case, the respective ranking scenario implies the following stepwise hierarchization algorithm (Avise and Johns 1999; Holt and Jønsson 2014; Jønsson et al. 2016; Kraichak et al. 2017). A particular taxonomic hierarchy is first established by any appropriate algorithm applied to any kind of data, after which it is aligned according to a particular timing ruler. The latter is shaped as a sequential *temporal banding* with the conventionally established *band widths* as specific 'windows of time'. The provisional hierarchy is projected onto the temporal banding sequence, so that the taxa falling within the same temporal band are grouped into a distinct 'age class' and are ascribed the same taxonomic rank. The American zoologist Graham

Griffiths (1937–2009) suggested that the taxa belonging to the respective 'age classes' be specifically denoted: for instance, the terms *Paleotaxon*, *Eotaxon*, *Oligotaxon*, *Miotaxon*, and *Pliotaxon* would refer to the groups having originated during respective epochs of the Tertiary (Griffiths 1976).

The 'absolute' indirect timing ruler has a great potential for the taxonomic ranking, as it can be uniformly calibrated and applied to any fragments to the overall Tree of Life. This approach, which was initially based on the neutral model of molecular evolution (Kumar 2005), provides a possibility for any taxon, for which the respective molecular data are available, to be ranked independently of all others, and all taxa ranked in such a fashion are potentially comparable with each other with respect to their timing across that Tree. However, it faces the fundamental challenge of an uneven run of molecular clock, with the latter's speed being specific to different evolving groups and molecular structures, whereas certain 'technical' obstacles make synchronization of molecular clock in different lineages rather uncertain (Ayala 1999; Marshall et al. 2016). It should be highlighted that these results agree with the general idea of heterogeneity of time (in the sense of Prigogine 1981; Levich 1996). Besides, the notable discrepancies between timing estimates by molecular clock and paleontological data, which cause the so-called *ghost ranges* on phylogenetic trees (Donoghue and Smith 2003; Wills 2007), also contribute to the ranking uncertainty based on such a timing ruler. All these facts, though they do not violate the fundamental principle of ranking uniformity altogether, make its implementation rather problematic because of an approximate and fuzzy character of any timing estimates.

* * *

The macroholistic approach to the organismal diversification is based on a systemic worldview and implies far more complex hierarchization scenarios regarding macroscale structuration of the overall biological (functional and structural) organization, including complexity, body plans, and ontogenetic patterns (Schmalhausen 1968; Corning 1983, 2022; Nazarov 1991; Iordansky 1994; Rasnitsyn 2020; Ochoa 2025). These scenarios can be reasonably divided into two major groups: *macroevolutionary* ones accentuates historical changes of organisms, whereas *structural* scenarios are of stationary kind.

The macroevolutionary scenarios consider evolutionary process at a macroscale and correlate major evolutionary events not with the local acts of speciation, but rather with the global development of the living. Accordingly to this viewpoint, the respective transformations, although they take up to several millions years, are treated as *quasi-discrete* events within the several milliard years of the historical development of biota. This general model suggests an ordered character of macroevolution that proceeds by regular parallel trends and involves successive stepwise alternations of the *aromorphous* (arogenetic, anagenetic) and *idioadaptational* phases of organismal transformations (Figure 3.4). The first of these involves emergence of *key innovations* as relatively quick significant changes in biological organization, whereas the second involves relatively slow and wide diversification within the achieved new capacities (Sewertzoff 1929; Schmalhausen 1939; Simpson 1944; Nitecki 1990; Nazarov 1991; Iordansky 1994, 2001; Heard and Hauser 1995; Vamosi

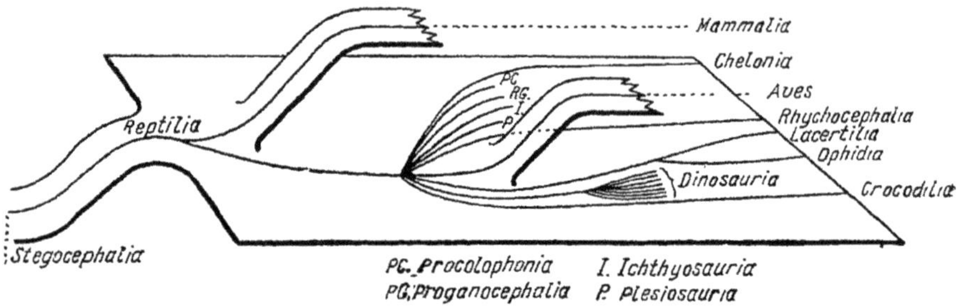

FIGURE 3.4 Relation between aromorphous and idioadaptational phases in the tetrapod macroevolution (Sewertzoff 1929, p. 332).

and Vamosi 2011; Wagner 2011). It is assumed that the aromorphous transformations in biological organization may be interpreted as *phase transitions* between qualitatively specific states of a complex evolving system (Scott et al. 2022; Muro et al. 2025). This classical model is currently being refined by the *evo–devo* one, which highlights epigenetic mechanisms for the qualitative changes in ontogenetic patterns entailing respective more or less drastic transformations of various scales (Løvtrup 1989; Shatalkin 1993, 2012; Minelli 2003, 2015; Shishkin 2006; Laubichler and Maienschein 2008; Lewens 2009; Riegner 2013; Minelli and Pradeu 2014; Ochoa 2025). The ordered character of macroevolution is formalized by the conception of *phylocreod* endowed with a *canalized evolutionary potential* (Waddington 1962; Meyen 1984; Amundson 1998; Wagner and Stadler 2003; Ochoa 2024, 2025), and its stepwise character is formalized by the conception of *quantum evolution* (Simpson 1953; Grant 1963; Nazarov 1991; Vasilieva 1998; Gould 2002; Rasnitsyn 2020; not in the sense of McFadden 2001).

The respective general ranking scenario takes into consideration all manifestations of evolutionary transformations suggested by this macroevolutionary model, including both divergent and parallel trends, emergence of key innovations making the general trend stepwise, and complex epigenetic regulation of evolutionary transformations. Its assumed stepwise character implies that the inclusive hierarchy of the resulting phylogenetic pattern (a) is structured by the quasi-discrete levels, which can be treated as ranks, and (b) encompasses all broadly understood (Haeckelian) morpho-biologically prominent monophyletic groups, which may be both holo- and paraphyletic.[15] According to such a scenario, a specific ranking ruler is elaborated, in which the sequences of aromorphous phases within phylogenetic lineages define ranking step sequences. It is further assumed that the biologically meaningful scaling of organismal transformations defines the linear hierarchy of rank levels, with the larger transformations corresponding to the higher levels. The groups of organisms which are diagnosed by the transformations of (nearly) the same scale are allocated to the same taxonomic rank (category). Within each taxon thus defined, its own lower-level ranked structure is determined by its diversification corresponding to the idioadaptational phase of its evolution. This interior ranking ruler, formed by the organismal features involved in the evolutionary transformations, can be supplemented by the exterior ones. One of them is formed by a hierarchy of specific adaptive zones, the penetrations into which have stimulated those transformations (Simpson 1961; Van Valen 1971); another is timing ruler, which should be applied not instead of but in addition to other rulers (Lücking 2019).

This macroevolutionary approach to the elaboration of ranking scenarios is usually criticized by the proponents of a positivist-style simplified reasoning, who believe that the macro-level hierarchies devoid of any signs of stepwise character (e.g., Bentham 1875; Kozo-Polyansky 1922; Michener 1963; Griffiths 1976; Løvtrup 1977; de Queiroz and Gauthier, 1992, 1994; de Queiroz and Cantino, 2001; Mishler 2009; Zachos 2011). Accordingly, they regard any judgments about stepwise structure of both evolutionary processes and resulting phylogenetic patterns as too subjective and referring to nothing real. Leaving aside the question of the validity of this criticism, it should be recalled that from the point of view of conceptualism, it proceeds from a certain onto-epistemic position, which looks no less speculative, although simpler (see Section 2.6), and, therefore, is equivalent in this respect to the one being criticized. What does actually provides a serious challenge to the implementation of the macroevolutionary ranking scenarios is the above-mentioned *principle of NP-completeness*, which inversely correlates the reliability of any empirical solutions to (in our case) the ranking issue with the complexity of the underlying evolutionary models (Felsenstein 2004).

The macroholistic structuralist-based hierarchization scenarios, in their 'pure' non-evolutionary guise, generally proceed from the seeing the Earth biota, considered from a morpho-biological perspective, as a specifically organized hierarchically structured whole. There are two major looks at the latter's nested hierarchy: organicism natural philosophy (Oken, etc.) likens it to a super-organism, whereas the classical typology (Cuvier, Geoffroy de Saint-Hilaire, Owen, etc.) treats it as a hierarchy of body plans of various levels of generality. In both cases, the overall hierarchical structure of the wholeness of biota is likened to the gross anatomical construction of real organisms, from which it is mentally inferred (Naef 1931; Troll 1951; Beklemishev 1994; Shatalkin 1994, 2012; Lyubarsky 1996; etc.). Besides, the pattern cladistic conception of the *levels of generality of defining characters*, which is directly linked by its authors to the Baerian developmental typology (Platnick 1979; Nelson and Platnick 1981), should be mentioned in this regard. Such a largely typological viewpoint implies partonomic partition of the whole into its parts (Lyubarsky 1996, 2018), and the resulting sequences of the hierarchy levels for particular archetypes (body plans) serve as the relevant 'meronomic rulers' for the taxonomic hierarchization. This entails a specific ranking scenario for the structure of systematic diversity, which implies an assumption that the basic constituents (merons) of an imaginary super-organism (an insect, a mammal, a flower plant, etc.) determine the respective structural units of the studied systematic diversity. Accordingly, the hierarchically arranged levels of generality of those constituents define ranks in the hierarchies of taxonomic systems elaborated for the real groups of respective organisms.

This general structuralist (primarily typological) conception is traditionally objected by evolutionists of various schools for its 'idealistic' nature and seemingly stationary world picture (see Mayr 1982 for a summary), and this standpoint is strengthened by a strange declaration that the 'structuralist thinking necessarily eliminates any reflection of macroevolution in classification' (Zander 2010a, p. 391). However, such criticism seems to be based on a simplified and biased view of typology and does not take into account the latter's quite organic embodiment into contemporary evolutionarily interpreted structuralism (Griffiths 1996; Amundson 1998, 2005; Gayon 1998; Vasilieva 1998; Brigandt 2007, 2009, 2017; Laubichler and Maienschein 2008; see also Section 2.8). In particular, as said above, this structural scenario is actively discussed within the scope of evo–devo concept (Laubichler and Maienschein 2008; Lewens 2009; Hall 2012; Riegner 2013; Brigandt 2017; Jaeger 2019; Villegas and Triviño 2023; Ochoa 2025).

However, as far as systematics is concerned, there still remains an important vacuity in the general structuralist argumentation scheme, namely, a vague substantiation of the ranking principles for taxonomic diversity with the reference to structural interpretation of the imaginary super-organisms. Within the frame of 'transcendental' typology, the respective rulers are based on the levels of generality of merons within generalized archetypes, and this is claimed to be the 'ranking objectivation' (Lyubarsky 1991, 1996, 2018). However, the latter suggestion seems to be largely unjustifiable because of a significant impact of epistemic consideration of concrete organismal structurations (Pavlinov 2018). It might be possible that at least some developmental patterns assumed by the evo–devo scenarios, such as the stepwise changes in the evolving ontogenetic patterns by *phylotypic stages* (Hall 1996; Slack et al. 1993; Duboule 1994; Richardson et al. 1998; Martynov 2012), will be helpful for a more correct formulating and resolving the structuralist ranking issue.

* * *

The ranking scenarios of basically epistemic kind are most usually elaborated as *pragmatic*: the conventionally ranked taxonomic hierarchies serve as a means for making the wide-scope taxonomic systems easier manageable (Stevens 1994; Lyubarsky 2018; Pavlinov 2018). Such scenarios and respective ranking rulers are generally regarded as something 'given by tradition', so their substantiation indirectly relies on a kind of 'case law': their practitioners just reproduce the ranking systems established by their predecessors for the given groups. The above-mentioned suggestions of phenetic distance-based ranking rulers seem to be of the same epistemic kind, to the degree they do not imply any ontic substantiation.

A curious outcome of such a pragmatic 'case law' ranking approach is the current practice of ranking newly discovered prokaryotic and earliest eukaryotic groups with an eye to those that have been delineated and ranked a century or more earlier. Indeed, regarding the overall Tree of Life, these new groups usually branch off at its base and are therefore ascribed the highest ranks, such as kingdoms or phyla (e.g., Whittaker 1969; Cavalier-Smith 1993, 2007; Kusakin and Drozdov 1994; Margulis and Schwartz 1997; Cracraft and Donoghue 2004; Shatalkin 2004, 2012). This results in a 'multi-regnal' structure of the global ranked taxonomic system, and Carl Linné probably observes it from heaven with a satisfied smile.

An epistemic consideration of the ranked taxonomic hierarchy carries out a kind of solid framework function for the overall classification of the living and thus makes its fragments comparable at leas in some cognitively significant respects. In this regard, it may be likened to the above mentioned ordinal numbers in set theory, which are natural numbers indicating order of particular inclusive hierarchy levels in relation to other levels, such as first, second, and third orders (Jardine and Sibson 1971). In particular, both ranks and ordinals are similar in that their sequences are finite, so their orders can be enumerated; in this important feature, any ranked hierarchy differs from the rankless one, as the latter is infinite and therefore more likens fractal hierarchy (see Section 3.4 on the latter).

Of great importance is that, as already noted above, such an ordering framework function of the ranked hierarchy likens it to the Cartesian coordinate system of graphical representation of the imaginary quasi-spatial relations between objects (Lyubarsky 2018). An important corollary of such a likening is that hardly anyone insists on the 'objectivity' of the coordinate system—but hardly anyone denies its meaningfulness as a special kind of descriptive metaphoric 'framework model' of the real physical world. The same seems to be justifiably and rightfully extended to the ranked hierarchy in systematics: even if it is more or less 'artificial', it nevertheless serves as an effective means for a 'framework' description of the hierarchical structure of systematic diversity. Another very significant analogy is interpretation of the ranking system as a specific 'cognitive map' (Atran et al. 2002).

It is to be reminded that the Cartesian coordinate system is widely used in numerical systematics for the construing multidimensional phenetic hyperspaces, which axes correspond to the classificatory characters (Sokal and Sneath 1963; Sneath and Sokal 1973; Abbot et al. 1985; Sokal and Rohlf 2009). Such a representation most often performs a technical function for a preliminary arrangements of organisms for their subsequent classifying, but it is sometimes considered as a full-fledged classification (Dupraw 1964, 1965).

* * *

The following conceptual propositions about dilemma of rankless *vs.* ranked hierarchies are useful to be briefly highlighted in order to summarize the above consideration.

First, the hierarchical structures of both systematic diversity and classifications representing it are fuzzy, i.e., uncertain to a degree. Applying this observation consistently, one have to assume that, among other things, the very boundary between the rankless and ranked hierarchy representations is fuzzy, so their 'absolute' opposition seems to be incorrect. Such a specific fuzziness is one of the multiple manifestations of the global taxonomic uncertainty that outlines some semantic area, within which it becomes unreasonable to insist upon an 'ultimately true' of either of these two representations.

Thus, instead of being likened to Jonathan Swift's Big-Endians and Little-Endians in their persistence in the endless and futile debates, it would be more beneficial to try to find other, more 'peaceful' and constructive solutions of the considered disagreement. For instance, why not think of ranks as fuzzy constructs being endowed with certain 'confidence intervals', so that if the contiguous ranks in a hierarchy overlap, they could be acknowledged as indistinguishable, and the respective fragment of a taxonomic hierarchy could be treated as a 'locally rankless'? It might also be reasonable to recognize a kind of 'hierarchization pluralism', implying that systematic diversity can be differently hierarchized in its particular fragments, which could be represented by rankless or ranked structures, accordingly (Reif 2003; Shatalkin 2012; Hamilton 2014).

Considering taxonomic ranks from such a 'fuzzy' perspective brings the *principle of ranking uncertainty*, as another manifestation of the just-mentioned taxonomic uncertainty, to the fore (Zarenkov 1988; Pavlinov 2018). This implies that (a) specific ranking rulers may imply somewhat different ranked hierarchies, (b) particular ranks cannot be unambiguously defined, (c) particular taxa cannot be unambiguously allocated to specific ranks, and (d) an equivalence of the equiranked taxa is not strict. Indeed, zoologists generally agreed that the orders of mammals were supposed to be equivalent to the families of insects in some biologically meaningful parameters, such as geological age and diversity/disparity. Such a nonequivalence of the allegedly equiranked taxa in the distant fragments of a global classification is traditionally taken for an irrefutable argument against ranking by the latter's opponents (Hennig 1950, 1966; Van Valen 1971; Shatalkin 1988; Rasnitsyn 1996; Minelli 2000), but the last of the just-listed uncertainty provisions seems to partly weaken this objection.

It should be reminded that at least some sort of ordering of taxa through their ranking is of use in the analyses of taxonomic diversity (Raup and Sepkoski 1984; Sepkoski 1996; Hallam and Wignall 1997; Markov and Neimark 1998; Contoli and Pignatti 2011). To the contrary, the application of rankless hierarchies entirely deprives the respective research of an important unified reference system of ranks as a basis for the broad and meaningful taxonomic comparisons (Vences et al. 2013; Lyubarsky 2018).

As intriguing as discussion about rank equivalence/nonequivalence may look, its conceptual consideration allows for seeing in this issue a new perspective. For this, let us consider a classification system as the one which is ordered by the information flows between its taxa, with these flows running either along the hierarchy paths or across them. The first option is implied by the application of interior hierarchization/ranking rulers, which provides a local relative leveling/ranking of taxa within particular hierarchy paths. The second option results from the application of exterior rulers, which provides a global 'absolute' leveling/ranking of taxa within the entire classification systems. To ensure a more fundamental background for such a consideration, it should be placed in the context of two basic models of causal interactions, *short-range* (proximity) and *long-range*, elaborated in cosmology (e.g., Hawking and Ellis 1975; Carroll 2004). For the purposes of our analysis, these models can aptly be formalized as the principles of the same name (short-range and long-range), which generalize the effects of application of the interior and exterior rulers, respectively, in a specific fashion (Pavlinov 1998, 2018). This analysis is facilitated by the usage of some important features of non-equilibrium systems and tree-like style representation of the hierarchical arrangements (see Sections 2.7 and 4.2 on these).

The application of interior ranking rulers is assumed to entail the short-range information interactions between tree nodes corresponding to taxa in a classification system. The information flows between them involve some amount of uncertainty in each transition between any two adjacent tree nodes, which means the lost of some information contained in the starting node. This evidently entails an accumulation of the information losses with increase of the distance between nodes along the hierarchy paths connecting them. Therefore, the more distantly any two taxa, corresponding to the respective tree nodes, are situated in a classification system, the less possibility for them to be meaningfully correlated in their biologically significant properties. With this, such a short-range effect seems to increase with the increase of hierarchy asymmetry, and this effect is manifested in a full capacity in the case of rankless hierarchies, whereas it is somehow smoothed by introducing discrete ranks.

Unlike this, the application of exterior hierarchization/ranking rulers is assumed to imply the long-range information interactions between taxa in a classification. This is because all its taxa are aligned with the reference to such a ruler (a) uniformly and (b) independently of each other, and this can be treated as if their information interactions are mediated by the exterior ruler. Accordingly, the respective information flows do not run by long hierarchy paths connecting the taxa but move across them through the ruler uniting them 'near-immediately' with a minimal information lost. Such a long-range effect makes taxa quasi-equivalent regardless of their position in a global classification system, which is most evident in the case of ranked hierarchies. It should be added that the alignment of a rankless hierarchy relative to a certain exterior ruler, such as the timing one, makes it 'near-ranked'.

NOTES

1 Unfortunately, the term 'laveling' is applied in quite a different sense to denote flatting surface, so these two usages should not be confused.
2 It is to be kept in mind that many taxonomists do not distinguish between the notions of character and modality.
3 Please note that the term 'typological hierarchy' is used in linguistics with quite another meaning (e.g., Cristofaro and Zúñiga 2018).
4 Not *polyhierarchy* in the technical sense of computer science, which refers to a possibility for a system constituent (notion, conception, etc.) to exist in more than one place in a hierarchy, or for a 'child' to have several 'parents' in a lineages network (e.g., Babikov et al. 2004; Hedden 2010; Morville et al. 2015; Laubheimer 2018).
5 The notion of levels of organization is usually applied in quite another meaning (e.g., Bertalanffy 1969; Bunge 1969; Auger 1989; Crumley 2005).
6 This pair of terms is probably close to those applied in the number theory of mathematics (e.g., Chattopadhyay et al. 2022).
7 Currently, N. Eldredge seemed to get back to the historical terminology by using Haeckelian 'economic' instead of now commonly known 'ecological' (Tëmkin and Eldredge 2015; Eldredge 2016).
8 As said above, this general 'causation' rule does not seem to be observed in the case of 'Chinese boxes' or Russian matryoshka dolls, which are integrated by spatial relation only (King 1997).
9 Not in the sense of 'category' introduced by E. Rosch (1973, 1988), see Section 1.2.
10 In this regard, it is worthy mentioning that the numerical classificatory methods based on fuzzy methodology have been already being developed (e.g., Buckley 1985; Bezdek et al. 1999; Bora and Gupta. 2014; Grover 2014; etc.).

11 Its predecessor was a *general meronomic-taxonomic relation* formulated by Georgy Lyubarsky within his version of 'new typology', with a special emphasize on the meronomic aspect of organismal diversity (Lyubarsky 1991, 1996, 2018).

12 W. Zimmermann's (1934) original term was *semophylesis* as 'phylogeny of single character' (p. 382).

13 It is to be noted that Beklemishev's book was actually prepared in the late 1920s.

14 It is very pertinent to recall the famous A. Einstein's allegory: 'When a blind beetle crawls over the surface of a curved branch, it doesn't notice that the track it has covered is indeed curved' (Isaacson 2008, p. 197).

15 It is to be reminded that the Copean evolutionary model pays much attention to the broadly paraphyletic (i.e., near-polyphyletic) groups (Cope 1868, 1887).

4 The Hierarchies Representations

What turns something into a scientific representation of something else?
R. Frigg and J. Nguyen (2016)

In order to effectively analyze any directly unobservable natural (epi)phenomena, including various aspects of biodiversity and their hierarchical structures, it is necessary to represent them in a meaningful and perceptible form. The latter is commonly referred to as *knowledge representation* (Bergman 2018), and its particular aspect of our prime interest is *scientific representation*. They are designed to reflect the objects' properties of interest, including interrelations between their constituents, while their perceptibility is ensured by certain technical (including conceptual) 'devices'. The latter can be regarded as specific *metaphors* allowing for the abstract conceptions to be expressed through some more familiar and simple items (Kittay 1987; Punter 2007; Gibbs 2008; Kövecses 2010). Both the conceptions and their metaphorical representations serve as cognitive models of various levels of generality, and they compose another deductive hierarchy as a part of the overall conceptual pyramid of cognitive situation: conceptions and notions constitute its first-order level, while their particular representations belong to its second-order level (Kittay 1987). Accordingly, to the extent that such technical 'devices' are well-designed by their creators and adequately understandable by their target users, they can facilitate comprehension of the interrelations between constituents of the complex systemic entities (Bertin 1983; Beghtol 1996).

The practical implementations of such 'technical metaphors' depend on the nature of the modeled entities and the questions addressing them. Being conceptual in their background, representations can actually take various forms, including textual, graphical, mathematical, and physical. Most of them emerged as conceptually loaded with certain hierarchization scenarios, although a few seem non-theoretical; their formats and popularity have changed over time.

In biological systematics, the most significant are two representation styles that differ markedly in both shaping and perceiving information they contain, namely, *textual* (including linguistic and 'formular') and *graphical* (including diagrammatical) (Messaris 1994). The representation formats of the latter were suggested to be denoted as *conceptual graphs* (Sowa 1976) and treated as special signs of a definite internationally understandable 'graphical language' (Brinton 1919). Of these, two principal kinds are important to be recognized for our purposes, *tree-like* and *plane-like* diagrams. The visually perceivable graphical representation style was the earliest in the history of humankind, as evidenced by the well-preserved petroglyphs, but the textual style seemed to be primary in the history of rationality.

DOI: 10.1201/9781003597094-4

There are two principal requirements for any representation style: (a) it should clearly characterize the hierarchization scenarios underlying respective classifications and (b) its structure should be easily readable and understandable. It is evident that both provisions require certain education and skills. This chapter reviews the most important representation styles and their diverse formats, which are implied in biological systematics.

4.1 THE TEXT-LIKE STYLE

A simple verbal expression of the knowledge about the surrounding worlds by the use of specific terms and their combinations had preceded a more developed text-like representation style. The latter could be traced back to the earliest written cultures, it dominated in Antique natural philosophy, and it permeated the entire history of systematics. Other text-like representation styles emerged as its technical add-ons and gained popularity with the progress of more formalized methods for representing scientific knowledge in general.

In the earliest history (or rather, late prehistory) of systematics, the textual representation of the diversity of wildlife was mostly pragmatically motivated, with the *alphabetic* arrangement of taxa based on the initial letters of their names having been quite popular among herbalists (Sachs 1906; Arber 1938; Kupriyanov 2005; Pavlinov 2021a; Minelli 2023b) (Figure 4.1). This format is usually characterized by modern scholars as more practical than theory-laden, though some influence from Platonic or Kabbalistic ideas might be suspected. The latter imply that the names of organisms are not just random letter combinations, but meaningful, so that their alphabetic arrangement may relate to their hidden meaning within an enigmatic 'world of words'.

The subsequent development of the textual representation style freed it of its previous mostly pragmatic motivation with the accompanying formats, including alphabetic one, and suggested a new, *systemic* textual arrangement of taxa. This new format should in some way reflect the idea of *Systema Naturae*, though it served also as a pragmatic 'search engine' (Winsor 2025.). Its most simple linear format appeared to be equally applied to both non-inclusive and inclusive hierarchies by arranging the named items into specific lists. The former is illustrated by a progression of the named organisms along the Scale of Being in C. Bonnet's *Traite d'insectologie* (Bonnet 1745; see Figure 1.8), and the latter is exemplified by the sequence of taxonomic ranks ordered as a 'scale of inclusiveness' in A. de Candólle's *Lois de la nomenclature botanique* (see Figure 1.4). It is of interest to note that (probably for the historical reasons), the representation of Scale of Being was of Aristotelian ascending format, whereas that of 'scale of inclusiveness' followed Platonic descending format. The Candólle's list illustrated one of popular textual arrangement implying a stepwise left-to-right indentation of the designations of named items (in his list, these were taxonomic categories), aligned according to their hierarchical levels (ranks); Marc Ereshefsky highlights the significance of this format (Ereshefsky 2004).

The more extended taxonomic systems, known as *synoptic classifications*, include the lists of particular taxa with some associated data, such as lists of synonyms, bibliographic data (publication dates and authors), distributional data (geographic and geological), and sometimes diagnostic characters. Their information contents may

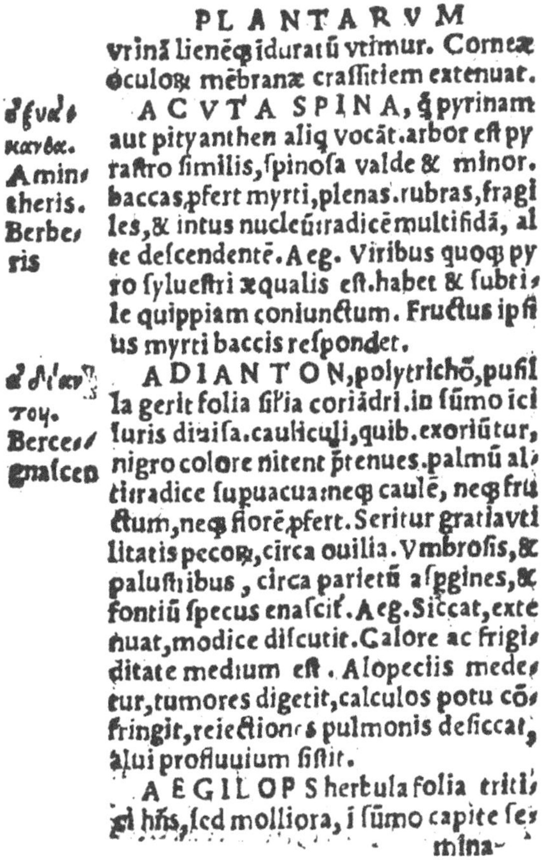

PLANTARVM
vrinã lienéq; induratũ vtimur. Corneæ
oculorũ mẽbranæ craſſitiem extenuat.
ὀξυά- ACVTA SPINA, q pyrinam
κανθα. aut pityanthen aliq vocãt. arbor eſt py
Amin- raſtro ſimilis, ſpinoſa valde & minor.
theris. baccas pfert myrti, plenas rubras, fragi
Berbe- les, & intus nucleũ radicẽ multifidã, al
ris te deſcendentẽ. Aeg. Viribus quoq; py
ro ſylueſtri æqualis eſt. habet & ſubti-
le quippiam coniunctum. Fructus ipſi
us myrti baccis reſpondet.
ἀδίαν- ADIANTON, polytrichõ, puſil
τον. la gerit folia ſilia coriãdri. in ſũmo ici
Berce- ſuris diuiſa. cauliculi, quib. exoriũtur,
gnaſcen nigro colore nitent ptenues. palmũ al-
ti radice ſupuacua neq; caulẽ, neq; fru
ctum, neq; florẽ pfert. Seritur gratia vti
litatis pecorũ, circa ouilia. Vmbroſis, &
paluſtribus, circa parietũ aſpgines, &
fontiũ ſpecus enaſcit. Aeg. Siccat, exte
nuat, modice diſcutit. Calore ac frigi-
ditate medium eſt. Alopeciis mede-
tur, tumores digerit, calculos potu cõ-
fringit, reiectiones pulmonis deſiccat,
alui profluuium ſiſtit.
AEGILOPS herbula folia triti-
ci hñs, ſed molliora, i ſũmo capite ſe-
mina-

FIGURE 4.1 Alphabetic arrangement of the list of taxa in C. Gesner's *Catalogus plantarum* (Gesner 1542, p. 1).

vary among different classifications, and their current more or less standardized textual representation style was inherited from pre-Linnaean times. The respective texts are most usually shaped (and sometimes denoted) as tables, in which the taxa of different ranks are centered or stepwise left-to-right indented, with their different position in taxonomic hierarchy being indicated by specific styles of their header designations (Figure 4.2).

It is evident that the more fractioned and 'deep' is hierarchy of such classifications and the more extensive is their coverage, the less operable become their text representations, especially if they employ exclusively indentation format. So it just remains to concur with David Hull's opinion that 'indentation is too feeble a mode of representation to permit any such systematic relationship to be established' (Hull 1968, p. 473). Certain 'classical' formatting (fonts, etc.) still remains relevant in standard textbooks and checklists, but they seem to loose their applicability in many fields of practical systematics, in which digital technologies becomes dominating (see below). Such a textual representation of ranked and rankless taxonomic systems differ only in that designated ranks are indicated in the former and absent in the latter.

ORDER PYGOPODES. DIVING BIRDS.

SUBORDER PODICIPEDES. GREBES.

FAMILY **PODICIPIDÆ.** GREBES.

GENUS **ÆCHMOPHORUS** COUES.

Æchmophorus COUES, Pr. Ac. Nat. Sci. Phila. April, 1862, 229. Type, *Podiceps occidentalis* LAWR.

1. **Æchmophorus occidentalis** (LAWR.).
Western Grebe.

Podiceps occidentalis LAWR. in BAIRD'S B. N. Am. 1858, 894.
Æchmophorus occidentalis COUES, Pr. Ac. Nat. Sci. Phila. 1862, 229.

[B 704, C 608, R 729, C 845.]

HABITAT. Western North America, eastward to Manitoba.

GENUS **COLYMBUS** LINNÆUS.

Colymbus LINN. S. N. ed. 10, I. 1758, 135. Type, by elimination, *Colymbus cristatus* LINN.

SUBGENUS **COLYMBUS.**

2. **Colymbus holbœllii** (REINH.).
Holbœll's Grebe.

Podiceps holbœllii REINH. Vid. Med. 1853, 76.
Colymbus holbœllii RIDGW. Water B. N. Am. II. 1884, 428.

(a)

Class MAMMALIA Linnaeus, 1758, p. 14.
Subclass PROTOTHERIA Gill, 1872, p. vi.
Order MONOTREMATA Bonaparte, 1838, p. 110 (=Ornithodelphia De Blainville, 1834, *fide* Palmer, 1904, p. 780). Monotremes.
Fam. Tachyglossidae Gill, 1872, p. 27 (=Echidnidae Burnett, 1830b, p. 365). Pleist.–R.; Aus. Spiny "anteaters," echidnas.
Tachyglossus Illiger, 1811 (=*Echidna* Cuvier, 1798, *nec* Forster, 1788). Pleist.–R.; Aus.
Zaglossus Gill, May 5, 1877 (=*Proechidna* Gervais, Nov. 30, 1877). R.; New Guinea.
Fam. Ornithorhynchidae Burnett, 1830b, p. 365 (=Ornithoryncina Gray, 1825, p. 343). Pleist.–R.; Aus. Duckbills, platypuses.
Ornithorhynchus Blumenbach, 1800 (=*Platypus* Shaw, 1799, *nec* Herbst, 1793). Pleist.–R.; Aus.

(b)

FIGURE 4.2 Fragments of synoptic classifications of (a) birds (Coues et al. 1886, p. 73) and (b) mammals (Simpson 1945, p. 39) represented as texts with two alignment modes, centered and stepwise left-to-right indented, respectively.

If a synoptic classification in a checklist includes lists of synonyms for particular taxa, they are formatted following certain specific rules elaborated in the 19th and the early 20th centuries. Generally speaking, synonyms are the taxonomic names that refer to the members of the same taxon of a specified level of generality (rank). Accordingly, the senior synonym is considered its valid name placed as a header in the beginning of taxon's account, and it is used as its unique *reference term* in the respective literature. All other names are junior synonyms, they are listed in a chronological or an alphabetic order, with the indication of their publication dates, authors, bibliographic references, and occasionally some other relevant information.

In the case of line arrangement of classifications as texts, the purely verbal construction of textual representation style can be complicated by the inclusion of certain symbols to indicate specific hierarchical relations between classification constituents. The most popular are two following mathematical symbols: '$A \Rightarrow B \Rightarrow C$' indicates the order of subordination of *A*, *B*, and *C* within a non-inclusive hierarchy, and '$A \supset B \supset C$' indicates the order of their inclusion within an inclusive hierarchy. These formalisms read (from left to right) that *A* command *B* and *C*, *B* command *C* in the first case; *A* contains *B* and *C*, *B* contains *C* in the second case. For instance, the former can refer to the military ranks (marshal, general, colonel, etc.), and the latter does to the respective army units under their command (army, brigade, corps, battalion, etc.).

A specific formalization of textual representation of hierarchies in biological systematics was suggested by the so-called *numericlature* with its special numerical coding of ranks in addition to the standard taxonomic names, as it was developed by Willi Hennig (Hennig 1969; see Section 6.4). For instance, the beginning of his hierarchical classification of the major groups of the ectognathous insects looks as follows:

2.1. Archaeognatha (Microcoryphia)
2.2. Dicondylia
2.2.1. Zygentoma
2.2.2. Pterygota
2.2.2.1. Palaeoptera
2.2.2.1.1. Ephemeroptera
2.2.2.1.2. Odonata

Such a 'numericalization' trend led to an invention of a 'near-formular' *nexus format* for the textual codifying inclusive hierarchy of taxa for the needs of numerical taxonomy (Maddison et al. 1997), which is now widely applied in the respective computer programs (Felsenstein 2004). This representation style writes the imaginary hierarchical relations between taxa *A*, *B*, *C*, *D*, *E* as follows: (*A*,(*B*,*C*), (*D*,*E*)) (see also Figure 4.19).

A specific text format is applied for compiling *identification keys*, being an important kind of special purpose pragmatic classifications. Unlike general purpose synoptic classifications, whose main idea is to reflect *Systema Naturae* in any of its versions (typological, phylogenetic, etc.), the keys implement algorithms for taxonomic identification of organisms, i.e., for their allocation to particular taxa. In the early history of biological systematics, taxonomic classifications functioned as both synoptic and identification (Bather 1927; Voss 1952; Mayr 1969, 1982; Nelson and Platnick 1981; Griffing 2011). These two basic functions were explicitly separated by J.-B. Lamarck in his *Flore Françoise* (Lamarck 1778), who applied a special stepwise hierarchical dichotomous structuration of the keys represented in a mixed textual-bracketed format, in which the identification steps were sequentially numbered (Figure 4.3).[1] An extensive literature on the theory and methodology of identification keys appeared in the second half of the 20th century, and several text and table formats were elaborated for them, including recent digital technologies (e.g., Lobanov 1972; Pankhurst 1975; Payne and Preece 1980; Tilling 1984; Sviridov 1994; Hagedorn et al. 2010; Wäldchen et al. 2022).

* * *

A special format of the text-like representation style is provided by 2D *tables* or *lattices*. In them, the columns and rows correspond to the particular parameters (variables, characters, etc.), and the cells correspond to the taxa of a one-level classification system. Such tables (lattices) represent parametric cognitive systems (see Section 2.7 on them), and the latter's proponents see their major advantage in their prognostication, usually with the reference to the periodic system in chemistry (Lubischev 1969, 1972; Zavarzin 1969, 1974; Tchaikovsky 1990; Popov 2008; Arkhangelsky 1996). Such a table is especially illustrative in the case of specific intercorrelations of parameters, which entail different probabilities of their particular combinations. This fact facilitates delineation of the higher-level taxa characterized by the most probable and stable combinations according to the above-mentioned *prohibition principle* (Stark 1966; Zavarzin 1969, 1974; Zarenkov 1976, 1988).

* * *

The recent advancements in digital technologies applied to biological systematics gave rise to the development of quite an extensive industry for *taxonomic databases* (TDB), which inevitably concern representation styles for the taxonomic systems they deal with. Of two major types of TDB architecture, *hierarchical* and *relational* (Hellerstein et al. 2007), the latter was found to be more suitable for the practical needs of systematics (Raguenaud et al. 1999). Although TDBs are 'technologically' digital, they nevertheless retain the general text-like style of representation of the relevant information about taxa, including their inclusive hierarchical arrangement. Hence, without going into technical details, TBDs seem to lack specific facilities for representing taxonomic hierarchies but rather manipulate the one-level tables, in which hierarchies are set by users in a standard textual fashion (Golovanov 2004; Tarasov and Burakov 2013).

The development of this innovative technological approach to gathering, maintaining, and retrieving data relevant to taxonomic issues faces certain challenges due to the inherently fuzzy character of systematic knowledge (Raguenaud et al. 1999; Pullan et al. 2000). The main issue is that it hinders the necessary strict links between

PRÉLIMINAIRE. lxix

ANALYSE.

Fleurs dont les étamines & piſtils peuvent aiſément ſe diſtinguer. 1.	Fleurs dont les étamines & piſtils ſont nuls, ou ne peuvent ſe diſtinguer. 16.

1. *Fleurs dont les étamines & piſtils peuvent aiſément ſe diſtinguer...*	Fleurettes nombreuſes, réunies dans un calice commun . . . . 2.
	Fleurs libres & non réunies dans un calice commun.. 9

2. *Fleurettes nombreuſes, réunies dans un calice commun. . . . .*	Fleurettes de même ſorte; elles ſont toutes en cornet, ou toutes en languette. 3
	Fleurettes de deux ſortes, les unes en cornet, & les autres en languette. 6

3. *Fleurettes de même ſorte. . . . .*	Fleurettes toutes en cornet. 4
	Fleurettes toutes en languette. 5

4. Fleurettes toutes en cornet.

Carduus marianus.

5. Fleurettes toutes en languette.

Hieracium murorum.

e iij

FIGURE 4.3 The first version of mixed textual-bracketed representation of identification key in J.-B. Lamarck's *Flore Françoise* with numbered identification steps (Lamarck 1778, p. 75).

different TDB records concerning taxonomic names and ranks, specimen allocation, etc. Among the most significant sources of such a troublesome situation are the lengthy and sometimes intricate history of taxonomic research and the diversity of their conceptual backgrounds, which entail a multitude of taxonomic systems for particular groups of organisms, as well as rather extensive lists of synonymy and homonymy.

In order to minimize the effect of fuzziness of the descriptive language of systematics, experts in TBD elaborate specific highly formalized unified *ontologies* (Smith 2004; Göpel 2024) and, in this regard, draw special attention to the linguistic (i.e., largely textual) aspect of their development (Vogt et al. 2010, 2013). Such ontologies are understood not philosophically but 'technologically', and they are designed to encompass strictly defined and rather formalized basic vocabularies for systematics, mainly for its meronomic branch (e.g., Ashburner et al. 2000; Bard et al. 2005; Mabee et al. 2007). However, all such attempts inevitably encounter the above-discussed inverse relation between the rigor and meaningfulness of notions (see Section 2.3), which puts certain limits on the strictness of TDB vocabularies. On this background, the occasional suggestions to incorporate somehow the fuzziness of the basic thesaurus of systematics in TDB structure by using non-strict notions, such as that of *potential taxon* (Berendsohn 1995), seem rather interesting.

4.2 THE TREE-LIKE STYLE

A tree-like schematic representation of the hierarchically arranged diversity of both things and ideas about them became standard since early scholasticism. This style evolved significantly over time with regard to the particular tree formats, some being schematic and others being very pictorial ('Baroque'), horizontal or vertical, ascending or descending, etc. With this, two basic sorts of them should be clearly distinguished for our purposes, which differ essentially in their meaning: they were initially designed to function as either/both *connection* or/and *categorization* trees. They became a part of the post-Antique philosophical discourses at approximately the same time, but their origins seemed to be different. The former was derived from a 'linking' style of a primordial idea of the World Tree, and the latter resulted from a 'divisive' thinking style rooted in the quite advanced Platonic natural philosophical idea of emanation of the *One*.

It is to be stipulated, however, that these two natural philosophical ideas were fundamentally similar in their supposed 'ideal' grounds for the real diversity of things. The latter were generated through the creative activity of a specific global root cause, either an emanation of the *One* or a fruiting of the World Tree. So their just-distinguished meanings should be attributed to the different cognitive intentions of early post-Antique thinkers, who chose the respective tree-like metaphoric representations as the most appropriate epistemic elements for their specific cognitive contexts.

* * *

As said in our historical overview, the earliest tree-like diagram has been drawn by the first commentators on Boëthius' *Isagogen Porphyrii* and was adopted as a standard illustration of the logical genus-species scheme in the majority of subsequent philosophical treatises since the 9th and 10th centuries (Barsanti 1992; Hacking 2007; Rosalinde 2010; Verboon 2014; Lima 2017; Dardağan 2018). It was called *Tree of Porphyry* (after the neo-Platonic philosopher Porphyry) in several treatises of the 13th century (de Rijk 1972), and this denotation would become later the common designation of all tree-like diagrams considered in the philosophical contexts regardless of their particular configurations. This tree varied in its formats, some were rather picturesque (as in Figure 1.2) and others applied a schematic *line* format (Figure 4.4). In any formats, it fulfilled both categorization and connection functions, i.e., illustrated (a) the sequences of divisions of higher genera into lower and ultimately into species and (b) the logical links between them. The line format appeared to be most suitable for the connection function and became the basic tree-like style in the later works of logicians, up to the contemporary, now in company with computer scientists (e.g., Aho and Ullman 1995; Huth and Ryan 2004; Hacking 2007). Besides, such a format would appear to be the most appropriate for the representing non-inclusive organizational hierarchies of decision trees (e.g., Bhushan and Rai 2004; Emrouznejad and Ho 2018).

The primary objective of such a tree-like figure, when dynamically interpreted, was to illustrate the sequential deductive inferences of notions of various levels of generality based on binary logic; statically considered, it functioned as a representation of the resulting logical genus-species hierarchy. The trees that appeared in the *Tractatus* by the above-mentioned Petrus Hispanus were characterized by two principal features (Barsanti 1992; Rosalinde 2010; Dardağan 2018). First, each of his trees was ordered by a scale of logical entities (genera), hierarchically ordered from the highest to the lowest along the tree's central axis; therefore, the Tree of Porphyry in this implementation was sometimes denoted as *Scala praedicamentalis* (scale of predicaments) (Peirce 1920). Second, the pairs of side offshoots, branched off from the nodes of central scale signifying these substances, demonstrated binary oppositions of logical species, into which the respective genera were divided. This tree most usually took a vertical position with a top-down sequence of genus-species divisions to visualize descending order of deduction, but an opposite, down-top (ascending) sequence could also be seen in some of *Tractatus* copies and in many of later treatises (Dardağan 2018). It is of interest that in some of its drawings, Tree of Porphyry was represented in a standard descending mode, which is evident from the top-down sequence of logical genera from *Substantia* to *Homo*, but had roots at its base, so it looked as if it had grown upward out of an imaginary 'ground' (Migne 1847; Peirce 1920; Ilg 2014; Bintley and Salonius 2024; see Figure 1.2). One can suppose that such a 'rooting' was reminiscent of the natural philosophical idea of the World Tree (see below on it).

The further graphical development of the tree-like representation style has led to its important modification to become known as *Tree of Ramus* named after also already mentioned Petrus Ramus (Ong 1958; Archibald 2014; Dardağan 2018). It replaced the previous line format with a new *bracket* one to illustrate more expressly the logical subordination of genera and species of various levels of generality (see Figure 1.3). One of the most important features of Tree of Ramus was that its bracket structure turned out to be appropriate for an implementation of classification

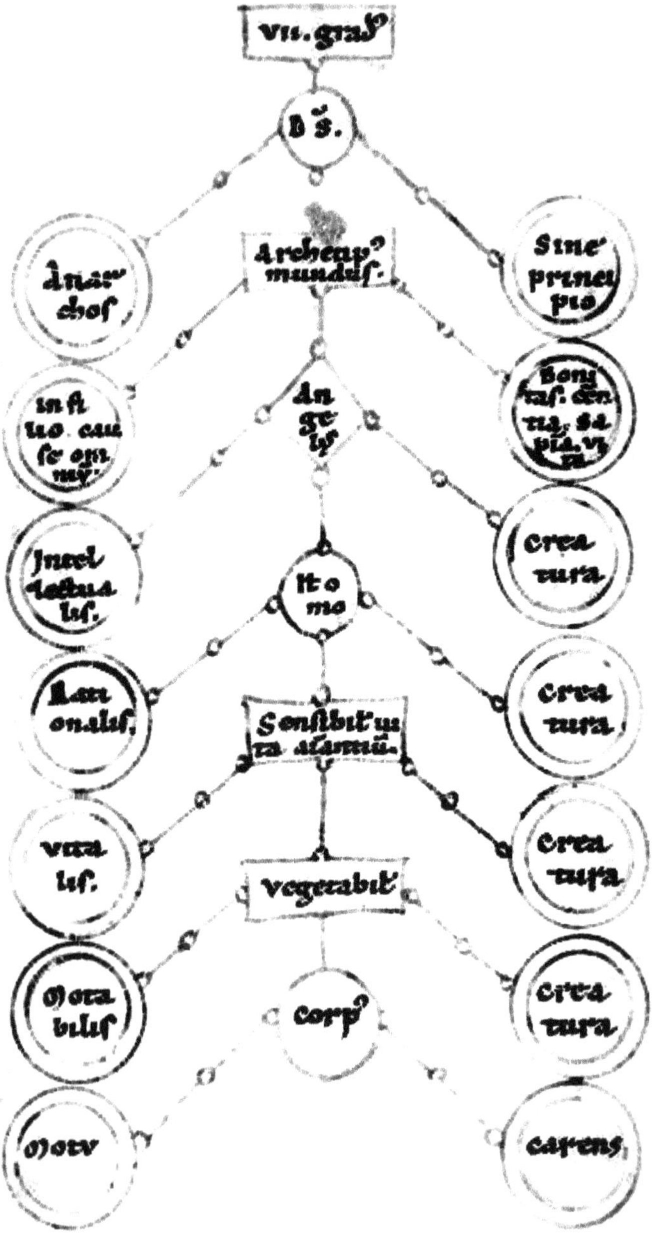

FIGURE 4.4 A very schematic representation of Tree of Porphyry drawn in the line format, with the central line corresponding to the basic *Scala praedicamentalis* (Honorius… 1001–1200, p. 3).

function, as compared to the standard Tree of Porphyry. The latter's structure was rigidly dictated by its scale-like central axis, which completely lacks in the Ramean tree. Therefore, this novel tree format provided a possibility for visualization of inclusive hierarchical classifications of the multi-aspectual diversity of numerous entities. In the same time, of importance was that the Ramean tree was fully compatible with the Porphyrian one, which provided a possibility for both these representation formats to be applied and interpreted in the same manner.

This Ramean tree format was applied in many of both scholastic and early post-scholastic systematic treatises up the first half of the 19th centuries. Its most substantive difference from scholastic one included ranked hierarchy of trees, which branching orderliness corresponded to the sequence of taxonomic ranks in classifications according to the already established tradition. Their most typical implications could be seen in *Prodrome d'une nouvelle distribution* by the above-mentioned H. de Blainville, who illustrated all major divisions of the animal kingdom by such bracket tree diagrams (de Blainville 1816). One of the last instances of this kind was the appearance of Tree of Ramus in *Taxonomic Units and Groups* by the Russian zoologist Sergei Usov (1827–1886), who represented in such a fashion taxonomic hierarchy of the major vertebrate groups (Usov 1867; Figure 4.5). Such late Ramean trees were sometimes denoted as 'quasi-genealogical' (Gill 1872; Patterson 1977; Panchen 1992), although in fact they have nothing in common with phylogeny.

Afterward, phylogenetic and then phenetic trees became dominating in biological systematics. These two major interpretations of tree-like diagrams did not seem to have derived from the logical categorization tree of Porphyry/Rame kinds. Indeed, the phylogenetic tree (considered below) is a special kind of connection tree, whereas

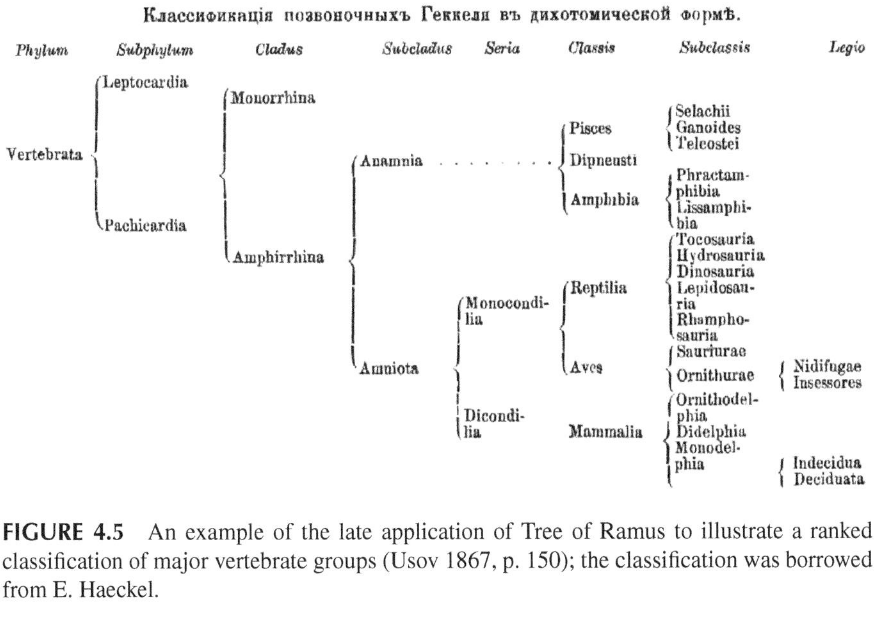

FIGURE 4.5 An example of the late application of Tree of Ramus to illustrate a ranked classification of major vertebrate groups (Usov 1867, p. 150); the classification was borrowed from E. Haeckel.

phenetic tree, commonly known as *phenogram*, is certainly of classificatory kind, and yet it cannot be linked directly to the earlier logical trees either, as it does not implies any logical syllogistics.

The basic function of phenetic tree (phenogram) is to illustrate hierarchy of the multi-character similarity relations between organisms, providing that its basal level corresponds to the level of minimal similarity of compared entities (e.g., Sokal and Sneath 1963; Sneath and Sokal 1973; Abbot et al. 1985; Sokal and Rohlf 2009). The phenograms can either be dividing or agglomerating depending on the inference algorithms applied. Their branching sequences manifest rankless hierarchies of the groups (clusters). By their orientation they can be either vertical or horizontal, with the former being either ascending or descending. All compared units take terminal position in them and are situated at the same level to indicate the absence of their reflexive relations. These units can correspond either to organisms (taxonomic aspect) or their characters (meronomic aspect), which is formalized as *Q* and *R* modes of the analyses of similarity relations, respectively.

* * *

Another tree-like representation format was being developed alongside the categorization format of Tree of Porphyry/Ramus, denoted here as *Generation Tree*. Its personified designation should probably be given after Spanish philosopher and theologian Raymund Lull (1235–1315) for his famous *Arbor Scientiae* (Lull 1515), which he explained as 'Generation [being] a general principle of the elemental tree' (cited after Dambergs 2003, p. 27). When considered dynamically, it illustrated supposed sequential origins of manifold entities of certain kinds, whether 'physical' (bodies) or 'ideal' (conceptions); regarded statically, it indicated their relationships ('affinities'); in both cases, their origins/affinities were indicated by the lines connecting them. It is to be reminded that Porphyrean Tree also fulfilled originally such a connecting function, along with a categorization one, as indicated by the connecting lines between sequentially emanated entities. However, as far as natural history is concerned, it was just the categorization function that appeared to be mostly fixed by the subsequent development of Tree of Porphyry. Unlike the latter, in the case of *Tree of Lull*, its connecting function of a special sort was initially and remained subsequently the only one.

Connection tree was apparently a specific derivative of the universal archetypal image of the so-called *World* (*Cosmic*) *Tree* inherent to the mythological world pictures of many ancient cultures.[2] This was an unimaginable colossal tree, which grew from its root and generated all material and ideal entities in Cosmos; being treated as something 'alive' that gave birth to all that existed, it was called also *Tree of Life* (*Lignum vitae*) to denote not only livings but all beings (Cook 1988; Ilg 2014). As a global *Axis mundi*, the World Tree was believed to provide harmonic integrity of the Cosmos by connecting the heavens (crown), the terrestrial world (trunk), and the underworld (roots). Due to a complete realization of its generative power, this Tree was thought of as a source not only of Cosmos but also of the knowledge about it, so it was at the same time the *Tree of Knowledge*, which became famous after the *Genesis* legend of the Fall of the first humans (Hammand 1918).

One of the most important (for us) implementations of the general idea of Generation Tree became *genealogical tree* (*Arbor genealogiae*) appeared in post-Antique culture. It initially illustrated pedigrees of selected Biblical personages and then of numerous Royal and Noble families, which gained a special popularity since Late Medieval (Klapisch-Zuber 1991; Barsanti 1992; Alexander and Baker 2003; Worm 2014; Lima 2017). Each genealogical tree was designed to represent the history of lineages within a family and the kinship relationships among its members. Both the history and kinship were demonstrated by the overall branching pattern of such a tree, which general orderliness was determined by the position of the fixed root corresponding to the initial event in the genealogical history. The genealogical trees were very diverse in their formats, with one of the important differences being their vertical orientation: some of them were descending and some ascending. Due to fact that such a genealogical tree referred to the historical events, it became a prerequisite for its subsequent transformation into evolutionary tree (see below).

Among the most impressive implications of the general idea of Generation Tree were tree-like representations of interrelations between human activities and virtues, sciences, languages, etc. (Gontier 2011; Reboiras et al. 2002; Pietsch 2012; Lima 2017). The best known of them were probably those appeared in *Arbor Scientiae* by already mentioned R. Lull, which, besides the title generation Tree of Knowledge (*Arbor Scientiae*, Figure 4.6), contained 16 annotated trees showing, among other matters, *Ars mechanica*, *Ars moralis*, *Theologia*, *Philosophia*, *Medicina*, etc.[3] The relevance of tree metaphor to such interpretation was witnessed by the term *florilegium*, which denoted anthologies of the ideas of Church Fathers as 'flowers of wisdom' gathered in the hand-written incunables and (since the 16th century) printed books (Bintley and Salonius 2024).

Regarding natural history, the first connection tree-like diagram, vaguely reminiscent of a genealogical tree, or *tree of affinities*, appeared probably in *Plantarum umbelliferarum* by the Scottish botanist Robert Morison (1620–1683). He outlined four such trees, each called by him '*Tab. Cognationis & Affinitatis*' (Table of Kinship and Affinity), they had no evident vertical orientation, so (in contemporary terms) they were more unrooted networks rather than typically rooted genealogical trees (Morison 1672; Figure 4.7). These trees displayed not so much particular genera and species of umbellifer plants as anatomical features of their flowers, i.e., these were not taxonomic but meronomic trees. The taxa *per se* were characterized by Morison in the main part of his book, and their hierarchical relations were illustrated by numerous typical Ramean trees.

Thereafter, the connection tree-like diagrams illustrating affinities between members of various plant and animal groups became popular in systematic literature with the emergence of post-scholastic school of thought at the end of the 18th and the early 19th centuries (Barsanti 1992; Stevens 1994; Rieppel 2010; Hellström 2012; Pietsch 2012). These diagrammatic figures varied from rather simple to very complex, which evidently depended on how particular systematicians viewed the relations of the studied organisms. They could roughly be divided among two major

FIGURE 4.6 The Tree of Knowledge in R. Lull's *Arbor Scientiae*, being a typical example of the generation (Lullian) tree format (Lull 1515, title).

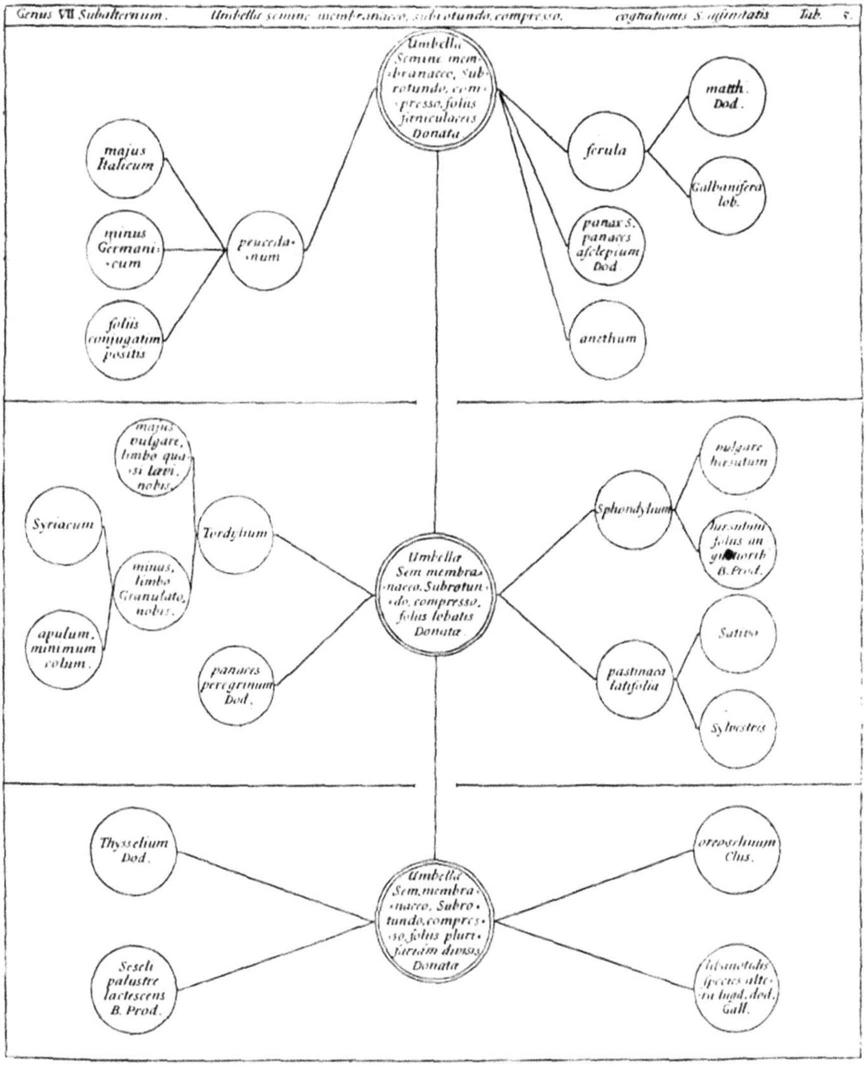

FIGURE 4.7 The unrooted network-like *Tabula Cognationis & Affinitatis* No. 5 in R. Morison's *Plantarum umbelliferarum* (Morison 1672, unnumbered page), instantiating probably the first real tree of affinities in natural history.

categories, networks and proper trees, of course with numerous transitional variants. The early rooted trees of connecting kind might incorporate certain features of categorization kind, borrowed from Ramean tree format, so they looked as specific combinations of these two (e.g., Rüling 1766). Many authors combined connecting (genealogical) trees with the plane-like style of representation of systematic affinities (see below).

The network representation format implies an unordered tree graph, and if the latter refers to the reticulate many-to-many relations between taxa, it takes form of a *web* (e.g., Eichwald 1821; Figure 4.8). Such networks became popular in the 20th century with the development of numerical phenetics (Lam 1936; Sokal and Sneath 1963; Smirnov 1969; Sneath and Sokal 1973; Figure 4.9).

One of the earliest tree-like representations of actually genealogical relations applied in systematics was *Arbre botanique* by the French botanist Augustine Augier (1758–1825) in his *Essai d'une nouvelle classification* (Augier 1801; Figure 4.10). This tree in its shape, on the one hand, evidently ascends to the Generation Tree and, on the other hand, could be taken for the prototype of future evolutionary trees. However, notwithstanding its upward rising character, it seemed to lack an explicit time dimension (Stevens 1983, 1994; Barsanti 1992;

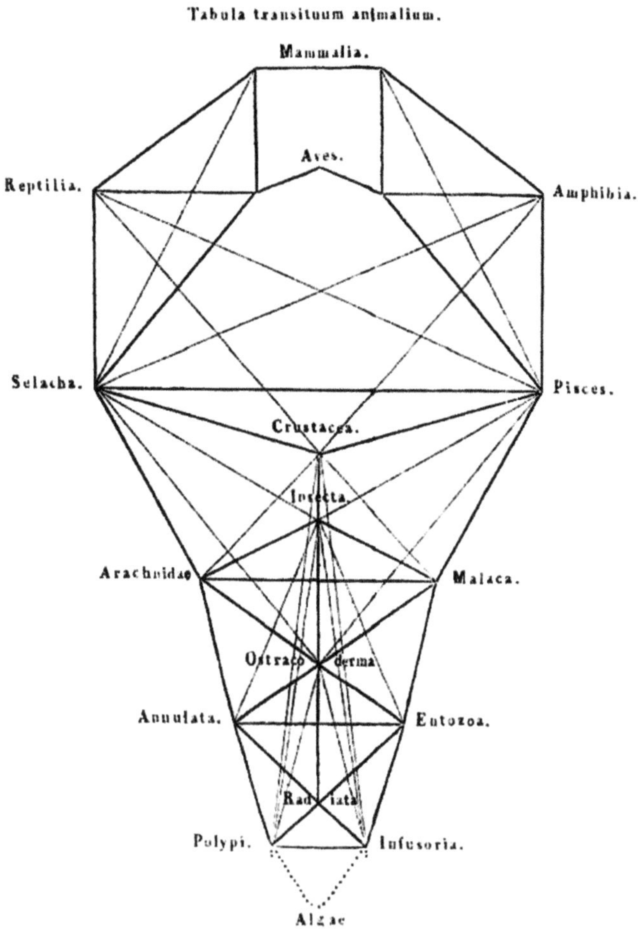

FIGURE 4.8 The web-like representation of many-to-many relations between major animal groups in E. Eichwald's *De Regni Animalis* (Eichwald 1821, pre-title page).

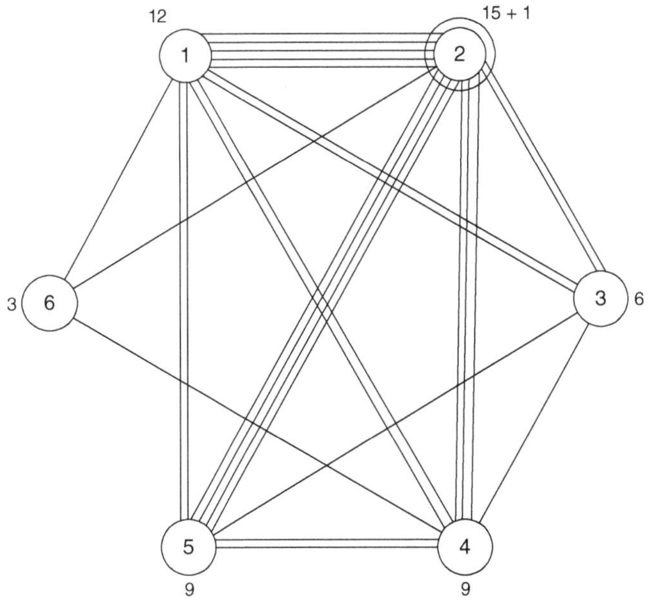

FIGURE 4.9 The network of similarity relations between congeneric species in H. Lam's *Phylogenetic symbols* (Lam 1936, p. 183). The number of connecting lines indicates the number of shared characters.

Pavlinov 2005; Ragan 2009; Rieppel 2010; Tassy 2011; Pietsch 2012). The first tree-like figure of just the evolutionary relations between some major animal groups appeared in already-mentioned J.-B. Lamarck's *Philosophie zoologique* (see Lamarck 1963). However, his tree was rather specific in two respects. First, its evolutionary meaning seemed to be implied rather than clearly depicted, as its connecting lines were dotted and looked as referring to something uncertain and speculative (Figure 4.11).[4] Second, it was descending, like Tree of Porphyry, and not ascending, which would become characteristic of all future evolutionary trees. And yet, the Lamarckian drawing was very important for its being with no doubts the first evolutionarily interpreted and therefore implying time dimension.

The next wave of implementation of the idea of Generation Tree in systematics was initiated by Darwin's and Haeckel's epochal treatises (Darwin 1859; Haeckel 1866) that laid foundation for the evolutionary interpretation of both the organismal diversity and its conceptual representations. This interpretation added a crucially important time dimension to the consideration of hierarchies, including tree-like style of their representation, now called *evolutionary* or *phylogenetic* tree or *tree of history*. C. Darwin only once depicted a 'local' genealogical scheme to illustrated his idea of gradual transformations of local forms into species (see Archibald 2014). E. Haeckel considered the Tree of Life encompassing all wildlife, so his tree-like diagrams became most influential and fateful in this regard. However, an important

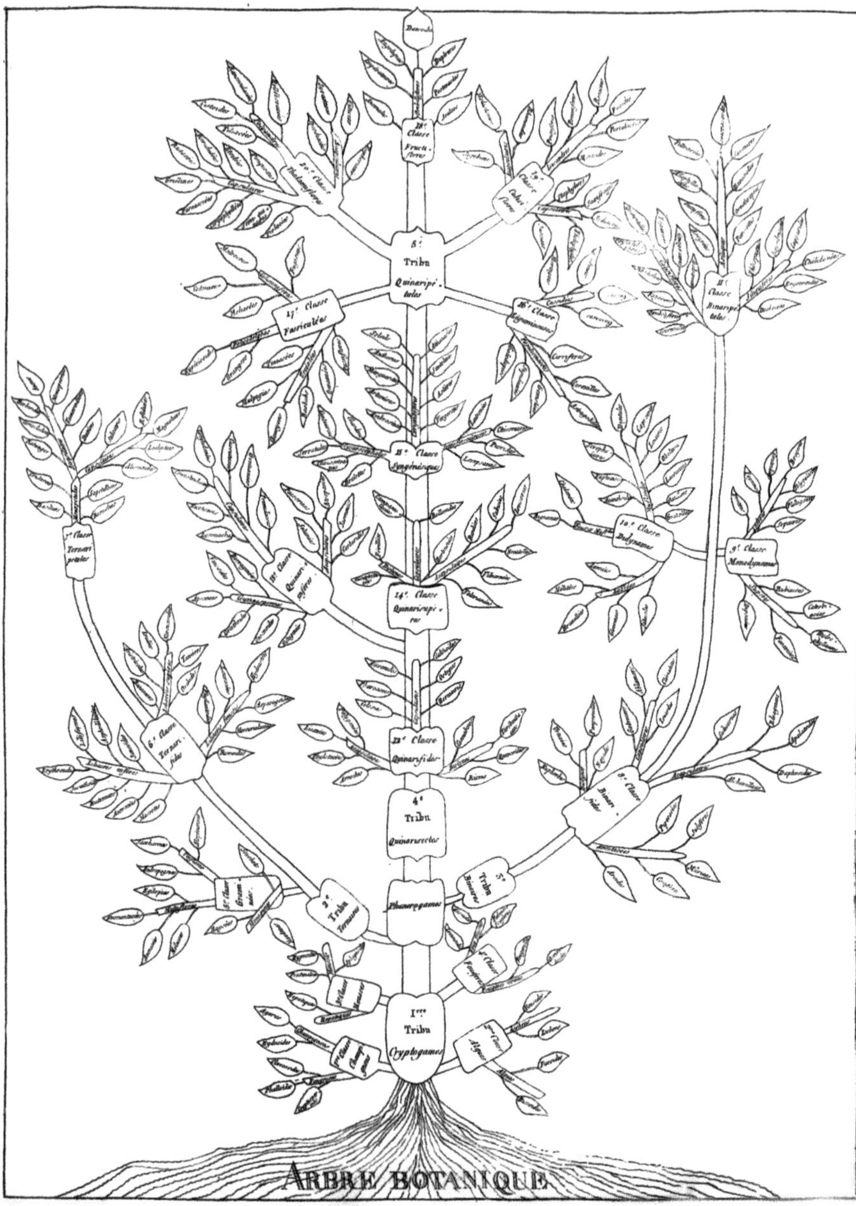

FIGURE 4.10 Probably the first ascending genealogical tree for plants (*Arbre botanique*) in A. Augier's *Essai d'une nouvelle classification* (Augier 1801, insertion after p. viij); it seemed to lack real time dimension.

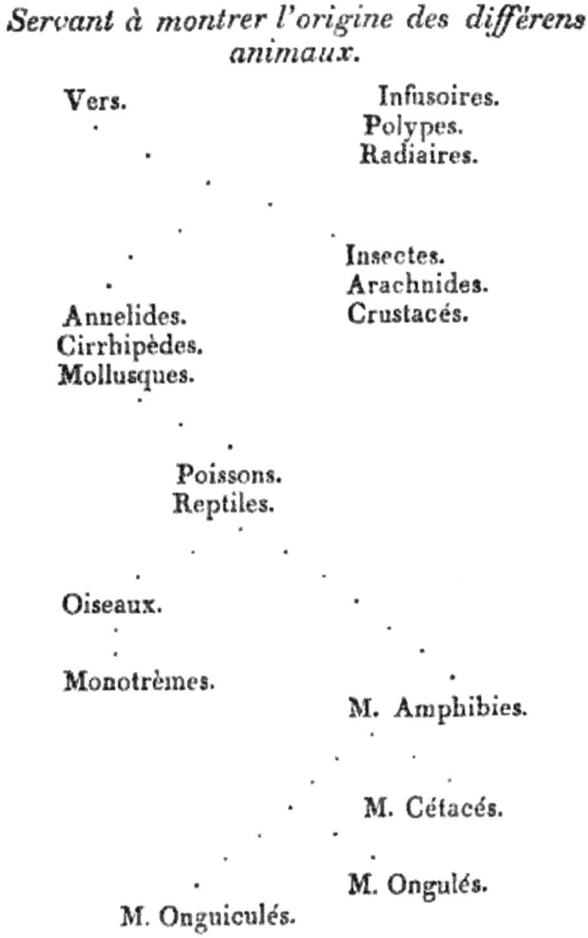

FIGURE 4.11 The first evolutionary tree for animals in J.-B. Lamarck's *Philosophie zoologique* (Lamarck 1809, p. 463). See text for comments.

conceptual shift landmarked by Haeckelian trees appeared to be not so radically new, if viewed from a broader natural philosophical perspective. In fact, his evolutionary branching trees, especially in a picturesque format, were typically depicted as growing upward with their central stems resembling the Ladder of Perfection, to which numerous side branches were added.

Regardless of their standard upward orientation and branching pattern, the evolutionary trees are very diverse in many other respects, including their contents and representation formats, so they might differ significantly in their appearances, just as their human genealogical predecessors (Lam 1936; Kozo-Polyansky 1949; Voss 1952; O'Hara 1991, 1996; Barsanti 1992; Pietsch 2012). The extremities of such differences were evident already from the Haeckel's trees, some of which were very schematic (see Figure 1.5) and others quite picturesque (Figure 4.12). The basic

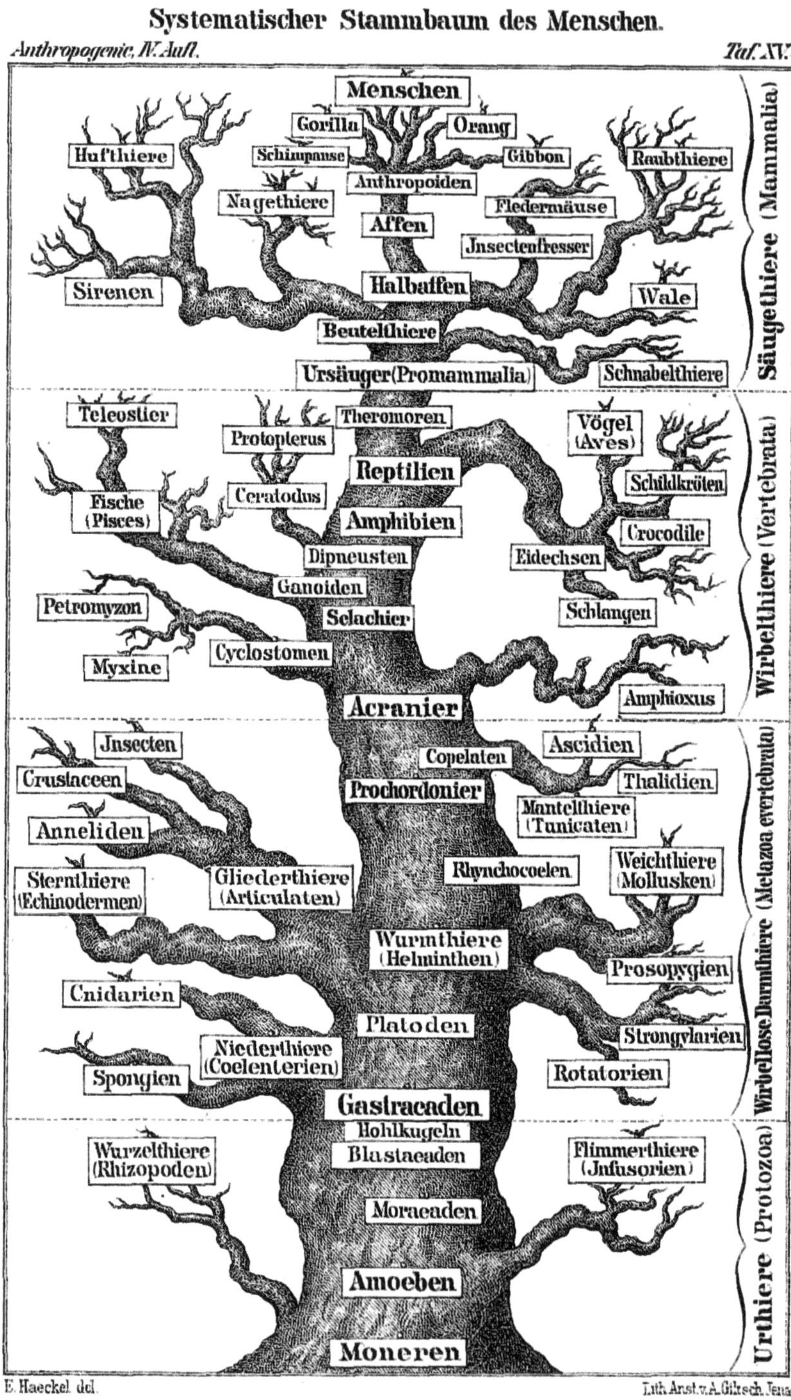

FIGURE 4.12 The most 'naturalistic' variant of E. Haeckel's phylogenetic trees appeared in his *Anthropogenie* (Haeckel 1891, p. 556).

evolutionary information contained by traditional phylogenetic trees is defined by (a) position of tree nodes relative to its root and (b) branch lengths and directions, whereas their exceptional decorativeness added nearly nothing to their informativeness. The node sequences represent the relative timing of phylogenetic events, the branch lengths correspond to the levels of advancement and duration of the respective groups, and branch directions indicate specifics of evolutionary specializations. Besides, the most informative evolutionary trees may indicate degrees of the divergence and advancement, geographic distribution, diversity, etc., of these groups (Figure 4.13). John Gregg called *truncated* hierarchies represented by the trees with different branch lengths and corresponding different position of terminal nodes relative to the tree roots (Gregg 1967; Griffiths 1973). Another format of phylogenetic tree, which illustrates the hypothetical ancestor-descendant relations in a specific stylized fashion, is referred to as *caulogram* (Zander 2010b, 2024, 2025).

The cladistic version of phylogenetic systematics suggests an extremely simplified overall understanding of the evolutionary history by analyzing only sequences of cladistic events in it (Hennig 1950, 1966; Nelson and Platnick 1981; Wiley 1981; Pavlinov 2005; Wägele 2005; etc.). Accordingly, all other just-listed features of evolutionary history, very important from the biological perspective, appear to be obviated from consideration. As a consequence, the stylized phylogenetic trees took shape of *cladograms* to represent branching patterns of phylogenetic lineages (see Figure 3.3). Such trees are considered sufficient for the correct representations of cladistic relations between elementary groups taking terminal positions on tree branches, and their branching points are sometimes treated as the principal cladogram components (Nelson 1979). Such formalized trees, which branch lengths differ according to the calculated patristic distances between nodes, are denoted as *phylograms* or *patrocladograms*; these lengths can also reflect timing and transform phylograms in time-calibrated *chronograms* (Pavlinov 1990a; Stuessy and König 2008; Scott and Baum 2016; Wilson et al. 2022; Bjornson et al. 2024). Besides, cladograms can occasionally indicate characters that are shared by the members of particular clades, with distinguishing graphically between their synapomorphic and non-significant symplesiomorphic modalities (Hennig 1966; Wiley 1981; Kitching et al. 1998; Wägele 2005). Cladograms of the last kind are called *synapomorphograms*; they are sometimes treated as 'the most parsimonious expression of character distribution without making any assumptions about the taxa' (Humphries 1983, p. 91).

One of the important and evident differences between phylograms (caulograms, chronograms, etc.) and proper cladograms is the interpretation of their internal nodes. The former identify them with the supposed ancestors and indicate ancestor-descendant relations, whereas the cladograms do not, with both the ancestors and their descendants being uniformly identified with the tree 'leaves' and called *terminal groups*, just as in the phenograms. Respectively, these major kinds of the stylized phylogenetic trees are called *stem-based* and *node-based* (Lee 1998; Martin et al. 2010; Wiley and Lieberman 2011). It is supposed that the refusal to consider ancestors deprives node-based cladograms of an explicit implying real (geological) temporary dimension (Tatarinov 1984; Wood 1994; Valentine and May 1996; Brummitt 2002; Upchurch and Hunn 2002; Pavlinov 2005; Rieppel 2010). Due to this, such trees can be positioned not only in a standard vertical ascending mode but also in

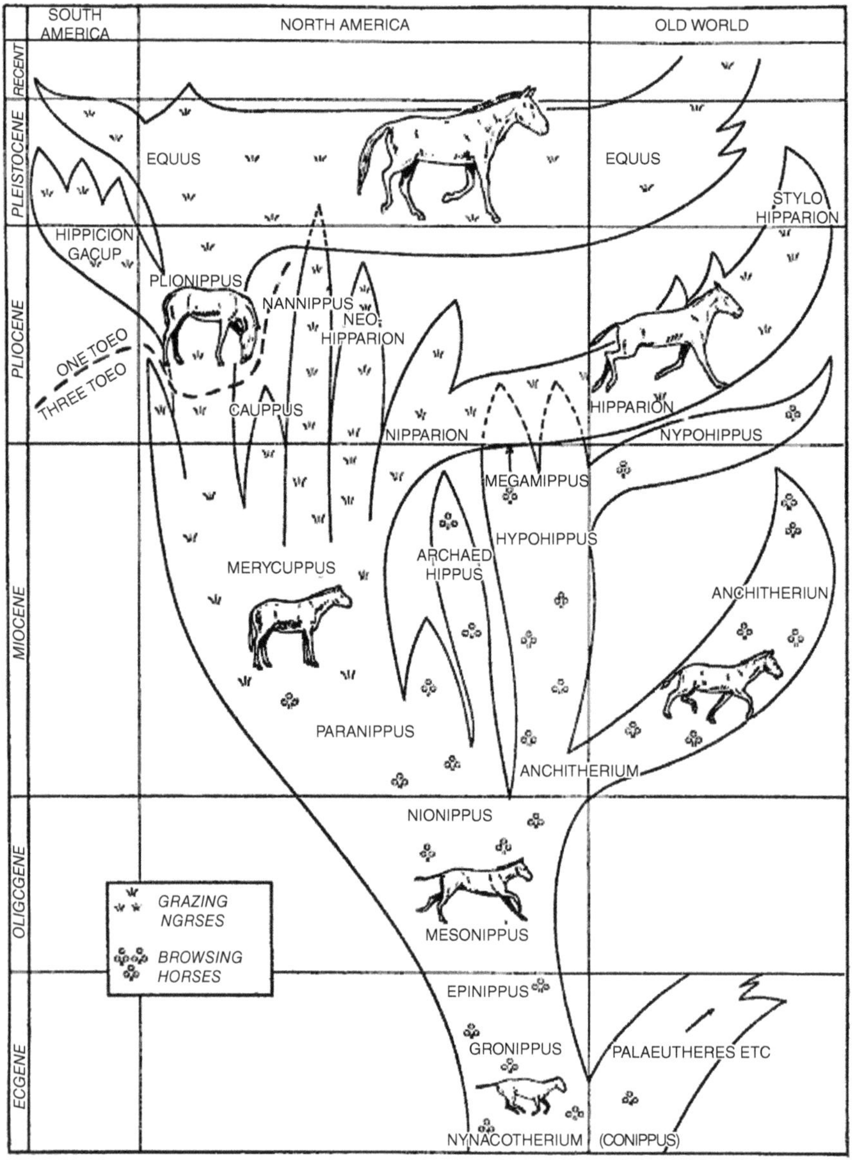

FIGURE 4.13 An informative representation of the equid phylogeny in G. Simpson's *The Story of the Horse Family* (Simpson 1951b, p. 28).

horizontal and even descending modes (the diversity of contemporary formats of genealogical trees is excellently illustrated in https://toffana.ru/phylogenetic/analysis).

Regardless of the declarative discarding of the anagenetic component of evolution by cladistics, it can be indirectly traced in most of cladograms. This is due to the fact that each of them is generally arranged by a specific interior hierarchization parameter defined by the sequence of appearance of apomorphic traits in the history of the studied group. With respect to their general configuration, cladograms can be of two particular formats. In one of these, the branches are of the same length, so a tree looks like as a 'ladder'. In another variant, all terminal groups are aligned along an imaginary line taking either vertical or horizontal position, and such trees are sometimes called aphoristically '*Hennigian combs*' (Shatalkin 1988; Pavlinov 2005; Yang et al. 2011). Their branches are formally of different lengths but bear no information concerning the 'amount of evolution'. Such 'combs' are usually asymmetrical, which is sometimes explained by specific character of evolution of the respective groups (Løvtrup 1977, 1989; Shatalkin 1988, 2012).

As noted above, the vertical ascending orientation of the evolutionary trees have several exclusions, one of which are the just-mentioned horizontal and descending cladograms. Another is a connection diagram, which represents something between typical tree and network. The dendrograms of the second kind are sometimes built into specific concentric circles that reflect major trends in the evolutionary diversification of the respective groups (Engler 1881; China 1933; Lam 1936; Pietsch 2012; Figure 4.14), and transitions from typical vertical to network-like phylogenetic trees can be found (compare China 1933, p. 195 and Engler 1881, p. 429). Similarly shaped near-unordered phylogenetic trees can be found in many recent molecular phylogenetic studies (e.g., Pace 2009; Hug et al. 2016; Kahana et al. 2024; see also: https://toffana.ru/phylogenetic/analysis). Another recent variant suggests to incorporate, wherever appropriate, network relations within rooted phylograms to show the inferred events of horizontal gene transfer (Heiss et al. 2025).

The active 'numericalization' and 'molecularization' of contemporary phylogenetics entail several important technical novelties in the tree-like representation style of the hierarchical patterns considered by phylogenetic systematics (Page and Holmes 1998; Nei and Kumar 2000; Felsenstein 2004). One of them includes indication of numerical estimates of the reliability of clades recognized within a tree; these estimates are calculated based on different methods (decay index, maximum likelihood, Bayesian or bootstrap support, etc.) and accordingly have specific particular interpretations (Cavalli-Sforza and Edwards 1967; Felsenstein 1982, 2004; Swofford et al. 1996; Kitching et al. 1998; Page and Holmes 1998; Nei and Kumar 2000; Nascimento et al. 2017; etc.). Another novelty involves incorporation of numerical estimates of the above-mentioned 'amount of evolution', which is visualized by different lengths of respective tree branches and thus transforms cladogram into phylogram (Pavlinov 1990a, 2005; Swofford et al. 1996; Felsenstein 2004). This 'amount' most usually reflects the differences in molecular characters and may be interpreted in the terms of already-mentioned molecular clock hypothesis, which provides a possibility for the time-calibrated molecular phylograms to be interpreted as chronograms (Page and Holmes 1998; Ayala 1999; Arbogast et al. 2002; Ho 2014; Bjornson et al. 2024; Duchêne et al. 2024; etc.). With some assumption, an evolutionary interpreted phenogram may be treated as a *gradogram* which represents not clades but grades in the phylogenetic history of a group (Shatalkin 2012). Some specific representation of microphylogenetic trees is suggested by phylogeography to illustrate the history of within-species spatio-temporal differentiation (Avise 2000). Reticulate evolution of prokaryotes, which is now vividly discussed, tends to turn the global

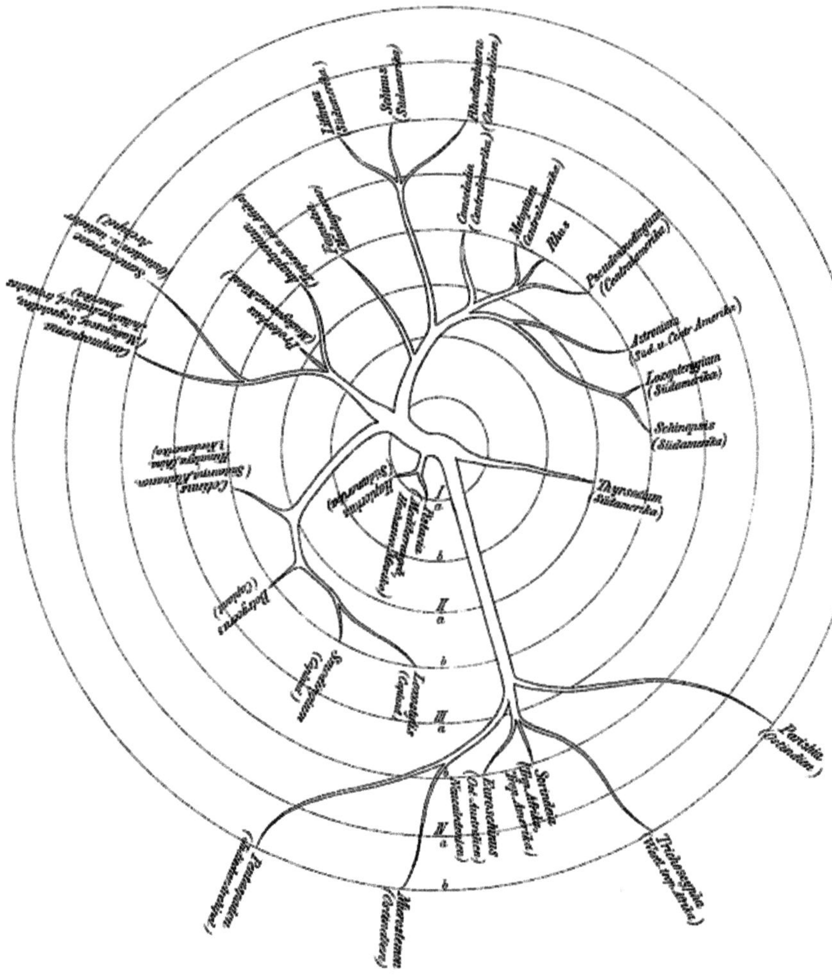

FIGURE 4.14 Affinity relations between species of the sumach (genus *Rhus*), imposed on the concentric circles showing divergence and advancement levels, in H. Engler's *Über die morphologische Verhältnisse* (Engler 1881, p. 429).

molecular Tree of Life at its base into (technically speaking) a network with circles (Sneath 2000; Bapteste et al. 2009; Gontier 2015a; Bjornson et al. 2024).

As noted above, the branching evolutionary trees of any kind, from their earliest Haeckelian versions to the most recent cladograms with all their modifications, are more or less compatible with the idea of linear Ladder of Perfection. Indeed, the traditional vertically positioned trees serve as an illustration of a steady progression from lower to higher organisms (Minelli 1993, 2023b; Gontier 2011; Jenner 2024). Besides, such a 'ladder' image is usually evident in the horizontal sequential arrangement of taxa according to the so-called *progression* (*deviation*) *rule* or *phyletic sequencing*. In the traditional trees the latter refer to the general gradual progression of organisms, whereas in cladistics it is defined by a succession of synapomorphies (Hennig 1966; Nelson 1974; Wiley 1981; Pavlinov 2005; Shatalkin 2012).

* * *

The revival of phylogenetic approach in research of taxonomic diversity in the second half of the 20th century inspired Robert O'Hara to declare 'Homage to Clio' by biological systematics and to designate the respective shift in systematicians' consciousness as gaining a *tree thinking* (O'Hara 1988, 1996, 1997). The latter neologism enjoyed a great attention from biologists who discussed various aspects of its meaning and applications (e.g., Omland et al. 2008; Novick et al. 2012; Baum and Smith 2013; Miller and Hartzog 2020; Evans et al. 2021).

O'Hara suggested this term to refer basically to the trees of histories and to oppose tree thinking thus understood to the so-called *population thinking* (the term coined by E. Mayr; see Mayr 1982, 1988) and *group thinking* (a kind of categorization thinking). However, as discussed above, application of the tree-like style for representation of the relations between both physical and logical entities (i.e., things and conceptions) was characteristic of both natural theology and natural philosophy from the very beginning of post-Antique conceptual history. This representation style initially combined two basic functions, categorization and connection—the former referred to the recognition of groups of entities, and the latter implied relations (affinity, etc.) between them. And there is very reason to suppose that such a representation style of whatever kind implied the same 'tree-like thinking', which impelled its bearers to see something of tree-like sort in the relations that structure the studied diversity.

Both tree-like representation and thinking styles were inherited from scholastics by naturalists of the 15th–16th centuries and afterward by systematicians as their 'intellectual descendants'. They applied this representation style first in its categorization format presuming no time dimension, and then in the connection format, with the latter meaning initially genealogy and subsequently time-scaled evolution (phylogeny)—and all this conceptual development was shaped by the tree-like (or simply tree) thinking in its general understanding. In contemporary systematics, tree-like representation is equally applied in both phenetic and phylogenetic studies. It should also be reminded that the tree notion is applied in graph theory to represent any partially ordered relations between entities of any kind (Bollobás 1998; Semple and Steel 2003; Bondy and Murty 2008), including those denoted as 'networks' by many members of numerical school of systematics (Sneath and Sokal 1973; Abbot et al. 1985; Swofford et al. 1996; Felsenstein 2004).

Taking all these considerations into account, it seems rather failed to follow O'Hara in denoting the narrow phylogenetic treatment of the respective thinking style as 'tree thinking', just because the latter had and still has much broader meaning. Therefore, the suggestion to refer to the style in question, as it has been outlined and defined by R. O'Hara, as *phylogenetic* (or *genealogical*) *thinking* (Pavlinov 2005, 2007b), seems to be quite reasonable. Another appropriate variant could be *lineage thinking* (Jenner 2024).

4.3 THE PLANE-LIKE STYLE

The plane-like diagrammatic representation style implies 2D planar graphical schemes, and it is as old and highly popular as the tree-like one. Just as the latter, the plane diagrams were initially designed to illustrate both the vertical hierarchical and horizontal network structure of the diversity of things. For this, such diagrams were composed as certain figures, which represented constituents of the systems of interest (in systematics these are taxa/merions), and their spatial arrangements represented constituents interrelations (kinship, similarity, etc.). Several diagrammatic formats were invented in order to achieve this goal, which could be roughly grouped into two major categories, namely, *circle*-like and *map*-like; they differ by some of their extreme variants and adjoin by others. Their implied hierarchical structures are fully compatible with those of tree-like schemes and easily interpretable in the same manner (Figure 4.15).

The probable original source of all such representation figures was the metaphor of *World Circle*, which occurred in the majority of primordial myths (Yardley 2014; Lima 2017). It was enjoyed by antique philosophers and flourished in Medieval and Renaissance natural philosophy and natural science. That metaphor was and is an omnipresent, universal symbol with very extensive and diverse meaning: without beginning and ending, it demonstrates perfection, wholeness, inclusivity, and unity of both the universe with all its manifestations and the knowledge about it.

Of numerous early graphical representations of the overall structure of World Circle, the nested *concentric* circles appeared to be the most developed format. Historically, it had several specific meanings. On the one hand, it illustrated a spatial non-hierarchical concentric structure of cosmic spheres, like in Ptolemaic geocentric model of the universe (Crowe 1990). On the other hand, such circle-like concentric diagrams represented non-inclusive hierarchies of both the universe and the knowledge about it. In this case, the physical or mental entities were arranged according to the degrees of their perfectness, so that the most perfect of them took the central position and the least perfect were spread along periphery (Barsanti 1992; Pietsch 2012; Yardley 2014). Such a configuration made such diagrams a kind of *Circle of Perfection* as a specific analogy of the Ladder of the same name. This analogy was evident from the fact that the above-mentioned theologian R. Lull represented the same *Scala Intellectus* both in the ladder and the concentric circle formats; with this, his

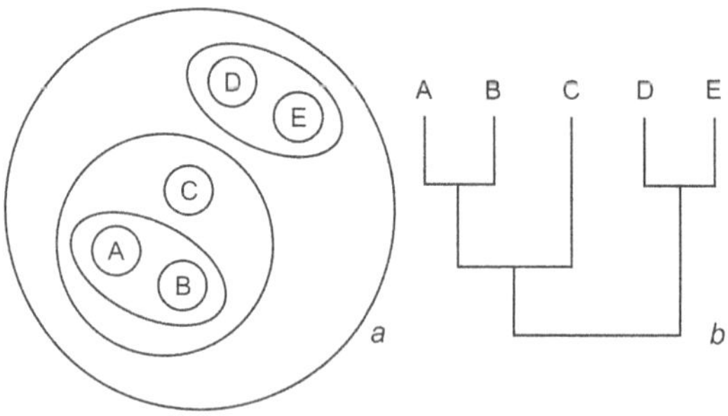

FIGURE 4.15 Isomorphism of circle-like (*a*) and tree-like (*b*) representations of the same hierarchical pattern.

representations illustrated both non-inclusive (by degrees of superiority) and inclusive (by levels of generality) hierarchies of concepts (the same references). In another variant, this format was applied by Lull to illustrate the inclusive hierarchical structure of knowledge (Gardner 1958). One more variant, probably most interesting to us, involved representation of the 'pedigrees of human races' (*Stemmata Stirpis Humanae*) in a concentric circle-like format by the Hispano-Roman scholar and theologian Isidore of Seville (Lat. *Isidorus Hispalensis*; ca. 560–636) (Lima 2017, p. 206).

With regard to biological systematics, this representation format seemed to be applied quite rarely by scientists. A well-known instance of its usage was *Lehrbuch der vergleichengen Zootomie* by the German naturalist and painter Carl Gustav Carus (1789–1869) (Barsanti 1992; Pietsch 2012). He pictured a hierarchical taxonomic system of the animal kingdom in a concentric circle-like manner and placed the humans (as Classe Mensch) in the center of this figure (Carus 1834; Figure 4.16).

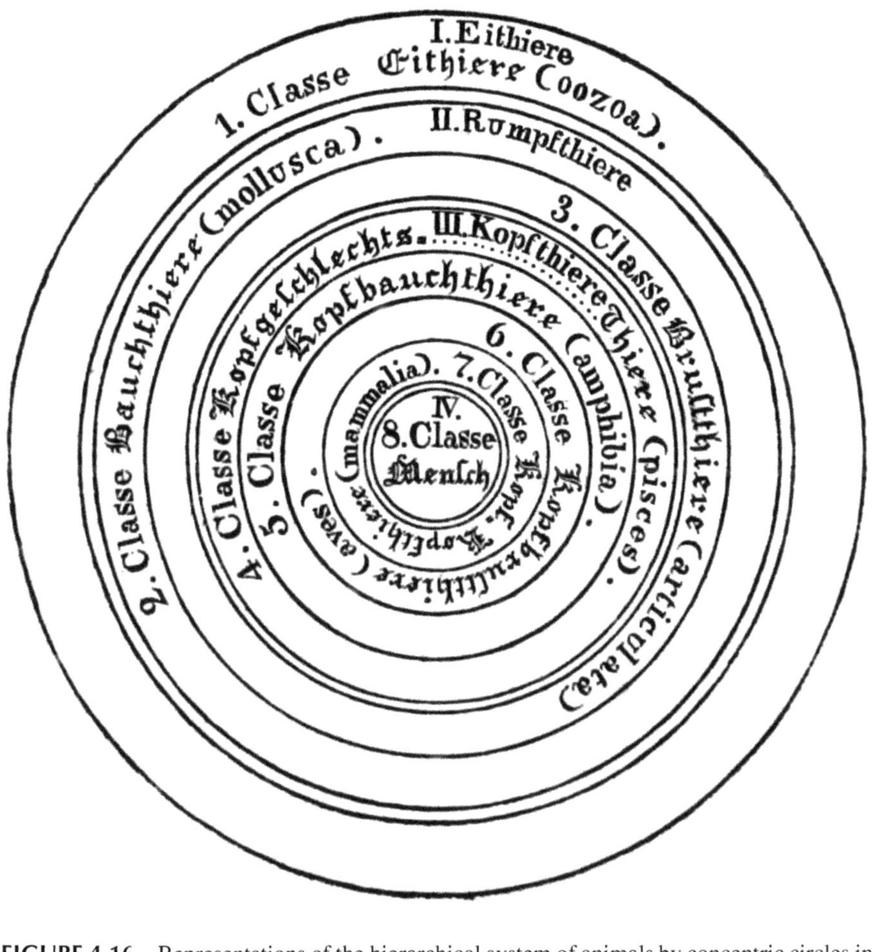

FIGURE 4.16 Representations of the hierarchical system of animals by concentric circles in C. Carus' *Lehrbuch der vergleichenden Zootomie* (Carus 1834, p. 20). Each circle corresponds to a particular level of generality and perfectness.

His system was intended to illustrate both inclusive (by generality) and non-inclusive (by superiority) hierarchies of living creatures, with the lowest forms taking peripheral position and the highest form (human) being placed in center. A similar circular figure could be found also in *Mémoire sur la famille des Mélastomacées* by A.P. de Candólle (1828; see also Pietsch 2012).

Another circle-like format, popular among both philosophers and naturalists, implies representation of inclusive hierarchical and some non-hierarchical relations by *non-concentric* circles, which can be strictly nested, overlapped, or adjoining. This format was also elaborated, along with the just-characterized, in order to demonstrate the overall structure of both the surrounding world and the knowledge about it (Baron 1969). In particular, the same R. Lull appeared to be probably the first to apply this format to illustrate the universality and plenitude of the logical structure of basic concepts he considered most relevant to the natural philosophical issues. The German polymath Gottfried Wilhelm Leibniz (1646–1716) actively applied this format in the same manner, usually limiting himself to three circles or ellipses. The English mathematician and logician John Venn (1834–1923) gave this format its final guise in 1880 and deserved that it would be named after him, the *Venn diagram* (also known as a *set* or *logic diagram*). He based his interpretation not on the logic of propositions but rather on the Cantor set theory that gained great popularity by the end of the 19th century; accordingly, the circles (or ellipses) of those diagrams represented graphically the particular sets, and their mutual position represented their logical interrelations.

The non-concentric format of Venn diagram is fully compatible with tree in representing inclusive hierarchical pattern of relations (see Figure 4.15), so it is applied in nearly all contemporary classificatory sciences, including systematics (Sneath and Sokal 1973; Abbot et al. 1985; Zarenkov 1988). An interesting early instance of the application of a variant of Venn diagram could be found in *Ueber die Entwicklungsstufen des Thieres* by the German zoologist Georg August Goldfuss (1782–1848) (Goldfuss 1817; Figure 4.17a). Another interesting modification of this plane-like format was elaborated by already-mentioned W. MacLeay, who applied it in the first volume of his *Horae Entomologicae* (MacLeay 1819; see Section 1.5). His quinarian system was illustrated by such a diagram, in which an inclusive group was represented by a large circle, and its five subgroups (called *series*) were represented by smaller circles distributed along circumference of the inclusive circle (Figure 4.17b). The German naturalist Johann Jakob Kaup (1803–1873) in his *Classification der Säugethiere und Vögel* brought quinarity to an extreme by representing the groups in question by five-pointed stars also arranged in circles (Kaup 1844).

* * *

The metaphor of *map* was borrowed by systematics from the cartographic practice to be used as a special representation format, illustrating inclusive hierarchies of taxonomic classifications; the American botanist Charles Edwin Bessey (1845–1915), when scarcely mentioning this metaphor, called it 'map theory' (Bessey 1909). In the respective diagrams, which could be properly denoted as *taxonomic maps* (Pavlinov 2018, 2021a), the groups (taxa) were analogized with the natural or political territorial

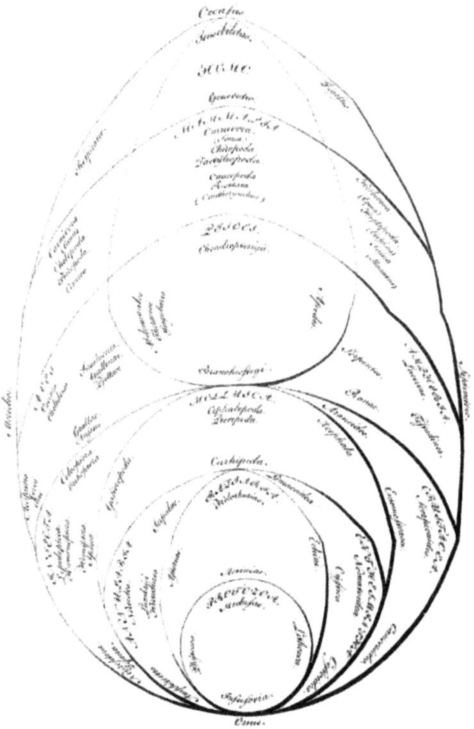

(a)

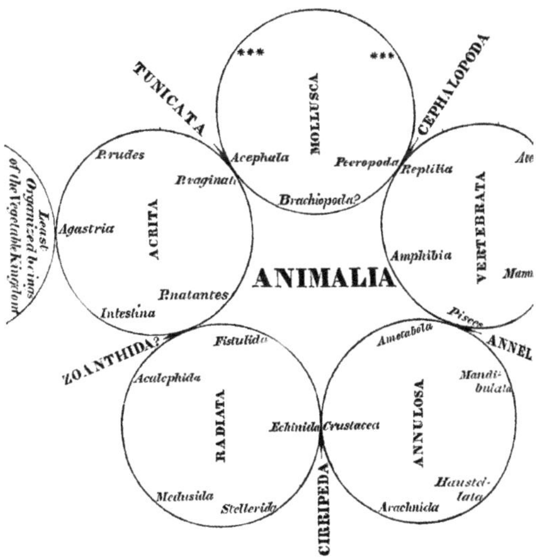

(b)

FIGURE 4.17 Two variants of non-concentric format of Venn diagram showing hierarchical pattern of animal diversity based on specific natural-philosophical ideas: (a) advancement stages of the animal Kingdom in G. Goldfuss' *Ueber die Entwicklungsstufen des Thieres* (Goldfuss 1817, insertion after p. 58); (b) quinarian system of the same Kingdom based on a neo-Pyphagorean worldview in W. MacLeay's *Horae Entomologicae* (MacLeay 1819, p. 318).

units of various levels of generality, the sizes of these units indicated diversity of the respective groups, and their arrangements indicated their affinities: close 'kins' were placed nearer to each other on such 'maps'. By the implied arrangements of the geometric figures (of arbitrary shapes) corresponding to the groups, this metaphor was consistent both with the vertical inclusive hierarchy and the horizontal network structure of the diversity of organisms (Stevens 1984, 1994; Lesch 1990; Barsanti 1992; Panchen 1992; Pietsch 2012; Pavlinov 2018).

In systematics, the map analogy was mentioned probably for the first time by C. Linné in his *Philosophia botanica* (Linnaeus 1751). The first diagram of this kind, which was vaguely reminiscent of a map, appeared in the treatise *Praelectiones in Ordines Naturales Plantarum* by Linné's German disciple Paul Dietrich von Giseke (1741–1796), who denoted his drawing as *Tabula genealogico—geographica Affinitatum Plantarum* (The table of genealogical geographic affinity) (Giseke 1792). A half of century later, the English naturalist Hugh Edwin Strickland (1811–1853) pointed it out that 'the true order of affinity can only be <…> illustrated by a series of *maps*' (Strickland 1841, p. 192; italics in the original). A more recent implementation of this idea by the Soviet botanist Boris Kozo-Polyansky (1890–1957) in his article *The Styles of Representations* illustrated a combination of stylized taxonomic map with phylogenetic tree (Kozo-Polyansky 1949; Figure 4.18).

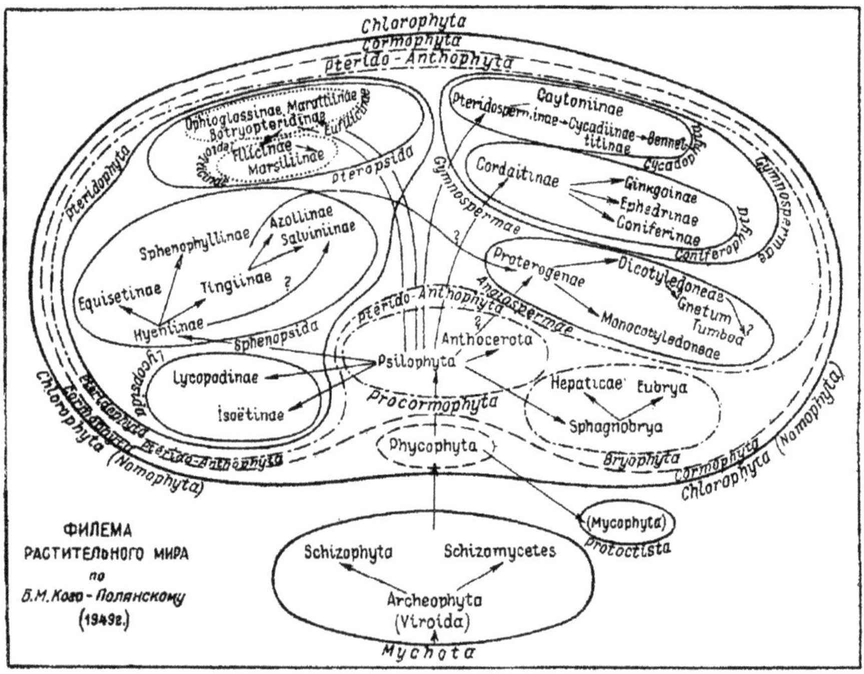

FIGURE 4.18 A combination of stylized 'taxonomic map' with tree-like graph showing affinities of major groups of plants in B. Kozo-Polyansky's *The Styles of Representation*, denoted as 'Phyleme of the vegetable world' by its author (legend in the lower left corner) (Kozo-Polyansky 1949, p. 252).

4.4 FROM CLASSIFICATIONS TO REPRESENTATIONS AND BACK AGAIN

The three major representation styles for hierarchical classifications differ in their relation to the latter and to each other. In general, this depends on how clearly the hierarchization scenarios underlying the respective classifications can be read in these representations, which in turn depends on the contents (especially complexity) of those scenarios.

The texts containing lists of taxa with (desirably) their characteristics and formatted according to certain rules are most usually considered as *the* classifications. Indeed, when someone says 'classification', the latter is most usually imagined just as such an ordered text (Chebanov and Martynenko 1998). Therefore, in the case of simple enough classifications, their text representations are usually considered sufficient, especially for non-professional tasks. However, if underlying hierarchization scenarios are quite intricate and are not evident from the respective texts, the latter are supplemented with the graphical schemes—of course, providing that they themselves meet certain necessary conditions.

The tree style is traditionally considered the most appropriate for the representing inclusive hierarchies: this idea has been embedded in it at the very beginning of scholasticism and inherited from it by early systematics. Therefore, in its scholastic (basically logical) account, tree graph is usually considered self-sufficient as a schematic representation of the logical relations between notions, including their ordering (upon a need) in formal classifications. However, if a specific substantive interpretations of tree graphs are assumed, they are usually accompanied by certain explanatory texts. An example of such a combined representation is provided by the Ramean tree most usually applied by early (scholastic) systematics. In such a case, tree topology defines taxa extensionally by indicating their hierarchical subordination, while the lists of taxonomic characters placed at the nodes, at which the taxa diverge, define them intensionally. Such trees may be denoted as *annotated* (in the sense of Wiley 1979, 1981), and they either complement pure text representations of taxonomic classifications or are considered their appropriate equivalents. Such annotated could be found in many treatises of the 17th–18th centuries.

In phylogenetic systematics, genealogical trees are primary relative to taxonomic classifications based on them; in the terms applied here, this means that the graphical representations are primary relative to the textual ones. Transformation of trees into classifications implies identification of tree branching patterns with the respective hierarchical arrangement of classifications, with identifying tree branches with taxa of various level of generality. Such transformation often faces certain challenges caused by the multiplicity of possible ways, in which a tree can be chop up into pieces (in the sense of Brummitt 2002). This seemingly simple technical procedure becomes more intricate with the complication of the phylogenetic patterns and respective tree topologies, which can be anticipated keeping in mind the principle of *NP-completeness* (see Section 2.1 on it). The most controversial are anagenetic ('vertical') and cladogenetic ('horizontal') chopping approaches, which usually result in the essentially different phylogenetic classifications (Hennig 1950, 1966; Simpson 1961; Wiley 1981; Shatalkin 1988, 2012; Pavlinov 1990a, 2018; Wägele 2005) (Figure 4.19).

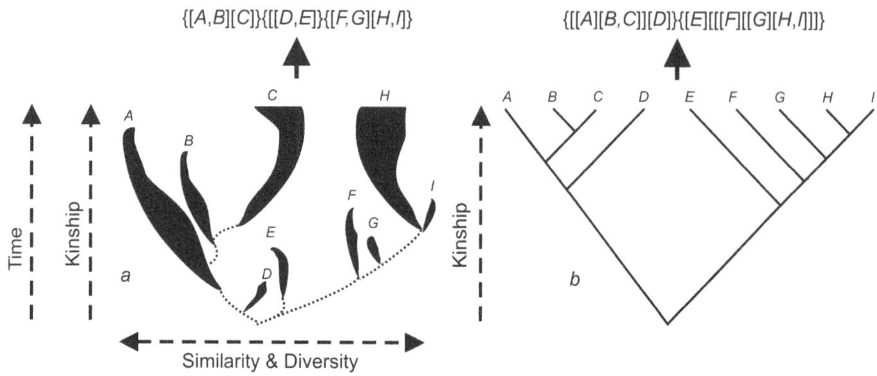

FIGURE 4.19 An artificial scheme showing two modes of representation of kinship relations of taxa *A–I*: (a) classical phylogenetic approach and (b) cladistic approach; the corresponding taxonomic systems are shown in the nexus format above trees. Based on Pavlinov 2021b, p. 11 (modified); courtesy of Phiton XXI Publ.

Such a double representation of taxonomic systems seems to be preferable also for some of the plane-like style formats, if the latter are based on the complex hierarchization scenarios. This is particularly true for the concentric circle representations, such as 'circles of perfection', as well as for non-concentric figures of MacLeay's quinarian system. Respectively, in all such cases, the standard textual representations of classifications must be accompanied by the graphical ones, without which they would appear to be essentially incomplete with regard to the underlying hierachization scenario. In contrast, Venn diagrams are fully compatible with both text- and tree-like styles in the case of simple hierarchical patterns, so any one of them would suffice for representing the respective hierarchical classifications.

At any rate, the potential (and real) multitude of representation styles is responsible for a part of the above-mentioned taxonomic uncertainty, as it means impossibility to reach a one-to-one corresponding between graphic-style and text-style representations of classifications based on the same hierarchization scenario for a given group. This kind of uncertainty can be partly 'circumvented' or minimized through several ways. George Simpson proposed that phylogenetic classifications should be merely *consistent* with the inferred phylogenies (phylogenetic trees), i.e., the former should not contradict the latter (Simpson 1961). W. Zimmermann and W. Hennig suggested reducing phylogeny to cladogeny and phylogenetic tree to cladogram: the latter's simplified branching pattern is easier to represent textually by the hierarchical system of holophyletic groups (Hennig 1950, 1966; Wiley 1981). However, even in such a seemingly simple case, cladistic classifications are most usually published in two formats, textual and graphical, with the latter being of primary meaning (Hennig 1966; Wiley 1981; Stuessy 1996; Kitching et al. 1998; Wägele 2005).

NOTES

1 Lawrence Griffing (2011) rightly considers Ramean tree as a direct predecessor of such graphical representation of identification keys.
2 This mythological World Tree should not be mixed with the so-called *sacred trees*, which are particular (real) trees endowed with certain sacred properties (e.g., Philpot 1897; Cusack 2011).
3 This book has been written in the beginning of the 1300s but published two centuries later (Pietsch 2012). Its complete translation in English was published recently by Y. Dambergs (2003).
4 Because of this, an evolutionary meaning of Lamarckian tree was actually presumed, rather than positively asserted, based on the general evolutionary attitude of its author.

5 Two Important Conceptual Pyramids: Species and Homology

Without the species notion, all [biological] science turns into nothing.
Friedrich Engels (1883)[1]

Homology [is] an unsolved problem [in biology].
Gevin de Beer (1971)

The basic thesaurus of biological systematics includes two important notions of very general meaning, namely, those of species and homology. They were defined above as the structural constituents of the object-based manifestations of systematic diversity, with species belonging to its taxonomic aspect and homologues to meronomic aspect (see Section 2.5). Such their understanding makes them mutually similar with regard to the general patterns of multitude of their interpretations, which suggests a possibility for them to be analyzed within the same conceptual framework.

In a full agreement with the scenario of conceptual development of the cognitive situation in systematics suggested by evolutionary epistemology, the history of these general concepts involved diversification of their particular interpretations. The latter diverged due to specific emphases placed on their either ontic or epistemic consideration, structural or dynamic explanation, accentuation of specific mechanisms of their origin and persistence, etc. With this, there was a very significant common trend in the historical development of both concepts, which can be designated as *biologization*, i.e., the recognition of a close connection between their essential features, on the one hand, and the natural history (for species) and structural-functional constitution (for homology) of organisms, on the other hand. It is this connection that appeared to be one of the most fundamental causes of the multiplicity of their particular manifestations and respective interpretations.

Such a diversification of these concepts gave rise to the general problems of species and homology, with two key words characterizing them, *pluralism* and *fuzziness*. Both problems are currently being actively debated not only within systematics but also in other branches of biology dealing with biodiversity, and the participants of these debates are not only scientists but also philosophers. The viewpoints vary from an unconditional acknowledgment of the great significance of these concepts in their monistic treatment to their categorical negation because of the impossibility for each of them to be defined in a unified manner. There is no end in sight to these discussions yet, and the only success in their resolving reached so far seems to be that 'the points of disagreement are much clearer now than they have been previously'

DOI: 10.1201/9781003597094-5

(Hull 2002, p. 119). And one of the most important points, philosophically speaking, concerns the relevance of the general conceptual approach to these problems (Pavlinov 2012, 2013, 2022b, 2023). In particular, such an approach implies an idea that may look like an 'intellectual chimera' from other perspectives: the respective entities 'are *conventional objects*. They are conventional, yet they are *real*' (Cosetta 2010, p. 271; italics in the original). Such a 'chimera' provides another evident manifestation of the repeatedly mentioned multifaceted onto-epistemic dualism of the basic thesaurus systematics develops and deals with.

This chapter briefly considers these two problematic issues, with the emphasis being placed on their hierarchical representation in the form of respective conceptual pyramids. The latter seem to provide an opportunity for a Mario Bunge's idea of *integrated pluralism* (Bunge 1969) to be implemented in order to preserve conceptual pluralism while satisfying an aspiration for a classical monism.

5.1 THE SPECIES PROBLEM

The species notion provides a typical instance of problematic concept due to its plurality, which gives it a 'nomadic' character (Minelli 2022). The latter means that it has a plethora of very different particular interpretations and applications in various disciplines, which were accumulating during 2,000 years of its active use in the humans cognitive activity addressing any kinds of entities, both material, supramaterial, and mental. It is this substantive plurality that was and still is a major source of the *species problem* that has been persisting in various guises throughout the long-lasting history of the notion in question.

The species notion has been originally rooted in Platonic *eidology* with its core concept of the overall hierarchy of *genê* and *eidê*, which was both linear (levels of perfectness) and inclusive (levels of generality) (see Section 1.3). In contemporary terminology, these two categories are denoted as *genus* and *species*, respectively (derived from the scholastic Latin). Their all-encompassing hierarchy was multi-aspectual in that it was composed of several tightly interrelated partial manifestations, which were either internested or complementary. On the one hand, this overall hierarchy referred to both the things (aggregates thereof) and the quiddities embodied in them, and such their two-dimensional character was denoted above as *taxonomic-meronomic dualism*. On the other hand, this hierarchy consisted of the interconnected ontic (the things with their quiddities) and epistemic (the respective conceptions and notions) aspects structured by the same categories of *genê* (genera) and *eidê* (species), which implied their *onto-epistemic dualism*. In each of these partial hierarchies, the status of these categories as either *genê* or *eidê* was defined contextually, and they were interchangeable. This implied one more manifestation of the complexly structured hierarchical Platonic world, which was denoted as *genos-eidos* (*genus-species*) *dualism* entailing *genos-eidos uncertainty*. This implied that the problem in question was originally shaped as the *genus-species* one. Such its treatment is correct from the contemporary perspective (Pavlinov 2018, 2023), but there seem to be nothing 'problematic' in the conceptual polysemy of *genê* and *eidê* for antique natural philosophers, as such their 'multi-dualistic' character manifested the same 'multi-dualistic' nature of both the cognizable universe and the knowledge about it.

The subsequent development of such a complex genus-species problem involved its more definite structuring by the disunion of basic categories as components of each of these dualities. The first and most important (for our consideration) step in this direction was made by Aristotle, who in his (*On the Generation of Animals*) (*Περὶ ζῴων γενέσεως*) distinguished between ontic and epistemic hypostases of *genos* and *eidos*, as applied to the living (see Aristotle 1953). With regard to the *eidos*, which is of our concern here, it was referred to as *phyleis* if considered as an ontic (natural) entity, and *homoeidos* if regarded in its epistemic (logical) account. Clarifying natural mechanisms that assumingly provided for the integrity and persistence of the *phyleis*, Aristotle expressly pointed to the *generation* (begetting) of organisms, and this idea is currently formalized as the *generative* species conception (Wilkins 2009a, b; Pavlinov 2023). By this, he anticipated an important contemporary idea of the close relation of the nature of species of living organisms to some of their biological features.

The refinement of the genus-species problem by scholasticism preserved all initial dualities of basic categories, with paying much attention to the epistemic aspect of the respective dualism. This accentuation resulted in the elaboration of the hierarchical classificatory *genus-species scheme*, which would play an exceptional role in the emergence of biological systematics.

However, such an epistemic bias did not prevent the further progress in the biologization of the species notion, which led to an important demarcation between universal (logical) and biological accounts of the species category. This trend was consolidated in the 13th century by Albertus Magnus, who expressly approved in his *De Animalibus* the Aristotelian idea that the persistence of species as a natural unit is ensured by 'reproducing one individual from another' (see Albertus 1999, Vol. I, p. 72). It is evident that such a species could in no way become 'genus' still understood logically: they were acknowledged as essentially different entities. Thus, the antique ontic genus-species dualism was gradually becoming obsolete.

The next very decisive step in this direction was made by Buffon, who fixed the 'generative' account of (in future terms) biological species by declaring in his *Histoire naturelle* that the ability to produce fertile offspring 'constitutes reality and unity of what should be called species, both in animals and in plants' (Buffon 1755a, p. 64), and this crucial idea was repeated by him in the fifth volume of D. Diderot's *Encyclopedie* (Buffon 1755b). Hence, it was Buffon who seemed to provide the first explicit formulation of what would be called *reproductive criterion* of biological species two centuries later (Sloan 1970, 1979, 1986; Wilkins 2009; Pavlinov 2023). The effect of this idea was evident, *inter alia*, from the fact that the *Elements of Logic* by the English philosopher Richard Whately (1787–1863) distinguished between logical species as a universal and species of living ('organized') beings that 'may have descended from a common stock' (Whately 1826, p. 166).

Considering the refinement of the old Aristotle's idea about biological (in its most general sense) species as a generative lineage by scholastic and Renaissance natural philosophers, one should not omit a decisive impact of the so-called biblical biology (in the sense of Berndt 2000) on them. Indeed, chapter I of the *Book of Genesis* says that God, after creating 'vegetation [and] living creatures of every kind' and 'humankind', blessed them, saying, 'Be fruitful and multiply' (*Gen.* 1: 22, 24, 28; cited after: The Holy 2025). By the way, taking into consideration that that Book was dated by the 13th–12th centuries BC (Shifman 2007), these were its authors who could be count to have been the first to express the generative idea of species.

This general biologization trend involved decomposing of each of the three antique dualisms, which made the species notion a highly fuzzy conceptual entity. The credit for this belonged to the most advanced actors of the mature phase of scholastic systematics, namely, J. Pitton de Tournefort and C. Linné. First, they decomposed genus-species dualism by introducing fixed ranks for these (and several other) categories, which was a part of their project of the ranked taxonomic hierarchy. Second, they disunited taxonomic-meronomic dualism by concentrating on the taxonomic account of species category and considering organismal features (meronomic account) merely a means for the delineating particular species taxa. Third, they destroyed species onto-epistemic dualism by accentuating its natural status and ultimately acknowledging the 'generative' idea as the primary cause of the integrity and persistence of species of the living, and many of their followers (de Jussieu, de Candolle, Lindley, etc.) and even opponents (Adanson, Lamarck, etc.) concur with them in this respect.

All these conceptual changes resulted in that the species category from then on referred primarily to the particular groups of organisms and was ascribed a salient status. Accordingly, this category considered by biologists lost its status of a universal classificatory unit and turned out to be just the *bio*species. This transformation was indirectly supported by the recognition of biology as a distinct natural science branch in early 19th century (Lunkevich 1960). Due to this, the species became for a while a nearly uniformly recognized by biologists as a specific unit of the overall hierarchy of *Systema Naturae*, which took a fixed place in the latter and was characterized by specific biological mechanisms ensuring biospecies integrity and persistence. This put an end to the previous multifaceted problem inherited from Antiquity and thereby fashioned the prerequisites for its development toward its treatment as a distinctive species problem.

* * *

The Darwinian revolution of the second half of the 19th century became a pivotal point in the post-Linnaean development of the species notion. This pivot has been initiated by C. Darwin's microevolutionary model which implied a gradual transformation of local infraspecies forms (races, subspecies, etc.) into distinct species. Proceeding from this model, Darwin stated that 'varieties have the same general characters as species, for they cannot be distinguished from species' (Darwin 1859, p. 58). Such a standpoint implied rejection of the salient status of 'Linnaean' species, which would give subsequently a reason to accuse Darwin of his 'eliminating the species as a concrete natural unit' (Mayr 1963, p. 14). This verdict was an evident exaggeration (Stamos 1996, 2003; Ereshefsky 2009, 2011; Sloan 2009; Pavlinov 2018, 2023), and yet some adversaries of such a radical solution to the 'old' species problem agreed with it and characterized that situation as the 'species crisis' (Skvortsov 1967; Zavadsky 1968; Gilson 1984). They were largely correct in their assessment, because such a conceptual 'obliteration' of the species category was culminated in its rejection as an abstraction having no real prototype in nature (Kerner von Marilaun 1895; De Vries 1904; Timiryazev 1904; Bessey 1908; Johnson 1908; etc.), which reproduced the old idea of Ockham's nominalism (see Section 1.4).

From the perspective outlined above by two key words, one of the most significant outcomes of the conceptual shift from the traditional macro- to a new microsystematics became the replacement of old uncertainties, caused by the specific multifaceted dualisms, with a new one, which appeared to be analogous in some way to the previous genus-species dualism. Now this was the *species-infraspecies dualism* and the *ranking species uncertainty* associated with it, which shaped the focal problematic issue denoted as the *species question* (Johnson 1908; Greene 1910; Turrill 1925).

This crises situation has changed closer to the mid-20th century due to the development of the *synthetic theory of evolution* (STE): in it, speciation (formation of new species) was suggested to be regarded as a nodal event of the evolutionary process (Dobzhansky 1935). This lead to a revival of the salient status of species treated realistically as a natural unit taking a fixed position in the hierarchy of biota. Due to this, a special attention was attracted to the biological mechanisms of species integrity, which rather unexpectedly has resulted in the emergence of the contemporary species problem.

In such a move of the biologization trend in the general understanding of species, Darwinian revolution happened to have plaid a very significant role. The essentially ecological character of the basic ideas of the microevolutionary model (Petrusiewicz 1959; Schwartz 1980; Gould 2002) provided for an emphasis on the ecological mechanisms of the origin and persistence of the species. It became clear soon that different groups of organisms differ markedly in such mechanisms, and this resulted eventually in the recognition of various *kinds of species*, which diversity was supposed to be non-accidentally related to the diversity of certain biological features (Romanes 1895; Cowles 1908). Of special interest was a metaphor of the latter author, who called 'nonhomologous' the species in the groups of organisms with different natural history. As a result, the *species problem* emerged in its modern guise with its fundamental idea of the natural multiplicity of the 'kinds of species' and gained its 'official' denomination in the 1920s–1930s (Treueman 1924; Turrill 1925, 1938; Robson 1928; Hawkins 1935; Faegri 1937; Hogben 1940).

This general idea found its explicit shaping in the form of the key notion of *species conception*. E. Mayr (1982) denoted all particular ideas about species as 'concepts', and W. Miller (2016) counted 'Three Centuries of Species Concept' (p. 33). However, this notion was probably introduced explicitly for the first time by the Canadian-Scots evolutionary biologist George Romanes (1848–1894), who just mentioned 'pre-Darwinian conception of species' in his *An Exposition Of The Darwinian Theory* (Romanes 1895, p. 284). This terminology was soon adopted by many naturalist studying species diversity (e.g., Cowles 1908; Shull 1923; Turrill 1925; Dobzhansky 1935); though looking minute at the first glance, it appeared in fact to be very important for a new configuration of the species problem, as it opened a possibility for its proper conceptualization. As said above, this made the analysis of species conceptions a focal point of the contemporary species problem.

* * *

The next several decades witnessed the multiplication of the particular understandings of the mechanisms responsible both for the statics and dynamics of species of plants and animals, which were fixed by different species conceptions. This gave Ernst Mayr a reason to declare 'that there are probably as many species concepts as there are thinking systematists and students of speciation' (Mayr 1942, p. 115). Such a pessimistic estimate of the seemingly enormous plurality of species conceptions was certainly another Mayrian exaggeration, and yet, their number grew up to two dozens or so by the end of the 20th century (Mayden 1997; Wilkins 2009a,b, 2011; Pavlinov 2013, 2022b; Zachos 2016). They differ in implying ontic (real, natural) or epistemic (nominal, classificatory) status of species, primarily ecological or genetic causes of integrity and specificity of natural species, emphasizing their stationary or dynamic (evolutionary) nature, etc. Heatedly discussed became the ontic status of natural species, which was treated as a set or a natural kind or a homeostatic property cluster or a quasi-individual. Such a biologically motivated multiple interpretation of the nature of species has finally destroyed the classical *species monism* and entailed the recognition of *species pluralism* with the conceptual *species uncertainty* accompanying it (Hey et al. 2003; Bartlett 2015; Pavlinov 2022b, 2023).

This diversification trend caused two opposite reactions. One of them, most reductionist and therefore quite attractive to positivistically minded scientists, revived the idea of negation of species category altogether and suggested developing a kind of *speciesless* (or 'post-species') biology (Kober 2008; Zachos 2016; Mishler 2021, 2022). Another was a specific integration trend, which involved the attempts to arranged particular species conceptions into certain conceptual systems to make their diversity more sensible and manageable (Minelli 2020, 2022). One of the first moves in this direction was made by the Ukrainian-American evolutionist Theodosius Dobzhansky (1900–1975), who recognized two major categories of species conceptions, static and dynamic (Dobzhansky 1935). The subsequent development of this idea resulted in several proposals to elaborate a conceptual pyramid of these conceptions according to their levels of generality (Kitcher 1984; Mayden 1997, 1999, 2002; Pavlinov 2009, 2013, 2018, 2022b, 2023; Richards 2010, 2013; Naomi 2011; Baetu 2012; Zachos 2016; Ereshefsky 2017).

Such a 'pyramidal' approach constitutes a significant part of the respective object-based partial taxonomic theory (see Sections 2.1 and 2.5). The latter, when implementing this approach, accentuates species definitions as the 'cores' of species conceptions. Therefore, the first issue to be considered in this regard is the proper elaboration of intensional definitions of species, which indicate its essential properties implied by the relevant conceptions. The general configuration of the sought pyramid of species definitions should basically follow a deductive principle, which is most consistent with standard deductive logic of definitions (Ivlev 2025; see Section 2.3). This means that a more general notion is to be defined first as a 'logical genus' and the less general ones are then defined in its context as its 'logical species'. To the contrary, as was emphasized above, the inductive definitions of a notion with the reference to some less general ones seems to be unsound.

With regard to the notion of 'species in general' of the living, the following important corollaries can be drawn from the preceding consideration. First, the currently accepted inductive definition of species through the interactions of its constituent organisms is inductive and therefore should be dismissed as logically inconsistent: it is similar to the definition of biological organism through its cells and tissues. Second, the customary placing of a particular species conception, which is treated as the 'most general' (evolutionary or else), at the top of conceptual pyramid (Mayden 1997, 1999, 2002; Naomi 2011; Baetu 2012; Zachos 2016) seems to be inappropriate, as it leaves the very notion of biological 'species in general' without any proper definition. Instead, the top position should be given to the notion of a certain encompassing natural phenomenon, within which 'species in general' could be properly defined as one of its structural units along with some others of the same level of generality ('of the same kind' in the sense of Urmantsev 1988).

* * *

The argumentation scheme developed by this author follows the general provisions of reduction cascade (see Section 2.2 on it) and suggests a specific hierarchization scenario for the species problem (Pavlinov 2013, 2018, 2022b, 2023). First, it places a certain general biologically sound concept at the top of pyramid, which can actually be treated as encompassing with regard to species. Second, the entire pyramid is constructed as cause-focused, which is essential for understanding the reasons why and how a distinct biological entity called 'species in general' arises (Mayr 1982; Barker 2010, 2019, 2022; Pavlinov 2013, 2022b, 2023). Next, it considers the particular kinds of species as particular 'modes of being' (hypostases) of that 'species in general' (Zavadsky 1968; Brigandt 2003a; Wilkins 2009a; Pavlinov 2013, 2022b, 2023), which entails a biologically sound substantiation of the respective species conceptions. All these make the latter, inferred in the substantive context of a single pyramid, conceptually interrelated and thus mutually interpretable.

This hierarchical conceptual construct was suggested to begin with the theory of life (Hey 2001), which seems to be too broad, or to consider evolutionary and ecological aspects of biota separately (Sluys 1991), which deprives the sought pyramid of its complex causality and desirable integrity. According to the author's opinion, the pyramidal top position should be given to the concept of biota as a self-developing non-equilibrium system (see Section 2.7 on this). The general idea presumes that its global structuration involves emergence of structural-functional units of various contents and levels of generality, including organisms and supra-organismal assemblages (Brooks and Wiley 1986).

The respective reduction cascade that underlies the pyramid in question may be shaped by the following chain of sequential deductive definitions: (a) *biota* is such a non-equilibrium system that <…>; (b) the *biotic structure* is such an immanent property of biota that <…>; (c) a supra-organismal structural-functional *biotic unit* is such a manifestation of the biotic structure that <…>; (d) a *species unit* is such a supra-organismal unit of the biotic structure that <…>; (e) a *local manifestation* of species unit is such its 'mode of being' that <…>. At each of these (and eventually possible other) steps, a certain natural entity is outlined and specified by the respective cognitive model in the form of a conception that comprises its definition by indication of its essential features. It is assumed that within such a definitional cascade, the 'species in general' is first conceptualized as one of the peculiar units of the biotic structure, and its local 'modes of being' are then delineated by particular species conceptions.

As can be seen from the immediate above, this conceptual pyramid implies a biologically meaningful interpretation of the species problem, so it is primarily ontically based. However, it should not be forgotten that, at the same time, it is inferred based on a specific reduction cascade, which is an attribute of the epistemic component of respective cognitive situation (see Section 2.2). This means that each species conception reflects not only a particular 'mode of being' of species unit but also a particular 'mode of view' of the latter, which accentuates its certain biological features regarded as essential within a certain cognitive task that implies a particular underlying theory. Accordingly, when discussing such an integrated conceptual construct, one should take into consideration its both ontic and epistemic constituents (see also below).

Regarding the ontic constituent, an understanding of species as a certain unit of the structure of biota accentuates the natural causes that operate at the level of biota and structure it, as it functions and develops. This implies that both the existence of species unit *per se* and the variety of its particular manifestations should be understood and explained as a non-accidental effect of the causation of biotic structure. Accordingly, the respective causal analysis should address those most fundamental properties of such units, to which the causes in question are supposedly applied.

Keeping in mind the basic properties of self-developing non-equilibrium systems and their constituents (see Section 2.7), it seems reasonable to assume that such a property of our interest, to be highlighted in this regard, is the *dynamic stability* of biotic units. Assuming that this property implies, before all, a certain unit integrity, it can be generalized by the notion of 'unit integrator', which in the case of species would be *species integrator* (Pavlinov 2023). Such an integrator is presumed to incorporate all biological mechanisms, which ensure collectively the dynamic stability of the respective biotic structural-functional units, including species. This approach provides a possibility for such units to be discussed within the same conceptual framework and, what is important to us, distinguish between species and 'nonspecies' units (biomorphs, guilds, etc.) as the ones that are characterized by different integrators.

It follows from the immediate above that the notion of species integrator may serve as the core of a presumed general species concept. Its supposedly multicausal nature implies that any biologically meaningful theoretical definition of species unit ('species in general') should be exhaustive and incorporating all factors that ensure its dynamic stability, including emergence, persistence, and diversification (Zavadsky 1968; Sluys 1991; Wilkins 2009a; Barker 2010, 2019, 2022; Baetu 2012; Pavlinov 2013, 2022b; Kendig 2014; Nathan 2017; Nathan and Cracraft 2020). Accordingly, the lower-level species conceptions are based on particular interpretations of the species integrator by highlighting those its constituents that are supposed to be most important in shaping the distinctive 'modes of being' of species unit in particular groups of organisms specific in this respect.

The basic principles of *new essentialism* (Ellis 2002; Devitt 2008; Austin 2018) suggest an opportunity to go further with this generalization and to supposes a hypothetical species essence that may be aptly denoted as *specieshood* (Pavlinov 1992b, 2009, 2013, 2018, 2021a, 2022b, 2023; Brooks and McLennan 1999; Griffiths 1999; Wilkins 2007; Kendig 2014; LaPorte 2017; Barker 2019). Its introduction proceeds from an assumption that the structure of biodiversity, including species diversity, is a constituent of the overall natural history of organisms, so it depends on some of its major properties. Accordingly, the specieshood is understood as an incorporated part of the overall group-specific natural history of organisms, which entails some significant ramifications.

First, this implies a possibility for a species integrator to be regarded as constituting the 'kernel' of specieshood, which is determined by the relevant biological features of particular groups of organisms. Second, this provides a general metaphysical substantiation of the diversity of kinds of species resulted from the biological specifics of the particular groups of organisms, which entails specifics of distinct manifestations of the overall specieshood. Third, this makes it reasonable to suppose that the global evolution of the overall biological organization of living beings involves the respective transformations of specieshood, which is thus an *evolving species quiddity*. In general, such an understanding of species unit and its distinctive quiddity (specieshood) appears to be a nowadays culmination of the biologization trend of both general notion and problem of species.

According to the above-outlined general conceptual framework for the contemporary species problem, the following major issues in it should be given the highest priority. One of them includes a careful examination of the interrelation between ontic and epistemic motivations for the recognition of particular species conceptions and their arrangement in a conceptual pyramid. Another issue concerns clarification of the concrete biological features of organisms that constitute and consolidate general specieshood as a characteristic species integrator; one of them is known since Aristotle's time to be reproduction system. As such features become known, the next major task would include an analysis of their diversity in different groups of organisms that provide for the variety of 'kinds of species', including an assessment of their actual spectrum. At last, considered pragmatically, the species problem may be interpreted as 'a problem of theory choice' (Otsuka, 2019, p. 1081). So the primary task becomes a proper choice between particular species conceptions, being specific models of particular manifestations of a certain unit of biodiversity structure, which are most adequate to the biological properties of particular groups of organisms.

One of the manifestations of the contemporary species problem includes the attempts to develop some universally applied operational conceptions to replace the traditional species concept, as far as it cannot be uniformly defined. Proceeding from the general ideas of reduction cascade and respective conceptual pyramid considered above, it should be emphasized that any such conceptions, if introduced *ad hoc* without reference to the higher-level substantive ones incorporating the notion of species integrator, remain biologically 'empty' (Ward and Stanley 2004; Pavlinov, 2018, 2023). The latter holds true for the notions of *paraspecies* or *surrogate species* (Nowak 1992; Oliver and Beattie 1996; Krell 2004; Oren 2004; Ward and Stanley 2004; Minelli 2019) and especially for the pure operational ones like that of a *least-inclusive taxonomic unit* (Pleijel and Rouse 2000).

* * *

When briefly discussing the plurality of species conceptions, David Hull asked humorously a rhetorical question: 'God created numerous different species, but did He create a *single sort* of species or *many different sorts* of species?' (Hull 1999b, p. 23; italics in the original). He did not dare answering it, but in fact, if forget its non-serious intonation and put it in another fashion, it becomes one of the most intriguing issues in the entire species problem. Indeed, considering enigmatic specieshood as a specific component of the natural history of organisms seems to allow for answering the first question: why are there different kind (sorts) of species unit? This is because there are different natural histories specific to particular groups of organisms. Now let us see whether it makes it reasonable to ask the next question: how many of these kinds might be actually there in nature? It seems that the meaningfulness of this very question is determined by the possibility to meaningfully answer it.

When placing the latter point in the general context of the concept of non-equilibrium self-developing biota, the above-mentioned metaphor of *evolving specieshood* comes to the fore, for which the following evolutionary scenario may be presumed (Pavlinov 2013, 2021a, 2022b, 2023). Its prerequisite basic assumption, suggested by the same open non-equilibrium system model, implies that biota has began its development as the least structured global entity, literally, as a 'living matter'. Its historical development involved its gradual structuration, which encompassed all its manifestations and levels, from organisms to supra-organismal units. In turn, this structuration involved a gradual elaboration of the integrators, specific to the particular units and ensuring their dynamic stability. This process implied an enhancement of those integrators to make them more effective in their basic functions: this is evident at the organismal level and reasonably assumable at the supra-organismal one.

This general metaphysical scenario looks plausible as applied to the species unit and its evolving specieshood. It is possible to suppose a 'pre-species' or 'non-species' form of the supra-organismal structure of the nascent biota at the very first stages of its evolution (Kuprevich 1949, 1950; Polyansky 1959; Morgan and Pitts 2008). Its progressive structuration and complication involved appearance of the first signs of more or less stable supra-organismal units, including some 'species-like' entities. This evolutionary stage is evidenced by the structure of systematic diversity of contemporary viruses and prokaryotes: species integrators in them are but weakly developed and cannot ensure discreteness of these 'species-like' entities due to horizontal gene transfer, so such units are aptly denoted as *pseudo-species* or *quasi-species* (Dobzhansky 1970; Eigen 1983; Eigen et al. 1988; Nowak 1992; Domingo 2002; Stamos 2003; Ereshefsky 2010a; Pavlinov 2013, 2021a, 2023; Andino and Domingo 2015; Domingo and Perales 2019); though, they are still sometimes referred to (probably by habit) as 'species' (Pringle 1991; Ochman et al. 2005; Doolittle 2012; Bobay and Ochman, 2018). The subsequent development of biotic structure involved an improvement of species integrators and led ultimately to the development of what can be denoted as *euspecies* (Pavlinov 2022b, 2023; not in the sense of Dillon 1966).

This scenario considers the evolutionary increase of the dynamic stability of species unit from the least to the most integrated stage due to the gradual enhancement of species integrator and thus explains a 'vertical' diversity of kinds of species. However, it does not explain a 'horizontal' structure of their diversity at each level of the evolutionary progression of specieshood. Therefore, the Hull's question of possible number of kinds (sorts) of species due to diversification of the mechanisms of species integrity still remains largely unanswered.

In this regard, a metaphoric *orbital model* of the variety of the kinds in question (Kokić 2009), looking rather strange at the first glance, should be mentioned. To the degree it is (hopefully) correctly outlined in the terms adopted here, it seems to imply the following scenario for the evolving specieshood. The entire diversity of the 'modes of being' of species unit can be imagined as a specific 'atom-like' entity. Its 'nucleus' is occupied by an idea of biological 'species in general' that maintains the integrity of the entire conceptual construct denoted by the notion of species unit. The emergence and diversification of specieshood resulted in the kinds of species, which variety may be supposed to be ordered by a limited number of biologically admissible distinctive combinations of mechanisms composing the general species integrator. Based on the tenets of biological structuralism (see Section 2.8 on it), one may speculate that the constrained diversity of those admissible combinations entails a limited amount of the biologically sound specieshood manifestations and respective 'modes of being' of species unit, which is a specific analogy of the naturally limited configurations of atomic orbitals. Thus, this metaphor implies that (a) the overall diversity of kinds of species is non-randomly structured, (b) it is actually constrained by a natural combinatorics of the causal mechanisms composing the general species integrator, and (c) this combinatorics entails a limited number of the realized 'modes of being' of species unit as the group-specific implementations of specieshood. Of course, this does not for now predict, even approximately, the number of kinds of species but at least allows for supposing their countable character.[2]

Such an evolutionary look at the diversity of the kinds of species as a result of the diversity of implementations of the general specieshood in particular groups of organisms seems to be relevant to the discussion of ontic status of species unit. As said above, this metaphysical issue is vividly debated in contemporary literature, and several versions are considered, namely, species as a *set/class*, or a *natural kind*, or a *homeostatic property cluster*, or a *quasi-individual* (see Pavlinov 2023 for detail). As is evident from this 'short-list', they differ basically in the levels of integrity of species unit, which can be easily explained in the general context of the theme of evolving specieshood. Indeed, the evolutionary scenario exposed above implies that the levels of integrity of the unit in question evolved together with the evolution of specieshood. Therefore, it is reasonable to suppose that various 'modes of being' of species unit, differing in their levels of integrity, are actually endowed with various ontic statuses. For instance, quasispecies can be treated as a natural kind, whereas euspecies as a quasi-individual. So we faces the species 'ontic plurality', which represents but another manifestation of species pluralism resulted from the variety of natural histories of organisms, with the characteristic patterns of their diversity being components of their natural histories.

* * *

The general metaphysics of evolutionary process supposed by the non-equilibrium thermodynamics supposes that the evolutionary increase in the integrity of species units due to the enhancement of species integrators does not completely eliminates a stochastic component of species diversity. It primarily occurs as a specific manifestation of the Darwinian model of microevolution, which emphasizes the random character of within-species variations. The latter serves as a source for the emergence of new species, which may proceed either regularly due to Darwinian selection or randomly, as suggested by the *stochastic* model of microevolution (Futuyama 1979; Stamos 2001; Lenormand et al. 2009; Malaterre and Merlin 2015).

One of the latter's effects allegedly involves a possibility for the local 'splash-outs' of the fluctuating intraspecific diversity on the species level, which results in the emergence of the so-called *virtual species* (Pavlinov 1992b, 2013, 2023). They arise spontaneously, exist for a more or less limited time, and may just as spontaneously vanish. It can be assumed that both the above-mentioned quasispecies of microorganisms and many cryptic euspecies that do not have obvious ecological and physiological specificity, belong to this category. If this is true, then the virtual species are quite widespread in the living nature.

5.2 THE HOMOLOGY PROBLEM

Any cognition, addressing the diversity of objects, implies their comparison, and the latter in turn implies that any two objects are comparable to the degree that they are of the *same kind*, i.e., the same in a sense. In one of scientific vocabularies most relevant to biology and systematics as its part, the *sameness* is denoted as *homology*, and the 'same' objects are denoted as *homologous*, i.e., they are *homologies*, or *homologues*. This general notion is 'enveloped' in the general *homology concept*, which is fundamental for science for its being crucial for the conceptual framing of all comparative approaches; the German zoologist Adolf Remane (1898–1976) believed that all comparative morphology—in fact, all comparison-based research (Rieppel 1988; Sluys 1996)—is the 'science of homology' (Remane 1956). It was recognized that understanding of the basic nature of this concept for any comparative studies forms a kind of *homology thinking* (Ereshefsky 2007b; Hall 2007; Matthen 2007). In systematics, the paramount role of homology is provided for by its leading role in the recognition of classificatory characters: it is traditionally assumed that only homologous characters can ensure the uncovering of actual affinities of organisms and the elaboration of natural systems (not in biomorphics; see Aleyev 1986; Pavlinov 2010).

As is typical for such general concepts, homology provides serious challenges regarding formulations of its unified all-encompassing and all-satisfying theoretical definition, and in fact it occurs in the form of several particular conceptions (Ghilarov 1964; de Beer 1971; Hall 1994a; Bock and Cardew 1999; Brigandt 2003b; Jamniczky 2005; Brigandt and Griffiths 2007; Kleisner 2007; Pavlinov 2012, 2018; Gonzáles 2015; Novick 2018; Ballego-Campos et al. 2023; etc.). This gave the British zoologist Gavin de Beer (1899–1972) a reason to conclude that 'homology [is] an unsolved problem' in biology (de Beer 1971, book title), so its current state was qualified as a 'crisis' (Wells and Nelson 1997). This resulted in the emergence of the *homology problem*, similar to the species problem in many respects (Jardine 1967; de Beer 1971; Ereshefsky 2010b; Pavlinov 2012; Gouvêa 2023).

The most fundamental features of homology to be realized first of all include its multi-aspectual *hierarchical* and *relative* nature (Woodger 1942; Van Valen 1982; Haszprunar 1991; Striedter and Northcutt 1991; Hall 1994a; Minelli 2003; Ghiselin 2005; Kleisner 2007; Pavlinov 2012; etc.). It is multifaceted and can be conceptualized as an array of the following relations.

First, an object is homologous not by itself but relative to some other objects, which are related as constituents of a certain whole: roughly speaking, no such a relation—no homology. Thus understood, homology in its most general meaning implies a certain *correspondence* between particular constituents of the whole (Brigandt 2002; Ghiselin 2005).

Second, the whole–parts relations that underlie homology constitute a specific, minimally *four-level* inclusive hierarchy (pyramid). The lowest level corresponds to the features of constituents, by which they are compared and homologized. The next level corresponds to the very constituents being compared, and it is the *focal level-0* (in the sense of Salthe 1985, 2012; see Section 3.4) in the studies of homologies. The higher level corresponds to the particular wholes, to which these constituents belong; in each comparison operation, homologies are related at this level of hierarchy. The highest level of this pyramid corresponds to a certain general whole encompassing the particular ones: it is evident that the latter cannot be compared without reference to something inclusive, which they themselves are constituents of.

Third, two or more compared objects can be assessed as homologous assuming that there might be their alternative, i.e., non-homology, commonly referred to as *analogy*. This means that the objects can be either homologous or *analogous*, though the boundary between these two is not absolute (see below). These two basic forms of correspondence belong to the same wholes and are allocated to the same level of hierarchy. A certain relation between homology and analogy is always implied, and their recognition and demarcation constitute the key issues in the homology problem.

Fourth, the above-indicated reservation 'in a sense' implies that in each case study, homology/analogy can be identified not universally but rather with the references to a particular *homologization basis*, relative to which the objects are considered homologous or not. Such a basis frames the context of comparison, and it is this context that provides a certain 'sense' for any judgments regarding homology. However, there seems to be a possibility for homology to be conceptualized in a quasi-absolute fashion: following terminology suggested by D. Hull (1980), a homologue can be treated as a specific organismal *replicator* that provides replication and continuity of biological information among the whole's constituents (Van Valen 1982; Roth 1991).

Fifth, a homologization basis is composed of certain *homologization principles* and *homology criteria*. The principles define what homology is both in general and particular meanings, whereas criteria ensure their proper implications. Both principles and respective criteria are multiple, which entails multiplicity of possible homologization bases.

* * *

A historical precursor to the distinguishing between homology and analogy is usually supposed (after Strickland 1846) to be the separation of *affinity* and *analogy* by already-mentioned William MacLeay in his *Horae Entomologicae* (MacLeay 1819). The considered notion in its contemporary understanding was first suggested by the English zoologist Richard Owen (1804–1892) in his *Lectures on the Comparative Anatomy* and treatise *On the Archetype and Homologies* (Owen 1843, 1848). He based his hierarchical typological (of a Platonic kind) conception on the Goethean idea of archetype with its basic constituents denoted as *homotypes* and recognized several important kinds of homology: *general* (unique parts of the generalized archetype), *serial* (repeated parts in the same organism), and *special* (the same parts in different organisms) (see Hall 1994b; Panchen 1994, 1999 for detail).

The next major steps in the conceptual history of homology involved its evolutionary (Haeckel, Gegenbaur, Lankaster, Cope, Osborn, etc.) and ontogenetic (Mivart, Bronn, etc.) interpretations. These entailed an initial fragmentation of the general concept of homology, and a suggestion to distinguish between *homogeny* (inherited from ancestor) and *homoplasy* (resulted from parallel evolution) by the English zoologist Edwin Ray Lankester (1847–1929) (Lankester 1870) would become of great significance for the formation of phylogenetics.[3] The subsequent development of this concept resulted in multiplication of its particular versions based on the diverse combinatorics of these two basic interpretations, so the number of 'homologisms' grew up to several dozens by the end of the 20th century (Voigt 1973; Blyakher 1976; Panchen 1999; Hall 2007). The elaboration of already-mentioned operational *homology criteria* (Remane 1956; Patterson 1982; Ballego-Campos et al. 2023) should also be mentioned. Some new categories of homologies are recently recognized in molecular studies, such as *orthology* and *paralogy* (Fitch and Margoliash 1967; Patterson 1988; Wheeler 2001; Sonnhammer and Koonin 2002; Theissen 2005), as well as *hemiplasy* (Avise and Robinson 2008) and *xenology* (Gogarten 1994).

An important step toward further generalization and biologization of the general homology concept involved recognition of *process* homologies in addition to the classical *structural* ones (Hall 1994b, 1996; Gilbert and Bolker 2001; Minelli 2003; Kleisner 2007). Such an extension was of two kinds. On the one hand, it encompassed the developmental processes that led to the formation of definitive structures of organisms. The correspondence of definitive structures resulting from either the same or different developmental trajectories was proposed to designate *syngeny* (or *generative* homology) and *allogeny*, respectively (Butler and Saidel 2000); the former was specifically denoted as *causal* homology (Otsuka 2017). On the other hand, it appeared productive to interpret certain functions of organisms in terms of the concept in question (e.g., Roth 1988, 1991; Wenzel 1992; Gilbert and Bolker 2001; Tetenyi 2013), which was denoted as *functional* homology (Love 2007). The latter includes, for instance, constituents of the Krebs cycle, viviparity in tertrapods, behavioral stereotypes, developmental psychology, gravity sensing, etc.

Rather unexpected and even discouraging appeared to be the results of the analyses of genetic foundations of 'classical' morphological homologies: no anticipated (and highly demanded) one-to-one congruity between the 'samenesses' at genotypic and phenotypic levels was revealed. It was shown that homologous genes can be responsible for non-homologous (in a traditional sense) morphological structures, while homologous (in the same sense) structures can be regulated by non-homologous genes (Striedter and Northcutt 1991; Wray and Abouheif 1998; Wray 1999).

Accordingly, *genetic* homology was proposed to be recognized as a special category considered in parallel to the old 'phenetic' one (Hossfeld and Olsson 2005), and the case, when morphological structures were treated as non-homologous in a traditional sense but proved to be regulated by the same genes, was denoted as *homocracy* (Nielsen and Martinez 2003; Svensson 2004). In addition, the evolutionarily conditioned sharing of both genetic and developmental pathways underlying the emergence of apparently different phenotypic features was denoted as *deep* homology (Shubin et al. 2009; Scotland 2010).

Given such an intricate character of both theoretical definitions and empirical inferring of homologies, any judgments about them should be considered as hypotheses that are put forward and tested according to certain rules implied by hypothetico-deductive argumentation scheme (Patterson 1982, 1988; Mindell and Meyer 2001; Pavlinov 2005, 2012, 2018; Richter 2005; Szucsich and Wirkner 2007). This means that the respective judgments are introduced in the particular cognitive situations as probabilistic presumptions, which true status can change as a result of the particular studies (see Section 2.5 on this point).

* * *

Philosophically considered, the most significant manifestations of the above-highlighted relative character of homology are determined by its onto-epistemic dualism. This implies, generally speaking, that the homology/analogy dichotomy (in fact, a near-continuum between these two extremes; see Mivart 1870; Boyden 1943, 1947; Hubbs 1944; Hall 1994b) cannot be attributed to the relations of the objects as such but represents the context- and subject-dependent judgments about them. However, in particular research, an emphasis is most usually put on either ontic or epistemic consideration, depending on the personal cognitive standpoints (including thinking styles) of the scientists conducting it.

The ontically interpreted homologization basis may be though of as defined by specific aspects of the structural and functional patterning of organisms and the latter's both phylo- and ontogenetic transformations (Roth 1988, 1991; de Pinna 1991; Lyubarsky 1996; Kuznetsov 1999; Butler and Saidel 2000; Brigandt 2003b; Pavlinov 2012, 2018). Considered epistemically, it is shaped basically by particular consideration aspects and research tasks (Meyen and Shreyder 1976; Meyen 1978; Lyubarsky 1996). These two primary views of homology seem to approximately correspond to the recognition of *explanatory* (causal) or *descriptive* (phenomenological) conceptions of homology, respectively (Griffiths 2007). A combined onto-epistemic consideration makes the basis dependent on the contents of the higher-order hypotheses, within which the respective structures and functions are homologized: for instance, in phylogenetics, a distinction between homogenies and homoplasies depends on the contexts of particular phylogenetic hypotheses (Hennig 1966; Wiley 1981; Pavlinov 2005). At last, at an operational level, the homologization basis is largely shaped by the ways of individuation of particular features, by which the organisms are compared.

* * *

Just as in the case of the above-considered species problem, the current 'homology crisis' causes two opposite reactions to the multiplication of particular homology conceptions. One of these rejects the very idea of any 'general concept' of homology and insists on the incompatibility of some basic conceptions of correspondence, such as structural and processual or phylogenetic and developmental (Borhvardt 1988; Shatalkin 1990), which is sometimes supported by the reference to scientific pluralism (Jamniczky 2005; Kleisner 2007). An opposite intension stimulates monistically oriented efforts for the elaboration of certain unifying conceptual constructs that may pretends to be treated as such general concepts, or *metahomologies*. The latter were suggested to be formally denoted as homology *guarantors*, referring to homologization bases in a specific manner (Kleisner 2007). Again, just as in the case of the species problem, the efforts of the second kind result in the hierarchical (pyramidal) arrangements of particular conceptions.

In their most simple versions, such hierarchies involve the groupings of conceptions into certain higher-level categories, each accentuating specific causal factors that shape major homologization bases. One of the most important ideas of this kind involves distinguishing between two its categories, already-mentioned *structural* (stationary) and *transformational* (dynamic, generative, processual). They differ basically by the ontic interpretation of the nature of interrelations between homologous elements: they are united through either structural correspondence or continuous transformations, respectively, of structures. Their conceptual summation provides concept of *synthetic* (sometimes called *biological*) homology, which is interpreted as a canalized development of the structural correspondences (Wagner 1989; Griffiths 2007; Szucsich and Wirkner 2007). Another version is *organizational* homology as a correspondence between morphoprocesses and morphostructures (Müller 2003; Kleisner 2007), and such a synthesis may help to solve some issues regarding ontological status of homologues (Suzuki and Saigo 2025). Placing the entire homology issue in evo–devo concept entails an understanding of a homologue as a structural unit capable of separate evolutionary development due to its ontogenetic (quasi) autonomy (Wagner 1994, 2014; Laubichler 2000; Brigandt 2002, 2003b, 2007; Amundson 2005; Ereshefsky 2010b; Minelli and Fusco 2013; Suzuki and Tanaka 2017). In this regard, of interest is a suggestion to consider a rather old C. Waddington's concept of *epigenetic landscape* as the one that could help to unite phylogenetic and developmental ideas about homologies (Striedter 1998). A quite advanced version implies generalization of the multiplicity of homology conceptions in a kind of conceptual pyramid, which is suggested to be denoted as *hierarchical*, or *combinatorial* (*factorial*) homology (Minelli 1998, 2016; Scotland 2010; Minelli and Fusco 2013).

NOTES

1 Cited after Engels (1964, p. 189).
2 I sincerely thank Prof. Tonći Kokić for his discussion of this interpretation of the orbital model (*in lit.*, July 09, 2025).
3 It is to be noted that in cladistics (after Hennig 1950, 1966), homology is identified with homogeny. The latter was proposed to be denoted as *tokogenetic* homology (Assis 2013).

6 Taxonomic Hierarchy and Nomenclature

Things are so because they are called so.
Chuang-tzu

The most significant portion of empirical knowledge about systematic diversity takes the form of taxonomic (taxono-meronomic) classifications. The latter, under certain assumptions, may be regarded as the specific texts written in a definite professional language of biological systematics (Löther 1972; Chebanov and Martynenko 1998). According to the provisions of cognitive science, any language is not separable from cognitive activity and evolves with its goals and principles (Langacker 1990). Biological systematics, in line with these basic ideas, has been persistently developing its professional descriptive language, as its specific cognitive device, throughout the whole of its history (e.g., Heller 1964; Nicolson 1991; Dayrat 2010; Pavlinov 2015, 2022a).

One of the important components of this language is *taxonomic nomenclature*.[1] Its immense significance for systematics was recognized in the 18th century, when it was declared to be a part of 'systematic philosophy' by Linnaeus (1751). This point is emphasized by most of the contemporary taxonomic textbooks, in which nomenclature may take a significant part of their contents (e.g., Blackwelder 1967; Mayr 1969; Shipunov 1999), though there are exceptions (e.g., Simpson 1961).

This chapter deals with the taxonomic aspect of nomenclature, which has been being purposefully developed by biological systematics during all its long history. With regard to its meronomic aspect, the latter's terminology has been attempted to be schematized and codified very provisionally by some advanced herbalists, systematicians, and anatomists of the 16th–18th and early 19th centuries (e.g., Bock 1546; Linnaeus 1751; Barclay 1803). However, the subsequent development of systematics became focused on the taxa and their ranking and naming, thus remaining within the scope of just the taxonomic nomenclature. The meronomic one appeared to be left to the other branches of biology, specifically dealing with the respective manifestations of organismal diversity (anatomy, histology, molecular biology, cytology, etc.) and developing their own nomenclatural systems in the form of specific 'ontologies' (e.g., Ashburner et al. 2000; Bard et al. 2005; Mabee et al. 2007; Göpel 2024). It hardly makes any sense to get into their detail here.

Taxonomic nomenclature is composed of two basic parts (Pavlinov 2015, 2022a). One of them, called *nominative* nomenclature, encompasses an array of designations of taxa and ranks (names, symbols, etc.). Another is *regulative* nomenclature, which contains specific regulators (conceptions, principles, rules, etc.) controlling those designations, their application, as well as certain parameters of the taxonomic objects

 DOI: 10.1201/9781003597094-6

being designated. Thus, regulative nomenclature can be regarded as a specific *meta-language* relative to nominative one (partly in the sense of Takhtadjan 1966).

It is a common belief, supported by the official declarations in the contemporary Codes, that taxonomic nomenclature is theory-free, which is formalized by the *principle of taxonomic freedom*, or *deontology* of nomenclature (e.g., Pauly et al. 2009; Dubois 2005; Aescht 2018; Dubois et al. 2019). However, this opinion is not correct; instead, a careful analysis of the conceptual history of systematics has clearly indicated a certain relation between taxonomic theory and nomenclature (Pavlinov 2015, 2022a).

One of the significant characters of this relation is that it connects taxonomic nomenclature in several ways with the hierarchical structure of classifications elaborated by systematics. Such a connection has become noticeable from the beginning of the history of systematics, and it got even stronger, at least in some parts, with the invention of the so-called *rank-dependent* (or *rank-based*) nomenclature in the 18th and 19th centuries. This nomenclature conception was subjected to harsh criticisms from several systematic schools in the second half of the 20th century, which suggested a *rank-independent* (or *rank-free*) one based on their own conceptual backgrounds. As a consequence, the clash between the adherents of rank-dependent and rank-free nomenclature conceptions threatens today to schism practical systematics.

The relation between taxonomic hierarchy and nomenclature manifests itself in both its regulative and nominative parts, which is most evident in the case of its rank-dependent version. In the first case, particular nomenclatural regulators are specifically applied both to the taxa of different ranks and the ranks themselves. As to the nominative nomenclature, its rank-dependence is most evident in the case of terminological fixation of particular ranks, as well in different rules of the formation of names of taxa of different ranks.

This particular relation, which attracts the most attention of nomenclaturists, is but one aspect of the multifaceted hierarchical character of the professional language of systematics. As a matter of fact, the nomenclature itself possesses its own complex multi-level structure, which is typical of any descriptive language developed by natural science. At its highest level, the just-mentioned nominative and regulative parts are distinguished. At the latter's lower levels, it is hierarchically structured by formulating the nomenclature regulators of various levels of generality.

The chapter analyzes two issues pertaining to the hierarchical consideration of taxonomic nomenclature. One of them is the relations between the latter and the hierarchical arrangement of taxonomic classifications, and another concerns the hierarchical structure of nomenclature itself. A special attention is given to two its basic versions mentioned above, namely, rank-dependent (rank-based) and rank-independent (rank-free).

6.1 TAXONOMIC THEORY AND NOMENCLATURE

According to one of the basic ideas of cognitive linguistics, the structure of languages, with which science describes diversity of whatever natural (epi)phenomena it studies, should be adequate to the structure of that diversity (Rosch 1983; Chomsky 1987; Croft and Cruse 2004; Geeraerts 2006). Due to this, a specific *linguistic world picture* appears as a specific part of the onto-epistemic framework of cognitive situation, in which it reflects the cognized world by specific linguistic means (Gumperz and Levinson 1996; Devitt and Sterelny 1999; Uryson 2003). As stated in Chapter 2, the natural (epi)phenomena are imbedded in the cognitive situations not by themselves but in the form of respective cognitive models. The latter in turn are elaborated based on the substantive theories addressing the (epi)phenomena in question. Accordingly, a specific relation is established between a theory, which models a studied (epi)phenomenon, and a specific language, with which the latter is described.

When applying this general idea to biological systematics, it is to be reminded foremost that its research object, namely, taxonomic diversity, is rationally defined by the general taxonomic theory (GTT) (see Section 2.5). Furthermore, the particular aspects of this diversity are specified by the respective partial taxonomic theories (PTTs). As we descend with our generalizations along the steps of the conceptual pyramid, we reach the level of taxonomic classifications, which are elaborated as specific cognitive models of concrete fragments of the diversity in question. Of importance is to remind that all lower-level conceptions, being cognitive models of the respective manifestations of taxonomic diversity, are elaborated within the substantive contexts outlined by the higher-level ones. This important point holds true for taxonomic classifications that remain meaningfully (biologically) empty if elaborated based on some formal ground out of certain substantive theory or conception.

An extension of the just-above idea of cognitive linguistics to the descriptive language of systematics, with taxonomic nomenclature being its integral part, leads to the following important observations (Pavlinov 2015, 2022a). From a conceptualism perspective, this language describes not an 'abstract' ('objective') biodiversity as something existing out of the cognitive situation of systematics but rather its particular manifestations as specific conceptual taxonomic realities defined by respective PTTs. A certain connection between the latter and nomenclature is really fundamental, as it ensures that the structure of descriptive language of systematics is actually adequate to the structure of taxonomic diversity. This connection most clearly manifests itself in the 'macroevents' of the history of systematics: each conceptual revolution in its theoretical background entailed respective changes in taxonomic nomenclature. Indeed, scholastic systematics developed and applied principles of *essentialist* nomenclature; early post-scholastic systematics replaced the latter with its *nominalist* conception and considered mostly higher-level taxonomic hierarchy; Darwinian revolution drew a great attention to the within-species taxa and their naming; the most recent cladistic ideas produced phylonomenclature. Such a regularity was clearly expressed by E. Mayr, who stated that 'the rules [of nomenclature] will have to be adjusted to the conceptual development of taxonomy' (Mayr 1969, p. 300). This indicates that the notorious principle of taxonomic freedom, which is expressly promulgated by the Codes to affirm as-if theory-neutral nature of taxonomic nomenclature, is generally not valid.

Because of this theory-dependence of descriptive language of systematics, it differentiates into distinct 'dialects' following conceptual divergence of PTTs, which define specific aspects of taxonomic diversity to be described using adequate linguistic means. The multiplicity of such 'dialects' generally indicates the non-rigid (fuzzy) character of professional language of biological systematics, which is undoubtedly holds for its nomenclatural part.

* * *

The basic tenets of essentialist nomenclature demonstrate perhaps most evidently its dependence on the ontic premises of a specific kind. It was rooted in the Platonic worldview, according to which the *genuine* names of natural entities were intimately linked to their essences, so these names should by all means be descriptive and reflecting those essences. That is why the pioneers of systematics of the 17th–18th centuries attached a very great importance to the taxonomic names: J. Pitton de Tournefort wrote that 'knowing the plants is tantamount to knowing their names, [therefore] the study of plants should begin with their names' (de Tournefort 1694, p. 1), and C. Linné called nomenclature one of two 'foundations' of botany, alongside the 'arrangement' (classification) of plants (Linnaeus 1751, § 151).

The effect of ontic considerations upon regulative nomenclature of essentialist kind is evident from Linnaean rules for the formation of names of taxa of different ranks. For him, genera were the most fundamental classification units as the 'works of Nature', whereas classes and orders were not so significant for their being the 'works of Nature and Art'. Therefore, it was important for Linnaeus to make evident from the names of plants to which genera they belonged, so 'when the species [were] referred to genera, the species should take the name of the genus' (Linnaeus 1751, § 284); in contrast, 'the names of the class and the order must never be included in the name of the plant' (§ 212).

Another striking example of the elaboration of a system of theory-dependent nomenclature is provided by the contemporary cladistic school of systematics. Its most dedicated adherents proposed the so-called *phylogenetic nomenclature* as the one 'that is more concordant with evolutionary concepts of taxa [and] more closely conform to the manner in which they are conceptualized' (de Queiroz and Cantino 2001, p. 269).

It is important to note that the relation between taxonomic theory and taxonomic nomenclature is not rigid: there is no one-to-one correspondence between their particular versions (Pavlinov 2015, 2022a). This means that different PTTs can develop similar nomenclature systems and, conversely, different systems of nomenclature can get along within the same PTT. Such a specific 'fuzziness' is clearly illustrated by the rank-dependent nomenclature, which is characteristic of both later scholastic systematics and most of its post-scholastic schools. Besides, the relation in question can be complete or partial, i.e., encompassing the entire nomenclature system or just some of its principles. For instance, phylogenetic nomenclature is totally theory-dependent, whereas the Linnaeus' nomenclature system was dependent on his natural philosophy with respect to some of its naming rules only. At last, the particular nomenclature regulators can be evaluated differently as theory-dependent or not in particular nomenclature systems or according to their specific understanding by scientists. For instance, the ranked nominative nomenclature is basically theory-dependent in the essentialist system, but it is principally devoid of such a dependence in the nominalist systems, so it was preserved by the latter just as a kind of 'atavism'.

The principle (rule) of typification is sometimes considered theory-dependent as presumably being rooted in the natural philosophical type conception (Ogilby 1838; Simpson 1940, 1961; Mayr 1969; Rasnitsyn 2002), while other nomenclaturists treated the principle (rule) of diagnosing as a manifestation of the essentialist worldview (Cook 1898, 1900; Dubois 2008, 2011).

6.2 THE HIERARCHIES OF REGULATIVE NOMENCLATURE

As said above, the professional language of biological systematics is complexly structured. Its structuredness is most evidently expressed in the inclusive hierarchy of nomenclature regulators of various levels of generality. This hierarchy in turn is defined by non-inclusive hierarchies of specific hierarchization rulers that arrange those regulators according to their importance for the carrying out nomenclatural activity (Smith 1962; Dubois 2008; Pavlinov 2015, 2022a; Aescht 2018; Dubois et al. 2022).

The information flows, ordering the inclusive hierarchy of nomenclature regulators as an information system, can be both ascending and descending. In the first case, the practical needs for a stable and universal nominative nomenclature serve as the main ordering factor that stimulates the development of no less stable and universal primary regulators. In the second case, the most general norms and principles of organization of the professional language of biological systematics are decisive for the elaboration of specific nomenclature principles that would be most compatible with the basic premises of respective taxonomic theories. As far as the most general and ultimate goal of the whole of the taxonomic nomenclature is to provide stability and universality of its nominative part, these ascending and descending causality flows are widely crossed.

The hierarchization of the whole body of taxonomic nomenclature commences with its differentiation into the above-mentioned dialects, which are defined by particular *nomenclature conceptions* and correspond to the specific PTTs. These conceptions are theoretical and therefore basically declarative in their contents, such as descriptive and nominalist, essentialist, and pragmatic. They may be grouped based on specific criteria, one of which involves distinguishing between rank-dependent and rank-independent conceptions, and another refers to the traditional and non-traditional nomenclature systems. The *nomenclature norms* are introduced at the next level, which may be considered as rather informal precepts (meta-principles) that define some general approaches to naming taxa, such as rationality and adequacy, and organize the whole of nomenclatural activity (see below on the latter). The norms are specified by certain *nomenclature principles* of more practical meaning, such as wordiness and typification, which are usually detailed by the respective ready-to-use *nomenclature rules*. The latter can be supplemented with the most particular lowest-level regulators, such as *recommendations*, which suggest the preferable ways of dealing with certain non-standard issues.

The draft *Linz ZooCode* suggested another basic block structure of nomenclature principles, which is laid on a mainly empirical basis and corresponds to the major steps of nomenclature activity (Dubois 2005; Dubois et al. 2019).

> In this regard, these principles are grouped as follows: (a) general, (b) regulating the availability of names, (c) regulating the taxonomic allocation of names, (d) regulating the validity of names, and (e) regulating the registration of names (Aescht 2018).

At each level of generality, many nomenclature regulators admit several particular interpretations, which entails sequential multiplication of regulators with passing from higher to lower levels of their generality. This results in a specific 'pyramidal' logical structure of the whole of the regulative nomenclature, so that the nomenclature conceptions take the top position in this 'nomenclature pyramid', the next go the general norms, whereas the particular working rules and recommendations are placed at its base. As is seen, such a 'pyramid' is quite similar to that produced by the hierarchy of taxonomic theories. From this analogy, it is reasonable to regard that the differentiation of the overall professional language of systematics into dialects is not accidental, and it entails a kind of *nomenclature pluralism* as an immanent property of the regulative taxonomic nomenclature (Pavlinov 2015, 2022a).

The conceptual pyramids of nomenclature regulators are specified by the *nomenclatural systems* developed as the sets of mutually agreeable nomenclature principles and rules, which are suitable for their direct implementation. Such systems are actualized by the *Codes of nomenclature* as the working normative documents. Both systems and especially Codes are ordered in their respective inclusive hierarchies; in the case of Codes, their hierarchies are quite evident from their formal arrangement in chapters (sometimes also sections) and articles.

All nomenclatural systems are implemented by the *nomenclatural activity*, which in its practical realization is ruled by the respective Codes. This activity includes several dozens of concrete *nomenclatural tasks*, which are arranged hierarchically according to several particular scales. One of the latter recognizes the tasks as *primary* and *secondary* based on their importance and precedence in the just-named activity. According to another scale set by the contents of the tasks being solved within this activity, they are divided into three principal groups: *classificatory*, *nominative*, and *procedural*.

At a certain level of generality, the overall hierarchy of nomenclature regulators is shaped by their specific contents and functions they accomplish within the nomenclatural activity. Accordingly, they are categorized before all into five major blocks, namely, *general regulatory* principles, *cognitive* principles, *linguistic* principles, *juridical* principles, and *taxonomic* principles. Within each of these blocks, their respective hierarchies are arranged to reflect the usage of regulators as either conceptual principles or working rules, as well as in accordance to their involvement in the decisions of the nomenclatural tasks as either primary or secondary.

The scale of significance of the principles, rules, etc., is not something constant but varies among nomenclature conceptions and systems, and it changed with the history of systematics (Pavlinov 2015, 2022a). For instance, with regard to the selection of valid (correct) taxonomic names, two principles were competing during the 19th century, namely, those of *priority* and *usage*, with the former becoming the winner in the 20th century in zoology and botany (Anonymous 1934, 1964), whereas the usage being retained in virology (Siddell et al. 2020). Another example is provided by the principles, with the reference to which the newly described taxa should be defined, these are principles of *circumscription*, *diagnosing*, and *typification*. The first dominated up to the late 18th century, the second has been given the highest priority in some Codes of the 19th century, whereas the third began to prevail in the traditional nomenclature of the 20th century. With this, some particular principles (rules) were once given such high significance that they used to be called ‘laws’, the most known instance is the *law of priority*.

A specific non-inclusive scale of rating (significance) was elaborated in the 20th century for the *type specimens*, by which new species-group taxa were to be defined according to the principle of typification. Thus, in zoological nomenclature, these ‘defining’ (sometimes termed as ‘name bearing’) type specimens are categorized either as *primary* (most significant and therefore mandatory, such as holotypes, neotypes) and *secondary* (less significant and non-mandatory, such as paratypes).

As to the nominative nomenclature, its non-inclusive hierarchy is most demonstrable in the ranking of taxonomic names according to their *nomenclatural status*. The latter is determined by the respective provisions of the Codes, which changed with time and more or less specifically formulated in different nomenclature systems. The most significant is the categorization of names as *valid* (*correct*) according to the principles of priority, publication, and some other; consequently, all other names take a subsidiary position relative to them. Another ranking of taxonomic names involves distinguishing between *senior* and *junior* synonyms and homonyms.

Besides, a specific kind of categorization is characteristic of the entire system of ranks, which are categorized as either primary or secondary (see Section 3.7).

6.3 RANK-DEPENDENT NOMENCLATURE

The rank-dependent (rank-based) nature of traditional taxonomic nomenclature concerns both its parts, regulative and nominative. In the first case, this implies that the application of certain working principles and rules depends on the particular taxonomic ranks. In the second case, the same holds for the rank terminology and the names of taxa, whose formation is in part rank-dependent. All now acting traditional Codes are entirely rank-dependent in this regard.

The ranked hierarchical arrangement of taxonomic classifications is more or less strongly regulated by all contemporary Codes of nomenclature, which is formalized by the *principle of ranked hierarchization*. The assessment of this principle as a part of the traditional nomenclature varies from its recognition as one of the most significant achievements of the ‘Linnaean reform’ (Ride 1988; Schuh 2003; Dubois 2011) to its decisively negative estimate (Griffiths 1976; de Queiroz and Gauthier 1992; Ereshefsky 1997, 2004). It is implemented by an explicit and exhaustive listing of categories (ranks) in the taxonomic hierarchy, which are officially certified by the respective authorities. Such a codification began with Tournefort and Linnaeus in the 18th centuries, continued in the 19th centuries (de Candólle 1813, 1818, 1867, 1883; Dall 1877), and now it occurs in all traditional subject-area Codes. It is worth noting that zoological nomenclature codifies both sub- and supraordinate auxiliary ranks, whereas botanical and microbiological ones do so with regard to subordinate ranks only, as A. de Candólle has suggested in the mid-19th century (see Section 1.5).

The zoological Code regulates taxonomic ranks from species up to the family with all their sub- and super- 'auxiliaries' and does not consider the order and higher ranks at all; the latter limitation is deemed a serious flaw by a number of contemporary nomenclaturists, who suggested to expand the nomenclatorial regulation to the highest ranks, as well (Stenzel 1950; Rodendorf 1977; Rasnitsyn 1982; Brothers 1983; Starobogatov 1991; Dubois 2005, 2006, 2011). This Code modifies the traditional continuous ranking system by recognizing three distinct *rank-group*s in it, namely those of family, genus, and species (Anonymous 1964). Each of these groups encompasses its principal rank with its 'auxiliaries': for instance, the family-group includes ranks of family, superfamily, subfamily, and tribe, the latter with its own 'auxiliaries'. The *principle of rank coordination* implies that within the same rank-group, certain basic nomenclature principles (such as priority and homonymy; see below) are applied uniformly to the taxa and their names. The contemporary botanical nomenclature codifies taxonomic hierarchy in all its capacity from infraspecies categories up to the kingdom. It suggests a specific scaling of ranks by dividing them into *principal* (from kingdom to species), *secondary* (tribe, section, and series), and *intercalated* (not listed explicitly). The draft *BioCode* combines zoological and botanical codified hierarchization systems by adopting (a) zoological rank-groups, which are supplementing with one more group for the order and higher ranks and (b) botanical recognition of two categories of ranks within each rank-group, which are referred to as *primary* and *secondary*, respectively (Hawksworth et al. 1994; Greuter et al. 1998; Greuter 2004). The draft *Linz ZooCode* suggests to designate the rank-groups as *nominal-series* and adds a lot of intercalated ranks, which are distinguished by specific prefixes: *suprem-*, *ano-*, *hyper-*, etc., for superordinate and *hypo-*, *cato-*, etc., for subordinate ones (Dubois 2006; Dubois et al. 2019).

* * *

The most significant impact of taxonomic hierarchy on the codified ranking of the nominative nomenclature is that it makes the latter partly rank-dependent. This is generally formalized by the *principle of ranking nomenclature* (Pavlinov 2015, 2022a), and the ranked hierarchy endowed with such a property is called *flagged* (Stevens 2002). The just-named principle does not duplicate that of ranked hierarchization of classifications, as the latter's ranking does not necessarily entail the ranking of the formation of taxonomic names. Anyway, it means that (a) the rules for naming taxa are more or less specific for the particular ranks (rank-groups), (b) the naming rules are applied uniformly within each of the ranks and rank-groups (if recognized), and (c) the taxonyms that are formed according to these rules are partly rank-dependent. These provisions have been informally recognized in the mid-19th century by stating, 'The form, number and arrangement of the names are dependent on the nature of each group' (Dall 1877, § XIV), where the 'nature' corresponds approximately to the 'category' in its contemporary meaning; they both are proposed to be formalized by the *principle of rank homogeneity* (Dubois 2005).

The rank-dependence of nominative nomenclature in systematics begins with codification of the terminological designation of the fixed ranks. Its predecessor was designation of the particular steps of logical division within the sliding rankless hierarchy of the genus-species scheme by the terms *Genus summum*, *Genus intermedium*, etc. (see Section 1.4). These steps became fixed and designated individually in the 17th-18th century to become *Classis*, *Ordo* (or *Famila*), etc. (see Section 1.5). Subsequently, this initial four- or five-level ranking system became more and more fragmented, including the introduction of auxiliary ranks in addition to the principal ones, so the overall number of ranks steadily increased. This fragmented ranking system became being codified since the mid-19th century, with the number, subordination, and designation of ranks being currently strongly regulated; variants of this system are specific to the particular subject-area nomenclature Codes.

Another manifestation of the rank-dependence of the nominative nomenclature involves *taxonyms* (taxonomic names, nomens). Its importance in this regard is that it makes taxonomic names partially performing a classification function by indicating position of the named taxa in classification. Such specifically organized designations indicate for each taxon (a) its allocation to a certain rank, (b) its relation to the nearest super- and subordinated taxa (if any) within a particular hierarchy path (series), and (c) its position in the major blocks of the overall taxonomic classification of the living.

A partially ranked character of the nominative nomenclature was implicitly presumed by the standard logical formula 'generic common and species particular' for the definitions of any entities developed by scholasticism. With time, such two-part definitions were gradually transformed into descriptive taxonomic names to make them *binary*, i.e., consisting of two parts, generic and specific. This transformation was formalized as a special nomenclature rule in the 17th–18th centuries (Rivinus, Tournefort, and Linné), and its subsequent refining was resulted in the so-called *binomial* nomenclature. The latter implies that the complete name of a species should consist of the name of the inclusive genus and the specific epithet, both being one-word; unlike this, the complete name of a genus and any of the higher-level taxa should always be one-word. In the second half of the 19th century and later, this binomial nomenclature became formally *polynomial* (partly returned to scholastic variant) as the species binomen was augmented with the additional words (epithets) to designate infraspecies taxa. This version is *trinomial* in zoology and bacteriology, in which only subspecies category is officially acknowledged, and it is truly polynomial in botany with its highly detailed infrasubspecies ranking and naming. Nonetheless, this traditional nomenclature systems are still called 'binomial', because the complete species name remains two-word.[2] Such an evident remnant of the earliest rank-dependent nominative nomenclature is acknowledged even in phylogenetic nomenclature summarized by *PhyloCode* (see below on it).

With regard to the regulative nomenclature, some rules for naming taxa were suggested as rank-dependent in *Philosophia Botanica* by Linnaeus to reflect in part his view of different ontic status of taxa (see above). Its post-Linnaean development, though abandoning its natural philosophical background, inherited an idea of the rank-specific application of some nomenclature rules. The first of them was suggested by M. Adanson in his *Families of Plants* (Adanson 1763), they became refined and codified during the 19th century and are now acknowledged by all traditional Codes.[3] One rule of this group regulates the root parts of the names of a supra-generic taxa: according to the principle of *automatic typification*, this part must be a derivative of the root part of the name of a genus, which the respective higher taxon includes, so this genus becomes its name-bearing type. Another rule regulates the formation of the standard rank-specific ends of the names of supra-generic taxa, which are different in particular subject-area Codes.

The following nomenclature regulators are certainly rank-dependent in the contemporary Codes. The principles of *priority* and *prohibition of homonymy* are applied in zoology only to the names of taxa of the same rank-group. The principle of *prohibition of genus-species tautonymy* in botany is also rank-dependent by its application to the names of these two ranks only. The principle of *typification* is rank-dependent in that the taxa are typified by the specific classificatory elements of the next lower ranks, which are type specimens for the taxa of species rank-group, species for the taxa of genus rank-group, genera for the taxa of family rank-group, etc.

* * *

The predominant trend in the development of taxonomic nomenclature involved implication of its just-characterized classical version, which implied the designation of taxa by the words or word-like lexemes. However, several attempts were made during the 18th–20th centuries to apply certain elements of the *rational-logical* nomenclature, in which some variants of symbolic designation of taxa were employed (Pavlinov 2015, 2022a).

Their predecessor was the above-mentioned universal 'philosophical language' elaborated in the mid-17th century by John Wilkins in his *An essay towards a real character* (Wilkins 1668). It was based on an overwhelming ranked hierarchical classification of all material (and some mental) entities, including animals and plants, which included three basic levels, those of 'genera', 'differentiae', and 'species', and it was illustrated by the Tree of Ramus (Slaughter 1982; Maat 2004). The primary objective of this 'philosophical language' was to designate each entity in the classification by a unique standard ID consisting of either four letters or four digits. Their specific combinations indicated position of the respective entities within the overall hierarchy.

This Wilkins' nomenclature system was subsequently implemented, with some modification, several times in particular classifications (Bergeret 1783–1785; Petit-Thouars 1822; Siu and Reese 1955). The American zoologist Charles Michener (1918–2015) saw one of the advantages of such a non-classical nomenclature 'futuristically' in that it would allow for an unproblematic designation of any hypothetical exobiological objects (Michener 1963). The most advanced version of such a rational-logical nomenclature, to become known as the so-called *numericlature*, was developed in the mid-20th century; it contained certain elements of rank-independent nomenclature and will therefore be considered in the next section.

6.4 RANK-INDEPENDENT NOMENCLATURE

It would perhaps be easier and more reasonable to link the origin of the very idea of rank-independent (rank-free) taxonomic nomenclature to the emergence of the rankless hierarchy of classifications based on the genus-species logical scheme of scholastics. However, the latter's specific terminology implied certain 'hidden' prototypical elements of the ranked nomenclature, and it was these elements that became really actualized in the first nomenclature systems by the systematicians of the 17th and 18th centuries. Thus, the most consistent rank-independent nomenclature seems to be a specific product of systematics of the second half of the 20th century. Its common characteristic is that (a) it does not employ any unified ranking system, (b) hierarchy paths in a classification are arranged independently of each other, and (c) taxonomic designators are shaped following certain rank-independent rules.

* * *

The first versions of rank-independent nomenclature were proposed within its rational-logical conception in the form of the already-mentioned *numericlature* in the mid-20th century due to the mastering of positivist philosophy of science by biological systematics (Rabel 1940; Jahn 1961; Michener 1963; Little 1964; Rivas 1965; Hull 1966, 1968; Yochelson 1966; Griffiths 1981; Ereshefsky 1994, 2004). Its predecessor has been probably the so-called *Universal Decimal Classification* system (UDC) elaborated at the end of the 19th century as a hierarchically arranged indexing and retrieval system for all branches of scientific knowledge to be used in the bibliographic works (Wilson 1969; McIlwaine 2009).

This new nomenclature system was suggested to replace completely the traditional nomenclature with its verbal taxonomic designators. It is rank-independent in its regulative part, since it is based on the implication of all nomenclature regulators uniformly at all levels of taxonomic hierarchy. However, its nominative part seems to be partly rank-dependent, since the respective numerical designations of the hierarchy paths can equivocally refer both to the unfixed levels of generality and fixed ranks. This nomenclature implies that the particular hierarchical paths are fixed in some way and all steps in each of them are sequentially assigned respective numerical designations denoted as *position numbers* by the American philosopher of science David Hull (1935–2010). They are accumulated in a specific manner while descending along the respective paths, thus resulting in a hierarchy of digital identifiers (IDs) that refer both to the hierarchical levels (ranks) proper within the particular paths and to the particular taxa allocated to them. In some cases, the figures can be supplemented with the letters, which provides a complex alpha-numeric designation system. One of the proponents of this conception claims that it has a more advanced syntax than any of the verbal ones, and therefore it 'can be used to organize and impart knowledge more efficiently than the present system of names' (Hull 1968, p. 472).

This highly formalized version of nomenclature did not gain an appreciation by taxonomic community as the basis for a completely reformed taxonomic nomenclature. One of the causes of its ignoring was that the digital taxonomic designations appeared to be totally inappropriate as a means of human communication. Besides, contrary to a supposition of Michener (1963) about its stability, numericlature is potentially very unstable due to the fact that any changes in the higher-level taxonomic arrangement in a classification will inevitably entail a recoding of the entire system of numerical IDs assigned to the respective lower-level taxa. With this, however, the contemporary development of computer-based technologies makes numericlature a useful technical device to be applied not instead of but in addition to the traditional verbal naming of taxa (Kennedy et al. 2005; Page 2006; Patterson et al. 2006, 2010; Schindel and Miller 2010). In particular, it is implemented for the elaboration of the Internet-based *International Species Information System* (ISIS) (Flesness 2003).

* * *

Another rank-independent conception and system of taxonomic nomenclature, referred to as *phylogenetic nomenclature* or *phylonomenclature*, was recently proposed by some most devoted adherents of cladistic systematics. Its main objective was declared to bring the descriptive language of systematics in line with the content of contemporary phylogenetic theory, which assumes a hierarchically arranged rankless phylogenetic pattern (de Queiroz 1992, 1997, 2005; de Queiroz and Gauthier 1992, 1994; Minelli 1995; Ereshefsky 1997, 2001b, 2004; Knapp et al. 2004; Laurin 2007, 2024; Cantino and de Queiroz 2010, 2020; etc.). In order to 'divorce' terminologically traditional and phylogenetic conceptions of nomenclature, the former was designated as 'taxonomy' (not in the sense of theoretical part of systematics) and the latter as *cladonomy* (Brummitt 1997; Dubois et al. 2021).

In accordance with the basic ideas of cladistics, this nomenclature system implies the following basic propositions formalized by the *PhyloCode* (Cantino and de Queiroz 2010, 2020; Laurin 2024). First, it is declared that only cladistically consistent classifications should be considered eligible; second, for this, only clades (holophyletic groups) are considered candidate taxa to be recognized and named properly; third, these taxa are arranged in a strict hierarchical manner; forth, their inclusive hierarchy is rankless; consequently and finally, taxonomic names are rank-irrelevant. In order to minimize conflicts with the accepted binomial nomenclature, it is agreed that the species and their subunits are not clades, so the clauses formulated by the *PhyloCode* are not applied to them and their traditional naming rules are conserved (which means that certain traces of rank-dependent nomenclature are preserved).

The rules for both the recognition of clades and the formation of their names are regulated very strictly and rather sophisticatedly. These names are referred to as *phylonyms* (Cantino and de Queiroz 2010, 2020) or *phyloreferences* (Cellinese et al. 2021). In order to distinguish between clades with different phylogenetic statuses (holo-, para-, etc.), various prefixes (*Pan-* and *Apo-*) are inserted into their names. A possible homonymy of the names established under different subject-area codes (botanical, zoological, etc.) is removed by adding group-specific prefixes to the coinciding preexisting names (*Phyto-*, *Zoo-*, etc.). All these provisions make the entire cladistic hierarchy partly *annotated* (in the sense of Wiley 1979, 1981).

The nomenclature reform undertaken by phylocoders seemed to be initially stimulated by their conviction that cladistic theory of systematics is the most compatible with the contemporary evolutionary philosophy of entire biology, and therefore is destined to dominate in the future. Accordingly, its linguistic part is supposed by its most active proponents to provide a unified nomenclature system for all subject areas of biological systematics. However, the objections against phylonomenclature are quite extensive (e.g., Brummitt 1997, 2002; Benton 2000; Nixon and Carpenter 2000; Stuessy 2000; Carpenter 2003; Keller et al. 2003; Monsch 2006; Dyke and Sigwart 2007; Rieppel 2006c, 2008; Dubois 2007; Kraus 2008). Its most serious criticism addresses its rather strong and restrictive conceptualization, which implies a pretension to impose a particular taxonomic theory on the entire taxonomic community and, consequently, suppresses any personal freedom of systematicians to choose among different theories (Rasnitsyn 1996, 2002; Rieppel 2006c; Pavlinov 2015, 2022a). At any rate, it so happens that two major nomenclature conceptions, traditional and phylogenetic, are now acting in biological systematics in parallel, and this threatens another serious 'schism' in taxonomic nomenclature, causing its fundamental instability. Therefore, suggestions are being put forward to develop certain combinations of the most significant provisions of both nomenclature systems in order to avert such a collision (Kinman 1994; Barkley et al. 2004; Kuntner and Agnarsson 2006; Naomi 2014).

One more, quite peculiar version of the rank-independent phylogenetic nomenclature system was proposed by Nikita Kluge, who called it *hierarchical* (Kluge 1998a, 1999a, 1999b, 2000). His classification of Insecta lacks any fixed ranks, so taxonomic names lack any kind of the traditional rank-specific endings. Instead, special digital markers are inserted in the names to indicate the steps in the hierarchy paths, at which the respective taxa are recognized. With this, no equivalence of similarly designated steps in different parts of such a hierarchically arranged classification is supposed.

NOTES

1 It is commonly referred to as 'biological', but 'taxonomic' seems to be more accurate, as it is applied by systematics to describe just the taxonomic diversity. Unlike this, the term 'biological' encompasses all nomenclature systems applied by other biological disciplines, such as anatomy, genetics, and synecology (Pavlinov 2015, 2022b).
2 The virological Code of nomenclature does not acknowledge mandatory status of binomiality and suppresses any infraspecies categories (Anonymous 2025).
3 This means that contemporary traditional supra-generic taxonomic nomenclature is not 'Linnaean' but rather 'Adansonian'.

Bibliography

Abachiev, S. K. 2004. [*Evolutionary Theory of Knowledge. An Experience of Systemic Construction.*] Moscow: Editorial URSS. ISBN:9785354007240. (in Russian).

Abbot, L. A., F. A. Bisby, and D. J. Rogers 1985. *Taxonomic Analysis in Biology: Computers, Models, and Databases.* New York (NY): Columbia University Press. ISBN:9780231049269.

Abzhanov, A. 2013. Von Baer's law for the ages: Lost and found principles of developmental evolution. *Trends in Genetics* 29:712–22. DOI:10.1016/j.tig.2013.09.004

Adanson, M. 1763. *Familles des Plantes*, Pt. 1. Paris: Vincent. ISBN:9780913196250 (1964 reprint).

Aescht, E. 2018. Reflecting on a theory of biological nomenclature (onymology). *Alytes* 36:212–37. DOI:10.1099/ijsem.0.000778

Agassiz, L. 1859. *An Essay on Classification.* London: Longman. ISBN:9780486435800 (2004 reprint).

Ahl, V., and T. F. H. Allen. 1996. *Hierarchy Theory. A Vision, Vocabulary, and Epistemology.* New York (NY): Columbia University Press. ISBN:9780231084802.

Aho, A. V., and J. D. Ullman. 1995. *Foundations of Computer Science.* Hong Kong: Computer Science Press. ISBN:9780716782841.

Akhutin, A. V. 1988. [*The Concept of Nature in Antiquity and New Time.*] Moscow: Nauka. ISBN:9785020080164. (in Russian).

Albertson, D. 2014. *Mathematical Theologies. Nicholas of Cusa and the Legacy of Thierry of Chartres.* Oxford: Oxford University Press. ISBN:9780199989737.

Albertus, M. 1867. *Alberti Magni ex Ordine Praedicatorum. De Vegetabilibus Libri VII, Historiae Naturalis Pars XVIII.* Berlin: Georg Reimer. ISBN:9783112397619.

Albertus, M. 1999. *On Animals. A Medieval Summa Zoologica*, Vols. I, II. Baltimore (MD): Johns Hopkins University Press. ISBN:9780814254516.

Alcántara-Salinas, G., R. F. Ellen, L. Valiñas-Coalla, et al. 2013. Alternative ways of representing Zapotec and Cuicatec folk classification of birds: A multidimensional model and its implications for culturally-informed conservation in Oaxaca, México. *Journal of Ethnobiology & Ethnomedicine* 9:81. DOI:10.1186/1746-4269-9-81

Aldrovandi, U. 1618. *De Animalibus Insectis Libri Septem, Cum Singulorum Iconibus Ad Viuum Expressis.* [...]. Bononiae: Ioan. Bapt. Bellagambam. DOI:10.5962/bhl.title.147665

Alexander, T. D., and D. W. Baker. 2003. *Dictionary of the Old Testament: Pentateuch.* Downers Grove (IL): IVP Academic. ISBN:9780830817818.

Aleyev, Yu. G. 1986. [*Ecomorphology.*] Kïev: Naukova Dumka. (in Russian).

Allen, J., D. Massey, and A. Cochrane. 1998. *Rethinking the Region.* London: Routledge. ISBN:9780415168229.

Allen, T., and M. Giampietro. 2014. Holons, creaons, genons, environs in hierarchy theory: Where we have gone. *Ecological Modelling* 293:31–41. DOI:10.1016/j.ecolmodel.2014.06.017

Allen, T. F. H. 2001. A summary of the principles of hierarchy theory. *International Society for the Systems Sciences.* https://web.archive.org/web/20011218001638/http://isss.org/hierarchy.htm

Allen, T. F. H. 2008. Hierarchy theory in ecology. In *Encyclopedia of Ecology*, eds. S. E. Jørgensen, and B. D. Fath, 1852–7. ISBN:9780080454054.

Allen, T. F. H., P. C. Allen, and D. L. Wixon. 2009. Hierarchy theory in hydropedology. *Hydrology and Earth System Sciences Discussions* 6:2931–59. DOI:10.5194/hessd-6-2931-2009

Allen, T. F. H., and T. B. Starr. 1982. *Hierarchy: Perspectives for Ecological Complexity.* Chicago (IL): The University of Chicago Press. ISBN:9780226014319.

Amundson, R. 1998. Typology reconsidered: Two doctrines on the history of evolutionary biology. *Biology and Philosophy* 13:153–77. DOI:10.1023/A:1006599002775

Amundson, R. 2005. *The Changing Role of the Embryo in Evolutionary Thought Roots of Evo–Devo.* Cambridge (UK): Cambridge University Press. ISBN:9780521806992.

Anderson, L. 1976. Charles Bonnet's taxonomy and Chain of Being. *Journal of the History of Ideas* 37:45–58. PMID:11615700

Anderson, V., and L. Johnson. 1997. *Systems Thinking Basics: From Concepts to Causal Loops.* Waltham (MA): Pegasus Comm. ISBN:9781883823122.

Andino, R., and E. Domingo. 2015. Viral quasispecies. *Virology* 479:46–51. DOI:10.1016/j.virol.2015.03.022

Anonymous. 1934. *International Rules of Botanical Nomenclature Adopted by the Fifth International Botanical Congress, Cambridge, 1930.* London: Taylor & Francis Group. www.iapt-taxon.org/historic/Congress/IBC_1930/prepub.pdf

Anonymous. 1964. *International Code of Zoological Nomenclature Adopted by the XV International Congress of Zoology.* London: International Trust for Zoological Nomenclature. www.biodiversitylibrary.org/item/107127#page/5/mode/1up

Anonymous. 1969. Preface. In *Hierarchical Structures*, eds. L. L. Whyte, A. G. Wilson, and D. Wilson, vii–viii. New York (NY): American Elsevier Publ. ISBN:9780444000699.

Anonymous. 1999. *International Code of Zoological Nomenclature*, 4th ed. London: International Trust for Zoological Nomenclature. ISBN:9780853010067.

Anonymous. 2025. *The International Code of Virus Classification and Nomenclature.* https://ictv.global/about/code

Antipenko, L. G. 1986. [*Problem of the Theory Incompleteness and its Gnoseological Meaning.*] Moscow: Nauka. ISBN:9785971098539. (in Russian).

Antonov, A. S. 2006. [*Plant Genosystematics.*] Moscow: Akademkniga. ISBN:5946282719. (in Russian).

Aratyn, H., and A. Sorin. 2001. *Integrable Hierarchies and Modern Physical Theories.* Dodrecht: Springer Science+Business Media. ISBN:9780792369639.

Arber, A. 1938. *Herbals: Their Origin and Evolution. A Chapter in the History of Botany, 1470–1670.* Cambridge (UK): Cambridge University Press. ISBN:9781108016711.

Arbogast, B. S., S. V. Edwards, and J. Wakeley, et al. 2002. Estimating divergence times from molecular data on phylogenetic and population genetic timescales. *Annual Review of Ecology, Evolution, and Systematics* 33:707–40. DOI:10.1146/annurev.ecolsys.33.010802.150500

Archibald, J. D. 2014. *Aristotle's Ladder, Darwin's Tree. The Evolution of Visual Metaphors for Biological Order.* New York (NY): Columbia University Press. ISBN:9780231164122

Aristotle. 1953. *Generation of Animals.* London: William Heinemann. ISBN:9780434993666.

Arkani-Hamed, N., S. Dimopoulos, and G. Dvali. 2024. The hierarchy problem and new dimensions at a millimeter. *SLAC-PUB-7769/SU-ITP-98/13.* https://arxiv.org/pdf/hep-ph/9803315

Arkhangelsky, D. B. 1996. [*Orders as the Elements of Natural System of Contemporary Flowering Plants.*] St. Petersburg: Botanical Institute of RAS. (in Russian).

Armand, A. D. 2008. [*Two in One: A Law of Complementarity.*] Moscow: LKI Press. ISBN:9785382005263. (in Russian).

Asano, R., and U. Seifert. 2014. The evolution of human cognitive systems: Comparative approaches to language and music. In *The Evolution of Language: Proceedings of the 10th International Conference*, eds. E. A. Cartmill, S. Roberts, H. Lyn, et al., 377–8. Hackensack (NJ): World Scientific Publishing. ISBN:9789814603621.

Ash, R. B. 1990. *Information Theory.* Mineola (NY): Dover Publ. ISBN:9780486665214

Ashburner, M., C. A. Ball, J. A. Blake, et al. 2000. Gene ontology: Tool for the unification of biology. *Nature Genetics* 25:25–9. DOI:10.1038/75556

Assis, L. C. S. 2013. Species and tokogenetic homologies: 10 sutras. *Cladistics* 30:10. DOI:10.1111/cla.12031

Astolfi, P. and L. Zonta-Sgaramella. 1984. Phylogenetic trees: An analysis of the treeness test. *Systematic Zoology* 33:159–66. DOI:10.1093/sysbio/33.2.159

Atran, S. 1990. *The Cognitive Foundations of Natural History: Towards an Anthropology of Science*. Cambridge (UK): Cambridge University Press. ISBN:9780521438711.

Atran, S. 1998. Folk biology and the anthropology of science: Cognitive universals and cultural particulars. *Behavioral and Brain Sciences* 21:547–609. DOI:10.1017/S0140525X98001277

Atran, S. 1999a. The universal primacy of generic species in folk biological taxonomy: Implications for human biological, cultural, and scientific evolution. In *Species: New Interdisciplinary Essays*, ed. R. A. Wilson, 231–61. Cambridge (MA): The MIT Press. ISBN:9780262731232.

Atran, S. 1999b. Itzaj Maya folkbiological taxonomy: Cognitive universals and cultural particulars. In *Folkbiology*, eds. D. L. Medin, and S. Atran, 119–205. Cambridge (MA): The MIT Press. ISBN:9780262279611.

Atran, S. 2002. Modular and cultural factors in biological understanding: An experimental approach to the cognitive basis of science. In *The Cognitive Basis of Science*, eds. P. Carruthers, S. P. Stich, and M. Siegal, 41–72. Cambridge (UK): Cambridge University Press. ISBN:9780511613517.

Atran, S., and D. Medin. 2008. *The Native Mind and the Cultural Construction of Nature*. Cambridge (MA): The MIT Press. ISBN:9780262514088.

Atran, S., D. L. Medin, and N. Ross. 2002. Thinking about biology: Modular constraints on categorization and reasoning in the everyday life of americans, maya, and scientists. *Mind and Society* 6:31–63. DOI:10.1007/BF02513147

Auger, P. 1989. *Dynamics and Thermodynamics in Hierarchically Organized Systems. Applications in Physics, Biology and Economics*. Oxford: Pergamon Press. ISBN:0080401805.

Augier, A. 1801. *Essai d'une Nouvelle Classification des Végétux*. Lyon: Br. Ainé et Comp.

Austin, C. J. 2018. *Essence in the Age of Evolution. A New Theory of Natural Kinds*. New York (NY): Routledge. ISBN:9781351240857.

Avise, J. C. 2000. *Phylogeography: The History and Formation of Species*. Cambridge (MA): Harvard University Press. ISBN:9780674666382.

Avise, J. C., J. Arnold, R. M. Ball, et al. 1987. Intraspecific phylogeography: The mitochondrial DNA bridge between population genetics and systematics. *Annual Review of Ecology and Systematics* 18:489–522. https://www.jstor.org/stable/2097141

Avise, J. C., and G. C. Johns. 1999. Proposal for a standardized temporal scheme of biological classification for extant species. *Proceedings of the National Academy of Sciences* 96:7358–63. DOI:10.1073/pnas.96.13.7358

Avise, J. C., and J.-X. Liu. 2011. On the temporal inconsistencies of Linnean taxonomic ranks. *Biological Journal of the Linnean Society* 102:707–14. DOI:10.1111/j.1095-8312.2011.01624.x

Avise, J. C., and D. Mitchell. 2007. Time to standardize taxonomies. *Systematic Biology* 56:130–3. DOI:10.1080/10635150601145365

Avise, J. C., and T. J. Robinson 2008. Hemiplasy: A new term in the lexicon of phylogenetics. *Systematic Biology* 57:503–7. DOI:10.1080/10635150802164587

Ax, P. 1987. *The Phylogenetic System. The Systematization of Organisms on the Basis of Their Phylogenesis*. Chichester: John Wiley & Sons. ISBN:9780471907541.

Ayala, F. J. 1999. Molecular clock mirages. *BioEssays* 21:71–5. DOI:10.1002/(SICI)1521-1878(199901)21:1<71::AID-BIES9>3.0.CO;2-B

Azad, M., I. Chikalov, S. Hussain, et al. 2022. *Decision Trees with Hypotheses*. Cham (Switzerland): Springer Nature Switzerland. ISBN:9783031085840.

Babikov, P., O. Gontcharov, and M. Babikova. 2004. Polyhierarchical classifications induced by criteria polyhierarchies, and taxonomy algebra. In *Artificial Intelligence: Methodology, Systems, and Applications*, eds. C. Bussler, and D. Fensel, 411–9. Berlin: Springer. ISBN:9783540301066.

Baetu, T. M. 2012. Defining species: A multi-level approach. *Acta Biotheoretica* 60:239–55. DOI:10.1007/s10441-011-9143-z

Baghramian, M., and J. A. Carter. 2025. Relativism. In *The Stanford Encyclopedia of Philosophy*, eds. N. Zalta and U. Nodelman. https://plato.stanford.edu/archives/spr2025/entries/relativism/

Baier, G., and M. Klein (eds.). 1991. *A Chaotic Hierarchy*. Singapore: World Scientific Publ. ISBN:9810200234.

Bailey, L. H. 1894. Neo-lamarckism and Neo-darwinism. *The American Naturalist* 28:661–78. https://www.jstor.org/stable/2452518

Bailey, L. H. 1896. The philosophy of species-making. *Botanical Gazette* 22:454–62. DOI:10.1086/327442

Baker, R.J. and R. D. Bradley. 2006. Speciation in mammals and the genetic species concept. *Journal of Mammalogy* 87:643–62. DOI:10.1644/06-MAMM-F-038R2.1

Balaguer, M. 2025. Platonism in metaphysics. In *The Stanford Encyclopedia of Philosophy*, eds. E. N. Zalta, and U. Nodelman. https://plato.stanford.edu/archives/spr2025/entries/platonism/

Baldwin, J. T. (ed.). 1988. *Classification Theory*. Berlin: Springer-Verlag. ISBN:3540186743.

Ballego-Campos, I., S. K. V. Bonifácio, and L. C. S. Assis. 2023. A unified view of homology. *Cladistics* 39:398–417. DOI:10.1111/cla.12541

Balme, D. M. 1962. Γενοσ and ειδοσ in Aristotle's biology. *Classical Quarterly* 12:81–98. DOI:10.1017/S0009838800011642

Baofu, P. 2007. *The Future of Complexity. Conceiving a Better Way to Understand Order and Chaos*. Singapore: World Scientific Publ. ISBN:9789812708984

Bapteste, E., M. A. O'Malley, R. G. Beiko, et al. 2009. Prokaryotic evolution and the tree of life are two different things. *Biology Direct* 4:34. DOI:10.1186/1745-6150-4-34

Barclay, J. 1803. *New Anatomical Nomenclature, Relating to the Terms Which are Expressive of Position and Aspect in the Animal System*. Edinburgh: Ross and Blackwood, etc. ISBN:9781021503725 (2023 reprint).

Bard, J., S. Y. Rhee, and M. Ashburner. 2005. An ontology for cell types. *Genome Biology* 6:1–5. DOI:10.1186/gb-2005-6-2-r21

Barker, M. J. 2010. Specious intrinsicalism. *Philosophy of Science* 77:73–91. DOI:10.1086/650209

Barker, M. J. 2019. Eliminative pluralism and integrative alternatives: The case of species. *The British Journal for the Philosophy of Science* 70:657–81. DOI:10.1093/bjps/axx057

Barker, M. J. 2022. We are nearly ready to begin the species problem. In *Species Problems and Beyond. Contemporary Issues in Philosophy and Practice*, eds. J. S. Wilkins, F. E. Zachos, and I. Ya. Pavlinov, 3–38. Boca Raton (FL): CRC Press. ISBN:9780367855604.

Barkley, T., P. DePriest, V. Funk, et al. 2004. Linnaean nomenclature in the 21st century: A report from a workshop on integrating traditional nomenclature and phylogenetic classification. *Taxon* 53:153–8. DOI:10.2307/4135501

Baron, M. E. 1969. A note on the historical development of logic diagrams: Leibniz, Euler and Venn. *The Mathematical Gazette*, 53:113–25. DOI:10.2307/3614533

Barrett, J. A., and T. LaCroix. 2020. Epistemology and the structure of language. *Erkenntnis* 87:953–67. DOI:10.1007/s10670-020-00225-4

Barsanti, G. 1992. *La scala, la mappa, l'albero. Immagini e classificazioni della natura fra sei e ottocento*. Firenze: Sansoni. ISBN:9788838313844.

Bartlett, H. 1940. The concept of the genus. 1: History of the generic concept in botany. *Bulletin of the Torrey Botanical Club* 67:319–62. DOI:10.2307/2481068

Bartlett, S. J. 2015. *The Species Problem and its Logic*. http://philsci-archive.pitt.edu/11655/1/Bartlett_The_Species_Problem_and_Its_Logic.pdf

Bar-Yam, Y. 1997. *Dynamics of Complex Systems*. Cambridge (MA): Perseus Press. ISBN:9780201557480.

Bather, F.A. 1927. Biological classification: Past and future. *Quarterly Journal of the Geological Society of London* 83:lxii–civ. ISSN:0370-291X.

Bauhin, C. 1623. *ΠΙΝΑΞ Theatri botanici Caspari Bauhini [...]*. Basileae Helvet.: Ludovici Regii.

Baum, D. A., and S. D. Smith. 2013. *Tree Thinking. An Introduction to Phylogenetic Biology*. Greenwood Village (CO): Roberts and Co. Publ. ISBN:9781936221165.

Bazhanov, V. A. 2009. [*Vasil'ev and his Imaginary Logic. Resurrection of One Forgotten Idea*.] Moscow: Canon+. ISBN:9785883731968. (in Russian).

Beckett, S., and S. Lidgard. 2014. The normative structure of mathematization in systematic biology. *Studies in History and Philosophy of Science Part C: Studies inHistory and Philosophy of Biological and Biomedical Sciences* 46:44–54. DOI:10.1016/j.shpsc.2014.03.001

Beckner, M. 1959. *The Biological Way of Thought*. New York (NY): Columbia University Press. ISBN:9780231931403.

Beghtol, C. 1996. Graphic representations of hierarchical systems: Preliminary study. In *Proceedings of the 7th ASIS SIG/CR Classification Research Workshop*, ed. P. Solomon, 21–36. Baltimore (MD): American Society for Information Science.

Beghtol, C. 2000. A whole, its kinds, and its parts. In *Dynamism and Stability in Knowledge Organization*, eds. C. Beghtol, L. C. Howarth, and N. J. Williamson, 313–19. Würzburg: Ergon. ISBN:9783933563606.

Beklemishev, V. N. 1994. [*Methodology of Systematics*.] Moscow: KMK Scientific Press. ISBN:5873170053. (in Russian). [This book has been originally written in the 1920s.]

Bentham, G. 1827. *Outline of a New System of Logic: With a Critical Examination of Dr. Whately*. London: Hunt and Clarke. ISBN:9781855060296 (1990 reprint).

Bentham, G. 1875. On the recent progress and present state of systematic botany. *Report of the British Association for the Advancement of Science* (1874):27–54. DOI:10.2475/ajs.s3-9.53.346

Benton, M. J. 2000. Stems, nodes, crown clades. and rank-free lists: Is Linnaeus dead? *Biological Review* 5:633–45. DOI:10.1111/j.1469-185x.2000.tb00055.x

Berendsohn, W. G. 1995. The concept of "potential taxa" in databases. *Taxon* 44:207–12. DOI:10.2307/1222443

Berendsohn, W. G. and M. Geoffroy. 2007. Networking taxonomic concepts: uniting without "unitary-ism". In *Biodiversity Databases: From Cottage Industry to Industrial Networks*, eds. Curry, G., and C. Humphries, 13–22. Boca Raton (FL): Taylor & Francis Group. ISBN:978415332903.

Bergeret, J. P. 1783–1785. *Phytonomatotechnie Universelle, c'est- à- dire, l'art de Donner Aux Plantes Des Noms Tirés De Leurs Caractères [...]*, Vols. 1–3. Paris: Didot et Poisson. https://gallica.bnf.fr/ark:/12148/bpt6k10575844.image

Bergman, M. K. 2018. *A Knowledge Representation Practionary: Guidelines Based on Charles Sanders Peirce*. Cham (Switzerland): Springer International Publication. ISBN:9783319980911.

Bergman, M. K. 2022. Hierarchy in knowledge systems. *Knowledge Organization* 49:40–66. https://www.isko.org/cyclo/hierarchy

Berkov, V. F., and Ya. S. Yaskevich. 2001. [*History of Logic*] Minsk: Novoe Znanie. ISBN:984651648X. (in Russian).

Berlin, B. 1973. Folk systematics in relation to biological classification and nomenclature. *Annual Review of Ecology and Systematics* 4:259–71. DOI:10.1146/annurev.es.04.110173.001355

Berlin, B. 1992. *Ethnobiological Classification: Principles of Categorization of Plants and Animals in Traditional Societies*. Princeton (NJ): Princeton University Press. ISBN:9780691631004.

Berlin, B., D. E. Breedlove, and P. H. Raven. 1973. General principles of classification and nomenclature in folk biology. *American Anthropologist* 75:214–42. DOI:10.1525/aa.1973.75.1.02a00140

Berndt, C. 2000. *Biblical Classification of Life: A Framework and Reference for Authentic Biblical Biology*. Wildomar: Elihu Books. ISBN:9780967785004.

Bertalanffy, L. 1969. *General System Theory*. New York (NY): George Braziller Publ. ISBN:9780807604533.

Bertalanffy, L. 1972 The history and status of general systems theory. *The Academy of Management Journal* 15:407–26. DOI:10.2307/255139

Bertin, J. 1983. *Semiology of Graphics: Diagrams, Networks, Maps*. Madison (WI): University of Wisconsin Press. ISBN:9781589482616.

Bertrand, Y., F. Pleijel, and G. W. Rouse. 2006. Taxonomic surrogacy in biodiversity assessments, and the meaning of Linnaean ranks. *Systematics and Biodiversity* 4:149–59. DOI:10.1017/S1477200005001908

Bessey, C. E. 1908. The taxonomic aspect of the species question. *The American Naturalist* 42:218–24. ISSN:1537-5323.

Bessey, C. E. 1909. The phyletic idea in taxonomy. *Science* 29:91–100. ISBN:9780243041817 (2018 reprint).

Bezdek, J. C., D. Dubois, and H. Prade. (eds.). 1999. *Fuzzy Sets in Approximate Reasoning and Information Systems*. New York (NY): Springer Science+Business Media. ISBN: 9781461373902.

Bhushan, N. and K. Rai (eds.). 2004. *Strategic Decision Making: Applying the Analytic Hierarchy Process*. London: Springer Science+Business Media. ISBN:9781852337568.

Bianchi, V. 1916. [Species and its subordinate taxonomic forms.] *Russian Zoological Journal* 1:287–97. (in Russian).

Bininda-Emonds, O. R. P. 2004. The evolution of supertrees. *Trends in Ecology and Evolution* 19:315–22. DOI:10.1016/j.tree.2004.03.015

Bininda-Emonds, O. R. P., J. L. Gittleman, and M.A. Steel. 2002. The (super)tree of life: Procedures, problems, and prospects. *Annual Review of Ecology, Evolution, and Systematics* 33:265–89. DOI:10.1146/annurev.ecolsys.33.010802.150511

Bintley, M. D. J., and P. Salonius (eds.). 2024. *Trees as Symbol and Metaphor in the Middle Ages. Comparative Contexts*. Cambridge (UK): D.S. Brewer. ISBN:9781843846642.

Bjornson, S., H. Verbruggen, N. Upham, et al. 2024. Reticulate evolution: Detection and utility in the phylogenomics era. *Molecular Phylogenetics and Evolution* 201:108197. DOI:10.1016/j.ympev.2024.108197

Black, E., M. Raghavan, and S. Barocas. 2022. Model multiplicity: Opportunities, concerns, and solutions. In *Proceedings of the 2022 ACM Conference on Fairness, Accountability, and Transparency*, 850–63. New York (NY): Association for Computing Machinery. DOI:10.1145/3531146.3533149

Blackburn, P., M. de Rijke, and Y. Venema. 2001. *Modal Logic*. Cambridge (UK): Cambridge University Press. ISBN:9780521527149.

Blackwelder, R. E. 1967. *Taxonomy. A Text and Reference Book*. New York (NY): John Wiley & Sons. ISBN:9780471078005.

Blaxter, M., J. Mann, T. Chapman, et al. 2005. Defining operational taxonomic units using DNA barcode data. *Philosophical Transactions of the Royal Society, Ser. B* 360:1935–43. DOI:10.1098/rstb.2005.1725

Blyakher, L. Ya. 1976. [*Problems of Animal Morphology. Historical Essays.*] Moscow: Nauka. (in Russian).

Bobay, L.-M., and H. Ochman. 2018. Biological species in the viral world. *Proceedings of the National Academy of Sciences* 115:6040–5. DOI:10.1073/pnas.1717593115

Bobrov, E. G. 1970. [*Carl Linnaeus.*] Leningrad: Nauka. (in Russian).

Bock, G. R., and G. Cardew (eds.). 1999. *Homology*. Chichester: John Wiley & Sons. ISBN:9780470515662.

Bock, H. 1546 *Kreüter Buch, darin Underscheid, Würckung und Namen der Kreüter so in Deutschen Landen wachsen [...]*. Strasbourg: Wendel Rihel. DOI:10.5962/bhl.title.8043

Bock, W. J. 1973. Philosophical foundations of classical evolutionary classification. *Systematic Zoology* 11:375–92. DOI:10.2307/2412945

Bock, W. J. 1977. Foundations and methods of evolutionary classification. In *Major Patterns of Vertebrate Evolution*, ed. M. K. Hecht, P. C. Goody, and B. M. Hecht, 851–95. New York (NY): Plenum Press. ISBN:9780306356148.

Bock, W. J. 1981. Functional-adaptive analysis in evolutionary classification. *American Zoologist* 21:5–20. DOI:10.1093/icb/21.1.5

Bock, W. J. 1989. The homology concept: Its philosophical foundation and practical methodology. *Zoologische Beiträge* 32:327–53. ISSN:0044-5150.

Bodnar, I. 2025. Aristotle's natural philosophy. In *The Stanford Encyclopedia of Philosophy*, eds. E. N. Zalta, and U. Nodelman. https://plato.stanford.edu/archives/spr2025/entries/aristotle-natphil/

Boëthius. 1906. *Anicii Manlii Severini Boethii in Isagogen Porphyrii Commentorum Editionis Secundae*. Vienna: F. Tempsky/Leipzig: G. Freitag. http://www.logicmuseum.com/wiki/Authors/Boethius/isagoge/CPorII. (Russian edition of 1990, Moscow: Nauka).

Bokiy, G. B. 2000. [*Systematics of the Natural Oxides.*] Moscow: VINITI. (in Russian).

Bokulich, A., and P. Bokulich (eds.). 2011. *Scientific Structuralism*. Heidelberg: Springer Science+Business Media. ISBN:9789048195961.

Bollobás, B. 1998. *Modern Graph Theory*. Dodrecht: Springer Science+Business Media. ISBN:9780387984889.

Bolstad, W. M. 2007. *Introduction to Bayesian Statistics*. Hoboken (NJ): John Wiley & Sons. ISBN:9780470141151.

Bondarenko, D. M., L. E. Grinin, and A. V. Korotayev. 2002. Alternative pathways of social evolution. *Social Evolution & History* 1:54–79. https://www.sociostudies.org/journal/files/seh/2002_1/alternative_pathways_of_social_evolution.pdf

Bondy, A., and U. S. R. Murty. 2008. *Graph Theory*, 3d ed. Dodrecht: Springer Science+Business Media. ISBN:9781846289699.

Bonnet, C. 1745 *Traité d'insectologie ou Observations sur les pucerons [...], 1re Pt.* Paris: Durand & Fils. ASIN:B0BFKWZDPQ (2022 reprint).

Bora, D., and A. K. Gupta. 2014. A comparative study between fuzzy clustering algorithm and hard clustering algorithm. *International Journal of Emerging Trends & Technology in Computer Science* 10:108–13. DOI:10.14445/22312803/IJCTT-V10P119

Borhvardt, V. G. 1988. [Homology, live concept or dogma?] *Vestnik Liningradskogo Gosudarstvennogo Universiteta, Ser. Biol.* 3:3–7. ISSN:1025-8604. (in Russian).

Bornkessel-Schlesewsky, I., A. L. Malchukov, and M. D. Richards (eds.). 2015. *Scales and Hierarchies. A Cross-Disciplinary Perspective*. Berlin: Walter de Gruyter GmbH. ISBN:9783110344004.

Boster J., and D'Andrade R. 1989. Natural and human sources of cross-cultural ornithological classifications. *American Anthropologist* 91:132–42. DOI:10.1525/aa.1989.91.1.02a00080

Botha, M. E. 1989. Theory development in perspective: The role of conceptual frameworks and models in theory development. *Journal of Advanced Nursing* 14:49–55.

Boyden, A. 1943. Homology and analogy: A century after the definitions of "homologue" and "analogue" of Richard Owen. *The Quarterly Review of Biology* 18:228–41. DOI:10.1086/394676

Boyden, A. 1947. Homology and analogy. A critical review of the meanings and implications of these concepts in biology. *The American Midland Naturalist* 37:548–669. DOI:10.2307/2421470

Brenner, N. 2001. The limits to scale? Methodological reflections on scalar structuration. *Progress in Human Geography* 25:591–614. DOI:10.1191/030913201682688959

Brigandt, I. 2002. Homology and the origin of correspondence. *Biology and Philosophy* 17:389–407. DOI:10.1023/A:1020196124917

Brigandt, I. 2003a. Species pluralism does not imply species eliminativism. *Philosophy of Science* 70:1305–16. DOI:https://doi.org/10.1086/377409

Brigandt, I. 2003b. Homology in comparative, molecular, and evolutionary developmental biology: the radiation of a concept, *The Journal of Experimental Zoology, B Molecular and Developmental Evolution* 299B:9–17. DOI:10.1002/jez.b.36

Brigandt, I. 2007. Typology now: Homology and developmental constraints explain evolvability. *Biology and Philosophy* 22:709–25. DOI:10.7939/R33B5WP39

Brigandt, I. 2009. Natural kinds in evolution and systematics: Metaphysical and epistemological considerations. *Acta Biotheoretica* 57:77–97. DOI:10.1007/s10441-008-9056-7

Brigandt, I. 2010. Beyond reduction and pluralism: Toward an epistemology of explanatory integration in biology. *Erkenntnis* 73:295–311. DOI:10.1007/s10670-010-9233-3

Brigandt, I. 2017. Typology and natural kinds in evo-devo. In *Evolutionary Developmental Biology*, eds. L. Nuno de la Rosa, and G. Müller, 1–11. Cham (Switzerland): Springer Nature Switzerland Springer. DOI:10.1007/978-3-319-33038-9_100-1

Brigandt, I., and P. E. Griffiths. 2007. The importance of homology for biology and philosophy. *Biology and Philosophy* 22:633–41. DOI:10.1007/s10539-007-9094-6

Briggs, J. 1992. *Fractals: The Patterns of Chaos*. London: Thames and Hudson. ISBN: 9780500276938

Brillouin, L. 1962. *Science and Information Theory*, 2d ed. New York (NY): Academic Publ. ISBN: 9780121349509.

Brinton, W.C. 1919. *Graphic Methods for Presenting Facts*. New York (NY): The Engineering Magazine Co. ISBN:9781110356232 (2009 reprint).

Brisseau-Mirbel, C. F. 1802. *Histoire naturelle, Générale et Particulière, Des Plantes*, Vol. 1. Paris: F. Dufart. ISBN:9781013192203 (2019 reprint).

Broadfield, A. 1946. *The Philosophy of Classification*. London: Grafton.

Brooks, D. R., and D. A. McLennan. 1999. Species: Turning a conundrum into a research program. *Journal of Nematology* 31:117–33. PMID:19270882; PMCID:PMC2620355.

Brooks, D. R., and E. O. Wiley. 1986. *Evolution as Entropy*. Chicago (IL): The University of Chicago Press. ISBN:9780226075747.

Brooks, D. S., and M. Eronen. 2018. The significance of levels of organization for scientific research: A heuristic approach. *Studies in History and Philosophy of Biology & Biomedical Sciences* 68–69:34–41. DOI:10.1016/j.shpsc.2018.04.003

Brothers, D. J. 1983. Nomenclature at the ordinal and higher levels. *Systematic Zoology* 32:34–42. DOI:10.2307/2413218

Broughton, V. 2006. The need for a faceted classification as the basis of all methods of information retrieval. *Aslib Proceedings* 58:49– 72. DOI:10.1108/00012530610648671

Brower, A. V. Z., and R. T. Schuh. 2021. *Biological Systematics. Principles and Applications*, 3d ed. Ithaca (NY): Cornell University Press. ISBN:9781501752773.

Brown, C. H. 1976. General principles of human anatomical partonomy and speculations on the growth of partonomic nomenclature. *American Anthropologist* 78:400–24. DOI:10.1525/AE.1976.3.3.02A00020

Brown, C. H. 1984. *Language and Living Things: Uniformities in Folk Classification and Naming*. New Brunswick (NJ): Rutgers University Press. ISBN:9780813510088.

Brown, C. H. 1986. The growth of ethnobiological nomenclature. *Current Anthropology* 27: 1–19. DOI:10.1086/203375

Brown, D. E. 1991. *Human Universals*. Philadelphia (PA): Temple University Press. ISBN:9780070082090.

Brown, D. E. 2004. Human universals, human nature, human culture. *Daedalus* 133:47–54. DOI:10.1162/0011526042365645

Brummitt, R. K. 1996. In defence of paraphyletic taxa. In *The Biodiversity of African Plants*, eds. L. J. G. van der Maesen, X. M. van der Burg, and J. M. van Medenbach de Rooy, 371–84. Dordrecht: Kluwer Academic Publ. DOI:10.5962/bhl.title.3678

Brummitt, R. K. 1997. Taxonomy versus cladonomy: A fundamental controversy in biological systematics. *Taxon* 46:723–34. DOI:10.2307/1224478

Brummitt, R. K. 2002. How to chop up a tree. *Taxon* 51:31–41. DOI:10.2307/1554961

Brundin, L. 1972. Evolution, causal biology, and classification. *Zoologica Scripta* 1:107–20.

Brunelli, M. 2015. *Introduction to the Analytic Hierarchy Process*. Cham (Switzerland): SpringerBriefs in Operations Research. ISBN:9783319125015.

Brunfels, O. 1530–1536. *Herbarum Vivae Eicones Ad Naturae Imitationem Summa Cum Diligentia Et Artificio Effigiatse […]. T.* I–III. Argentorati [Strasbourg]: Ioannem Scottü.

Bryant, R. 2000. *Discovery and Decision: Exploring the Metaphysics and Epistemology of Scientific Classification*. Cranbury (NJ): Associated University Press. ISBN: 9780838638767.

Buck, R. C., and D. L. Hull. 1966. The logical structure of the Linnaean hierarchy. *Systematic Zoology* 15:97–111. DOI:10.2307/2411628

Buckland, M. 1991. *Information and Information Systems*. New York (NY): Greenwood Press. ISBN:9780313274633.

Buckley, J. J. 1985. Fuzzy hierarchical analysis. *Fuzzy Sets and Systems* 17:233–47. DOI:10.1016/0165-0114(85)90090-9

Buffon. 1835. Premier discours: De la manière d'étudier et de traiter l'histoire naturelle. In *Oeuvres Complètes de Buffon. Théorie de la Terre*, 37–79. Paris: Pourrat Frères. ISBN:978282064334 (2018 reprint).

Buffon, G.-L. 1755a. *Histoire Naturelle, Générale et Particulière, avec la Description du Cabinet du Roy, T. 5*. Paris: L'Imprimere Royale. ISBN:9782745320575 (2010 reprint).

Buffon, G.-L. 1755b. Espèce (Hist. Nat.). In *Encyclopedie, ou Dictionaaire raisonné, des Arts et Métiers*, T. 5, ed. D. Diderot, 956–7. Paris: Briasson, David, Le Breton & Durand. ISBN:9781295359714 (2013 reprint).

Bulmer, R. 1974. Folk biology in the New Guinea highlands. *Social Science Information* 13:9–28. DOI:10.1177/053901847401300402

Bulmer, R., and M. Tyler. 1968. Karam classification of frogs. *The Journal of the Polynesian Society* 77: 333–85. https://www.jstor.org/stable/20704577

Bunge, M. 1969. The metaphysics, epistemology and methodology of levels. In *Hierarchical Structures*, eds. L. L. Whyte, A. G. Wilson, and D. Wilson, 17–28. New York (NY): American Elsevier Publ. ISBN:9780444000699.

Bunge, M. 1979. *Treatise of Basic Philosophy, Vol. 4: A World of Systems*. Dordrecht: D. Reidel. ISBN:9789027709455.

Buol, S. W., R. J. Southard, R. C. Graham and P. A. McDaniel. 2011. *Soil genesis and classification*, 6th ed. Chichester: John Wiley & Sons.

Burgess, J. P. 2009. *Philosophical Logic*. Princeton (NJ): Princeton University Press. ISBN:9780691137896.

Burgin, M. 2015. *Theory of Knowledge. Structures and Processes*. Hackensack (NJ): World Scientific Publ. ISBN:9789814522670.

Burlando, B. 1990. The fractal dimension of taxonomic systems. *Journal of Theoretical Biology* 146:99–114. DOI:10.1016/S0022-5193(05)80046-3

Bursten, J. R. S. (ed.). 2020. *Perspectives on Classification in Synthetic Sciences. Unnatural Kinds*. Abingdon: Routledge. ISBN:9781351581301.

Burt, D. B. 2008. Evolutionary stasis, constraint and other terminology describing evolutionary patterns. *January Biological Journal of the Linnean Society* 72:509–17. DOI:10.1111/j.1095-8312.2001.tb01334.x

Butler, A. B., and W. M. Saidel. 2000. Defining sameness: Historical, biological, and generative homology. *BioEssays* 22:846–53. DOI:10.1002/1521-1878(200009)22:9<846::AID-BIES10>3.0.CO;2-R

Cain, A. J. 1994. Rank and sequence in Caspar Bauhin's Pinax. *Botanical Journal of the Linnean Society* 114:311–56.

Cain, A. J., and G. A. Harrison. 1958. An analysis of the taxonomist's judgment of affinity. *Proceedings of the Zoological Society of London* 131:85–98. DOI:10.1111/j.1096–3642.1958.tb00634.x

Calosi, C., and P. Graziani (eds.). 2014. *Mereology and the Sciences: Parts and Wholes in the Contemporary Scientific Context*. Dordrecht: Springer-Verlag. ISBN:9783319053561.

Cameron, M. 2007. Abelard (and Heloise?) on intention. *American Catholic Philosophical Quarterly* 81:323–38. DOI:10.5840/acpq200781243

Camp, W. H. 1951. Biosystematy. *Brittonia* 7:113–27. DOI:10.2307/2804701

Camp, W. H., and C. L. Gilly. 1943. The structure and origin of species, with a discussion of intraspecific variability and related nomenclatural problems. *Brittonia* 4:323–85. DOI:10.2307/2804896

Campbell, D. T. 1974. Downward causation in hierarchically organised biological systems. In *Studies in the Philosophy of Biology: Reduction and Related Problems*, eds. F. J. Ayala, and T. Dobzhansky, 179–86. London: Macmillan. ISBN:9780520026490.

Cantino, P. D., and K. de Queiroz. 2010. *International Code of Phylogenetic Nomenclature (PhyloCode), Version 4c*. www.ohio.edu/phylocode/PhyloCode4c.pdf

Cantino, P. D., and K. de Queiroz. 2020. *International Code of Phylogenetic Nomenclature (PhyloCode), Version 6*. Boca Raton (FL): CRC Press. ISBN:9781138332829.

Caponi, G. 2016. Lineages and systems: A conceptual discontinuity in biological hierarchies. In *Evolutionary Theory: A Hierarchical Perspective*, eds. N. Eldredge, T. Pievani, E. Serrelli, et al., 47–61. Chicago (IL): The University of Chicago Press. ISBN:9780226426051.

Capurro, R., and B. Hjørland. 2003. The concept of information. *Annual Review of Information Science and Technology* 37:343–411. DOI:10.1002/aris.1440370109

Carbonell-Nicolau, O. 2025. Measuring hierarchy. *Social Choice and Welfare* 64. DOI:10.1007/s00355-025-01582-1

Carpenter, J. M. 1994. Successive weighting, reliability, and evidence. *Cladistics* 19:215–20. DOI:10.1111/j.1096-0031.1994.tb00173.x

Carpenter, J. M. 2003. Critique of pure folly. *The Botanical Review* 69:79–92. DOI:10.1663/0006-8101(2003)069[0079:COPF]2.0.CO;2

Carr, D. C. 1923. *Linnaeus and Jussieu, or, the Rise and Progress of Systematic Botany*. Charleston (SC): Nabu Press. ISBN:9781287413240 (2013 reprint).

Carrier, M. 1994. *The Completeness of Scientific Theories*. Dordrecht: Kluwer Academic Publ. ISBN:9780792324751.

Carroll, S. M. 2004. *Spacetime and Geometry: An Introduction to General Relativity*. San Francisco (CA): Addison-Wesley. ISBN:9780805387322.

Carruthers, P., S. P. Stich, and M. Siegal (eds.). 2002. *The Cognitive Basis of Science*. Cambridge (UK): Cambridge University Press. ISBN:9780511613517.

Carter, J. A. 2025. Epistemology and relativism. In *Internet Encyclopedia of Philosophy*. https://iep.utm.edu/epis-rel/

Carter, J. G., C. R. Altaba, L. C. Anderson, et al. 2015. The paracladistic approach to phylogenetic taxonomy. *Paleontological Contributions* 12:1–9. DOI:10.17161/PC.1808.17551

Cartwright, N. 2016. Loose talk kills: What's worrying about unity of method. *Philosophy in Science* 83:768–78. DOI:10.1086/687862

Carus, C. G. 1834. *Lehrbuch der vergleichenden Zootomie*, Bd. 1. Leipzig : Fleischer Wien. ISBN:9780666658425 (2018 reprint).

Casacci, L. P., F. Barbero, and E. Balletto. 2014. The "evolutionary significant unit" concept and its applicability in biological conservation. *Italian Journal of Zoology* 81:182–93. DOI:10.1080/11250003.2013.870240

Cash, D. W., W. C. Clark, F. Alcock, et al. 2003. Knowledge systems for sustainable development. *The Proceedings of the National Academy of Sciences* 100:8086–91. DOI:10.1073/pnas.1231332100

Cavalier-Smith, T. 1993. Kingdom protozoa and its 18 phyla. *Microbiological Reviews* 57:953–94. DOI:10.1128/mmbr.57.4.953-994.1993

Cavalier-Smith, T. 2007. A revised six-kingdom system of life. *Biological Reviews* 73:203–66. DOI:10.1111/j.1469-185X.1998.tb00030.x

Cavalli-Sforza, L. L., and A. W. F. Edwards. 1967. Phylogenetic analysis: Models and estimating procedures. *Evolution* 21:550–70. DOI:10.2307/2406616

Cavalli-Sforza, L. L., and A. Piazza. 1975. Analysis of evolution: Evolutionary rates, independence and treeness. *Theoretical Population Biology* 8:127–65. DOI:10.1016/0040-5809(75)90029-5

Cellinese, N., S. Conix, and H. Lapp. 2021. Phyloreferences: Tree-native, reproducible, and machine-interpretable taxon concepts. *EcoEvoRxiv Preprintss*. https://ecoevorxiv.org/57yjs/

Cesalpino, A. 1583. *De plantis libri XVI Andreae Cesalpini Aretini [...]*. Florentiae: Georgium Marescottum.

Chakraborty, R. N., and E. Mohr. 2005. Coordination and motivation in flat hierarchies: The impact of the adjudication culture. *Economica, New Series* 72:563–76. DOI:10.1111/j.1468-0335.2005.00432.x

Chamberlain, J. R. 1992. Biolinguistic systematics and marking. In *Third International Symposium on Language and Linguistics*, 1279–93. Bangkok: Chulalongkorn University. http://sealang.net/sala/archives/pdf8/chamberlain1992biolinguistic.pdf

Chan, L. M., and A. Salaba. 2016. *Cataloging and Classification: An Introduction*, 4th ed. Lanham (MD): Rowman & Littlefield Publ. ISBN:9780810860001.

Chattopadhyay, A., U. Ghosal, and P. Mukhopadhyay. 2022. Robustly separating the arithmetic monotone hierarchy via graph inner-product. In *42nd IARCS Annual Conference on Foundations of Software Technology and Theoretical Computer Science*, eds. A. Dawar and V. Guruswami, 12:1–12. Madras. DOI:10.4230/LIPIcs.FSTTCS.2022.12

Chebanov, S. V. 2017. S. V. [Meyen's meronomy: On the 40th anniversary of its formulation.] *Lethaea Rossica* 14: 64–92. ISSN:2224-0446. (in Russian).

Chebanov, S. V., and G. Ya. Martynenko. 1998. [*Semiotics of descriptive texts (typological aspect).*] St. Petersburg: Saint Petersburg State University Publication. ISBN:9785288019715. (in Russian).

Checkland, P. 1988. *System Thinking, System Practice*. Chichester: John Wiley & Sons. ISBN:0471279110.

Chen, M.-H., L. Kuo, and P. O. Lewis (eds.). 2014. *Bayesian Phylogenetics. Methods, Algorithms, and Applications*. Boca Raton (FL): CRC Press. ISBN:9781466500792.

Chester, M. 1987. *Primer of Quantum Mechanics*. New York (NY): John Wiley & Sons. ISBN:0486428788.

Chiaradonna, R. C., and G. Galluzzo (eds.). 2013. *Universals in Ancient Philosophy*. Pisa: Scuola Normale Superiore. ISBN:9788876424847.

Child, J. 2019. *Hierarchy: A Key Idea for Business and Society*. New York (NY): Routledge. ISBN:9781138044388.

China, W. E. 1933. A new family of Hemiptera-Heteroptera with notes on the Phylogeny of the suborder. *Annals and Magazine of Natural History*, 12:180–96. DOI:10.1080/00222933308673767

Chippindale, P. T., and J. J. Wiens. 1994. Weighting, partitioning, and combining characters in phylogenetic analysis. *Systematic Biology* 43:278–87. DOI:10.1093/sysbio/43.2.278

Chislenko, L. L. 1977. [On the structure of taxa and taxonomic diversity.] *Zhurnal Obshchei Biologii* 38:348–58. ISSN:0044-4596. (in Russian).

Chomsky, N. 1987. *Language and Problems of Knowledge*. Cambridge (MA): The MIT Press. ISBN:9780262530705.

Christoffersen, M. L. 1995. Cladistic taxonomy, phylogenetic systematics, and evolutionary ranking. *Systematic Biology* 44:440–54. DOI:10.1093/sysbio/44.3.440

Clark, A. T., J.-F. Arnoldi, Y. R. Zelnik, et al. 2021. General statistical scaling laws for stability in ecological systems. *Ecology Letters* 24:1474–86. DOI:10.1111/ele.13760

Clayton, W. D. 1974. The logarithmic distribution of angiosperm families. *Kew Bulletin* 29:271–9. DOI:10.2307/4108541

Clements, F. E. 1908. An ecologic view of the species conception. *The American Naturalist* 191:253–64. DOI:10.1086/278930

Clusius, C. 1601. *Rariorum plantarum historia [...]*. Antverpiae: ex officina Plantiniana, Ioannem Moretum.

Colless, D. H. 1967. An examination of certain concepts in phenetic taxonomy. *Systematic Zoology* 16:6–27. DOI:10.2307/2411512

Colless, D. H. 1970. The phenogram as an estimation of phylogeny. *Systematic Zoology* 19:352–62. DOI:10.2307/2412276

Contessa, G. 2006. Constructive empiricism, observability and three kinds of ontological commitment. *Studies in History and Philosophy of Science Pt A* 37:454–68. DOI:10.1016/j.shpsa.2006.06.007

Contoli, L., and J. Pignatti. 2011. Phanerozoic biodiversity, macrotaxa and families. *Journal Mediterranean Earth Sciences* 3:73–80. DOI:10.3304/JMES.2011.005

Cook, O. F. 1898. Stability in generic nomenclature. *Science, New Series* 8:186–90. DOI:10.1126/science.8.189.186

Cook, O. F. 1900. The method of types in botanical nomenclature. *Science, New Series* 12:475–81. DOI:10.1126/science.12.300.475

Cook, R. 1988. *Tree of Life. Image for the Cosmos*, 2d ed. London: Thames & Hudson. ISBN:9780500810071.

Cooper, L., R. L. Walls, J. Elser, et al. 2013. The plant ontology as a tool for comparative plant anatomy and genomic analyses. *Plant and Cell Physiology* 54:e1–e1. DOI:10.1093/pcp/pcs163

Cope E. D. 1868. On the origin of genera. *Proceedings of the Academy of Natural Sciences of Philadelphia* 20:242–300.

Cope, E. D. 1887. *The origin of the fittest. Essays of evolution*. New York: Appleton. ISBN:9783337908492 (2020 reprint).

Coquand, T. 2022. Type theory. In *The Stanford Encyclopedia of Philosophy*, eds. E. N. Zalta, and U. Nodelman. https://plato.stanford.edu/archives/fall2022/entries/type-theory/

Corning, P. A. 1983. *The Synergism Hypothesis: A Theory of Progressive Evolution*. New York (NY): McGraw-Hill. ISBN:9780070131729.

Corning, P. A. 2022. A systems theory of biological evolution. *Biosystems* 214:104630. DOI:10.1016/j.biosystems.2022.104630

Corominas-Murtra, B., J. Goñid, R. V. Solé, et al. 2013. On the origins of hierarchy in complex networks. *Proceedings of the National Academy of Sciences* 110:13316–21. DOI:10.1073/pnas.1300832110

Corominas-Murtra, B., C. Rodriguez-Caso, and J. Goni. 2011. Measuring the hierarchy of feedforward networks. *Chaos* 21:016108. DOI:10.1063/1.3562548

Cosetta, E. 2010. Categories, taxa, and chimeras. In *New Essays in Logic and Philosophy of Science*, eds. M. D'Agostino, G. Giorello, F. Laudisa, et al., 264–78. London: College Publ. ISBN:9781848900035.

Coues, E., J. A. Allen, R. Ridgway, et al. 1886. *The Code of Nomenclature and Check- List of North American Birds [...]*. New York (NY): American Ornithologists' Union. www.biodiversitylibrary.org/item/16484#page/7/mode/1up

Cowles, H. C. 1908. An ecological aspect of the conception of species. *The American Naturalist* 42:265–71. DOI:10.1086/278931

Cracraft, J. 1989. Species as entities of biological theory. In *What the Philosophy of Biology is: Essays Dedicated to David Hull*, ed. M. Ruse, 31–52. Dordrecht: Kluwer Academic. Publ. ISBN:978-9401070201.

Cracraft, J., and M. J. Donoghue (eds.). 2004. *Assembling Tree of Life*. Oxford: Oxford University Press. ISBN:9780195172348.

Craske, A. J., and R. P. S. Jefferies. 1989. A new mitrate from the Upper Ordovician of Norway, and a new approach to subdividing a plesion. *Palaeontology* 32:69–99. https://www.palass.org/sites/default/files/media/publications/palaeontology/volume_32/vol32_part1_pp69-99.pdf

Crisci, J. V. 1984. Taxonomic congruence. *Taxon* 33:233–9. DOI: 10.2307/1221163

Cristofaro, S., and F. Zúñiga (eds.). 2018. *Typological Hierarchies in Synchrony and Diachrony*. Amsterdam: John Benjamins. 2018. p. 434. Hardbound: ISBN: 9789027200266.

Croft, W., and A. Cruse. 2004. *Cognitive Linguistics*. Cambridge (UK): Cambridge University Press. ISBN:9780511803864.

Cronk, Q. C. B. 1989. Measurement of biological and historical influences on plant classifications. *Taxon*, 38:357–70. DOI:10.2307/1222266

Cronquist, A. 1987. A botanical critique of cladism. *Botanical Review* 53:1–52. DOI:10.1007/BF02858181

Crowe, M. J. 1990. *Theories of the World from Antiquity to the Copernican Revolution*. Mineola (NY): Dover Publ. ISBN:0486261735.

Crowson, R. A. 1970. *Classification and Biology*. London: Heinemann. ISBN:9780435629847.

Crumley, C. L. 1987. A dialectical critique of hierarchy. In *Power Relations and State Formation*, eds. T. C. Patterson, and C. W. Gailey, 155–68. Washington (DC): American Anthropological Association. ISBN:9781879215122.

Crumley, C. L. 1995. Heterarchy and the analysis of complex societies. *Archeological Papers of the American Anthropological Association* 6:1–5. DOI:10.1525/ap3a.1995.6.1.1

Crumley, C. L. 2005. Remember how to organize: Heterarchy across disciplines. In *Nonlinear Models for Archaeology and Anthropology*, eds. C. S. Beekman, and W. S. Baden, 35–50. Aldershot (UK): Ashgate Press. ISBN:9781138266704.

Cusack, C. M. 2011. *The Sacred Tree: Ancient and Medieval Manifestations*. Newcastle: Cambridge Scholars Publ. ISBN:9781443828574.

Cuvier, G. 1801. *Leçons D'anatomie Comparée*, Vol. 1. Paris: Crochard. ISBN:9782329388854 (2020 reprint).

Dahdul, W. M., H. Cui, P. M. Mabee, et al. 2014. Nose to tail, roots to shoots: Spatial descriptors for phenotypic diversity in the biological spatial ontology. *Journal of Biomedical Semantics* 5:1–13. DOI:10.1186/2041-1480-5-34

Dahui, W., L. Menghui, and D. Zengru. 2005. True reason for Zipf's law in language. *Physica A: Statistical Mechanics and its Applications* 358:545–50. DOI:10.1016/j.physa.2005.04.021

Dall, W. H. 1877. *Nomenclature in Zoology and Botany [...]*. Salem (MA): Salem Press. ISBN:9781014841278 (2021 reprint).

Dambergs Y. 2003. *Mnemonic Arts of Blessed Raymond Lull*. https://archive.ph/Xk1aj

D'Amico, G., R. Rabadan, and M. Kleban. 2016. A theory of taxonomy. *ArXiv*:1611.03890 [physics.soc-ph]. https://arxiv.org/abs/1611.03890

Danser, B. 1929. Über die begriffe Komparium, Kommiskuum und Konvivium und üeber die Entstehungsweise der Konvivien. *Genetica* 11:399–450. ISSN:1573-6857.

Dardağan, A. 2018. *Neoplatonic "Tree of Life"* (*Arbor Porphyriana: A Diagram of Logic and Mystical Theology*). SocArXiv Papers. https://osf.io/preprints/socarxiv/g2qxe/

Darwin, C. 1859. *On the Origin of Species by Means of Natural Selection, or the Preservation of Favoured Races in the Struggle for Life*. London: John Murray. ISBN:9781615340378 (2009 reprint).

D'Atri, A., M. De Marco, and N. Casalino (eds.). 2008. *Interdisciplinary Aspects of Information Systems Studies*. Heidelberg: Physica-Verlag. ISBN:9783790820096.

Davies, B., and G. R. Evans (eds.). 1998. *Anselm of Canterbury: The Major Works*. Oxford: Oxford University Press. ISBN:9780199540082.

Dawson, M. N., R. S. Waples, and G. Bernardi. 2006. Phylogeography. In *The Ecology of Marine Fishes: California and Adjacent Waters*, eds. L. G. Allen, D. J. Pondella, and M. H. Horn, 26–54. Berkeley (CA): University of California Press. DOI:10.1525/california/9780520246539.003.0002

Dayrat, B. 2010. Celebrating 250 dynamic years of nomenclatural debates. In *Systema naturae 250—The Linnaean ark*, ed. A. Polaszek, 186–239. Boca Raton (FL): CRC Press. ISBN:9781420095012

de Beer, G. 1971. *Homology, an Unsolved Problem*. Oxford: Oxford University Press. ISBN:9780199141111.

de Blainville, H. 1816. Prodrome d'une nouvelle distribution du règne animal. *Bulletin de la Société Philomathique* 8:113–24. ISSN:1153-6470.

de Candólle, A. 1867. *Lois de la nomenclature botanique adoptées par le Congrès international de botanique tenu à Paris en août 1867 [...]*. Paris: J. B. Baillière. https://gallica.bnf.fr/ark:/12148/bpt6k981450.image

de Candólle, A. 1883. *Nouvelles remarques sur la nomenclature botanique. Supplément au commentaire du même auteur qui accompagnait le texte des lois*. Geneva: H. Georg. www.biodiversitylibrary.org/item/221146#page/7/mode/1up

de Candólle, A., and C. A. Cogniaux. 1876. Quelques points de nomenclature botanique. *Bulletin de la Société Royale de Botanique de Belgique* 15:477–85. ISSN:0037-9557.

de Candólle, A. P. 1813. *Théorie élémentaire de la botanique, oe exposition des principes de la classification naturelle [...]*. Paris: Deterville. www.biodiversitylibrary.org/item/88297#page/7/mode/1up. [The 2nd ed. of 1819; ISBN:9781142242220 (2010 reprint).]

de Candólle, A. P. 1818. *Regni Vegetabilis Systema Naturae [...]*, Vol. 1. Parisiis: Treuttel et Würtz. DOI:10.5962/bhl.title.59874

de Candólle, A. P. 1824. *Prodromus Systematis Naturalis Regni vegetabilis [...]*, Vol. 1. Parisiis: Treuttel et Würtz. DOI:10.5962/bhl.title.286

de Candólle, A. P. 1828. *Mémoire sur la famille des Mélastomacées*. Paris: Treuttel et Würtz.

de Pinna, M. C. C. 1991. Concepts and tests of homology in the cladistic paradigm. *Cladistics* 7:367–94. DOI:10.1111/j.1096-0031.1991.tb00045.x

de Queiroz, K. 1988. Systematics and the Darwinian revolution. *Philosophy of Science* 55:238–59. DOI:10.1086/289430
de Queiroz, K. 1992. Phylogenetic definitions and taxonomic philosophy. *Biology and Philosophy* 41:295–313. DOI:10.1007/BF00129972
de Queiroz, K. 1997. The Linnaean hierarchy and the evolutionization of taxonomy, with emphasis on the problem of nomenclature. *Aliso* 15:125–44. ISSN:0065-6275.
de Queiroz, K. 2005. Linnaean, rank-based, and phylogenetic nomenclature: Restoring primacy to the link between names and taxa. *Symbolae Botanicae Upsalienses* 33:127–40. https://repository.si.edu/bitstream/handle/10088/4506/VZ_2005deQueirozSymBotUps.pdf
de Queiroz, K., and P. D. Cantino. 2001. Phylogenetic nomenclature and the PhyloCode. *Bulletin of Zoological Nomenclature* 58:254–71. DOI:10.1201/9780429446320
de Queiroz, K., and M. J. Donoghue. 1990. Phylogenetic systematics and species revisited. *Cladistics* 6:83–90. DOI:10.1111/j.1096-0031.1990.tb00527.x
de Queiroz, K., and J. Gauthier. 1992. Phylogenetic taxonomy. *Annual Review of Ecology and Systematics* 23:449–80. DOI:10.1146/annurev.es.23.110192.002313
de Queiroz, K., and J. Gauthier. 1994. Toward a phylogenetic system of biological nomenclature. *Trends in Ecology and Evolution* 9:27–31. DOI:10.1016/0169-5347(94)90231-3
de Rijk, L. M. 1972. *Peter of Spain (Petrus Hispanus Portugalensis) Tractatus, called afterwards Summule Logicales*. Assen (The Netherlands): Koninklijke Van Gorcum & Comp. B.V. ISBN:9023209753.
de Tournefort, J. P. 1694 *Élémens de botanique, ou Methode pour connoître les plantes*, Vol. 1. Paris: l'Imprimerie Royale. ISBN:9781272619824 (2012 reprint).
de Vries, H. 1904. *Species and Varieties: Their Origin by Mutation*. Chicago (IL): Open Court Publ. ISBN:9781419148514 (2004 reprint).
Dean, J. 1979. Controversy over classification: A case study from the history of botany. In *Natural Order: Historical Studies of Scientific Culture*, eds. B. Barnes and S. Shapin, 211–30. London: Sage Publ. ISBN 9780803909595.
Decamp, D. 1971. Implicational scales and sociolinguistic linearity. *Linguistics* 9:30–43. DOI:10.1515/ling.1971.9.73.30
Dececchi, T. A., J. P. Balhoff, H. Lapp, et al. 2015. Toward synthesizing our knowledge of morphology: Using ontologies and machine reasoning to extract presence/absence evolutionary phenotypes across studies. *Systematic Biology* 64:936–52. DOI:10.1093/sysbio/syv031
Dekker, E., and P. Kuchař. 2021. Heterarchy. In: *Encyclopedia of Law and Economics*, eds. A. Marciano, and G.B. Ramello, H1–H6. New York (NY): Springer Science+Business Media. DOI:10.1007/978-1-4614-7883-6_640-2
Demirel, Y. 2002. *Nonequilibrium Thermodynamics. Transport and Rate Processes in Physical, Chemical and Biological Systems*. Amsterdam: Elsevier. ISBN:9780444641120
Denton, M. 1986. *Evolution: A Theory in Crisis*. Bethesda (MA): Adler & Adler Publ. ISBN:9780917561528.
Denton, M. 2016. *Evolution: Still a Theory in Crisis*. Seattle (WA): Discovery Institute Press. ISBN:9781936599325.
Dessalles, J.-L. 2001. Qualia and spandrels: an engineering perspective. Te*chnical Report ENST 2001-D-012*:hal-00614805f. https://hal.science/hal-00614805/document
Dessalles, J.-L., and D. Phan. 2005. Emergence in multi-agents systems: Cognitive hierarchy, detection, and complexity reduction, Part 1: Methological issues. In *Artificial Economics. Agent-Based Methods in Finance, Game Theory and Their Applications*, eds. M. Mathieu, O. Brandouy, and B. Beaufils, 147–59. Berlin: Springer. ISBN:9783540285786.
Devitt, M. 2008. Resurrecting biological essentialism. *Philosophy of Science* 75:344–82. DOI:10.1086/593566

Devitt, M., and K. Sterelny. 1999. *Language and Reality. An Introduction to the Philosophy of Language*, 2nd ed. Cambridge (MA): MIT Press. ISBN:9780262540995.
Dial, K. P., and J. M. Matzluff. 1989. Nonrandom diversification within taxonomic assemblages. *Systematic Zoology* 38:26–37. DOI:10.2307/2992433
Dillon, L. S. 1966. The life cycle of the species: an extension of current concepts. *Systematic Zoology* 15:112–26. DOI:10.2307/sysbio/15.2.112
Dobzhansky, T. 1935. A critique of the species concept in biology. *Philosophy of Science* 2:344–55. DOI:10.1086/286379
Dobzhansky, T. 1970. *Genetics of Evolutionary Process*. New York (NY): Columbia University Press. ISBN:9780231083065.
Domingo, E. 2002. Quasispecies theory in virology. *Journal of Virology* 76:463–5. DOI:10.1128/JVI.76.1.463-465.2002
Domingo, E., and C. Perales. 2019. Viral quasispecies. *PLoS Genetics* 15:e1008271. DOI:10.1371/journal.pgen.1008271
Dominguez, E., and Q. D. Wheeler. 1997. Taxonomic stability is ignorance. *Cladistics* 13:367–72.
Dompere, K. K. 2009a. *Fuzzy Rationality. A Critique and Methodological Unity of Classical, Bounded and Other Rationalities*. Berlin: Springer-Verlag. ISBN:9783540880837.
Dompere, K. K. 2009b. *Epistemic Foundations of Fuzziness. Unified Theories on Decision-Choice Processes*. Berlin: Springer-Verlag. ISBN:9783540880844.
Donoghue, P. C. J., and M. P. Smith (eds.). 2003. *Telling the Evolutionary Time: Molecular Clocks and the Fossil Record*. Boca Raton (FL): CRC Press. ISBN:0415275245.
Doolittle, W. F. 2012. Population genomics: How bacterial species form and why they don't exist. *Current Biology* 22:R451–3. DOI:10.1016/j.cub.2012.04.034
Dorofeeva, D. Yu. (ed.). 2001. *[The Book of Angels: An Anthology.] St.* Petersburg: Amphora. (in Russian). ISBN:5942780455.
Douven, I. 2021. Abduction. In *The Stanford Encyclopedia of Philosophy*, ed. E. N. Zalta. https://plato.stanford.edu/archives/sum2021/entries/abduction/
Dress, A., and A. von Haeseler. 1990. Introduction. In *Trees and Hierarchical Structures*, eds. A. Dress, and A. von Haeseler, 1–8. Berlin: Springer-Verlag. ISBN:9783540524533.
Driesch, H. 1908. *The Science and Philosophy of the Organism*. Aberdeen: University Print. ISBN:9781428640924 (2006 reprint).
Du Rietz, G. E. 1930. The fundamental units of biological taxonomy. *Svensk Botanisk Tidskrift* 24:333–428. ISSN:0039-646X.
Dubin, R. 1978. *Theory Building*, Rev. ed. New York (NY): Free Press. ISBN:9780029076200.
Dubois, A. 2005. Proposed rules for the incorporation of nomina of higher-ranked zoological taxa in the International Code of Zoological Nomenclature. 1. Some general questions, concepts and terms of biological nomenclature. *Zoosystema* 27:365–426. https://sciencepress.mnhn.fr/sites/default/files/articles/pdf/z2005n2a8.pdf
Dubois, A. 2006. Proposed rules for the incorporation of nomina of higher-ranked zoological taxa in the International Code of Zoological Nomenclature. 2. The proposed rules and their rationale. *Zoosystema* 28:165–258. https://sciencepress.mnhn.fr/sites/default/files/articles/pdf/z2006n1a9.pdf
Dubois, A. 2007. Phylogeny, taxonomy and nomenclature: The problem of taxonomic categories and of nomenclatural ranks. *Zootaxa* 1519:27–68. DOI:10.11646/zootaxa.1519.1.3
Dubois, A. 2008. Zoological nomenclature: Some urgent needs and problems. In *Future Trends of Taxonomy*, 15–18. Carvoeiro (Portugal): EDIT. https://www.academia.edu/11908654/FUTURE_TRENDS_OF_TAXONOMY
Dubois, A. 2011. The international code of zoological nomenclature must be drastically improved before it is too late. *Bionomina* 2:1–104. DOI:10.11646/bionomina.2.1.1
Dubois, A., E. Aescht, P. T. Aneesh, et al. 2022. The Linz Zoocode project: Sixth report of activities (2022). Nomenclatural availability. 5. *Bionomina* 31:24–45. DOI:10.11646/bionomina.31.1.2

Dubois, A., A. M. Bauer, L. M. P. Ceríaco, et al. 2019. The Linz ZooCode project: A set of new proposals regarding the terminology, the principles and rules of zoological nomenclature […]. *Bionomina* 17:1–111. DOI:10.11646/bionomina.17.1

Dubois, A., A. Ohler, and R. A. Pyron. 2021. New concepts and methods for phylogenetic taxonomy and nomenclature in zoology, exemplified by a new ranked cladonomy of recent amphibians (Lissamphibia). *Megataxa* 5:1–738. DOI:10.11646/MEGATAXA.5.1.1

Dubois, D., and H. Prade. 1980. *Fuzzy Sets and Systems. Theory and Applications*. New York (NY): Academic Press. ISBN:9780122227509.

Duboule, D. 1994. Temporal colinearity and the phylotypic progression: a basis for the stability of a vertebrate Bauplan and the evolution of morphologies through heterochrony. *Development (Supplement)*:135–42. DOI:10.1242/dev.1994.Supplement.135

Duchêne, D. A., S. Duchêne, J. Stiller, et al. 2024. ClockstaRX: Testing molecular clock hypotheses with genomic data. *Genome Biology and Evolution* 16:evae064. DOI:10.1093/gbe/evae064

Dunaev, V. 1984. [About rank distributions in classification.] In [*Scientific and Technical Information, Ser. 2*] 9:14–8. ISSN:0548-0027. (in Russian).

Dupraw, E. J. 1964. Non-Linnaean taxonomy. *Nature* 5:849–52. DOI:10.1038/202849a0

Dupraw, E. J. 1965. Non-Linnean taxonomy and the systematics of honeybees. *Systematic Zoology* 14:1–24. DOI:10.2307/2411899

Dupré, J. 1993. *The Disorder of Things. Metaphysical Foundations of the Disunity of Science*. Cambridge (MA): Harvard University Press. ISBN:9780674212619.

Dupuis, C. 1984. Willi Hennig's impact on taxonomic thought. *Annual Review of Ecology, Evolution and Systematics* 15:1–24. DOI:10.1146/annurev.es.15.110184.000245

Durkheim É., and M. Mauss. 1963. *Primitive Classification*. London: Cohen & West. ISBN:9780226173344.

Dvali, G. 2013. A lecture on the hierarchy problem and gravity. *CERN-Latin-American School of High-Energy Physics, CLASHEP 2011, Proceedings*, 145–56. https://cds.cern.ch/record/2120792/files/CERN-2013-003-p145.pdf

Dwyer, P. D. 2005. Ethnoclassification, ethnoecology and the imagination. *Le Journal de la Société des Océanistes* 120–21:11–25. DOI:10.4000/jso.321

Dyke, G. J., and J. D. Sigwart. 2007. A search for a 'smoking gun': No need for an alternative to the Linnaean system of classification. In *Animal Names*, eds. A. Minelli, G. Ortalli, and G. Sanga, 49–65. Venice: Istituto Veneto di Scienze, Lettere ed Arti. ISBN:8888143386.

Easterling, W. E., and K. Kok. 2002. Emergent properties of scale in global environmental modeling: Are there any? *Integrated Assessment* 3:233–46. DOI:10.1076/iaij.3.2.233.13576

Edgar, G. 2007. *Measure, Topology, and Fractal Geometry*. New York (NY): Springer Science & Business Media. ISBN:9780387747491.

Edwards, S. V., V. V. Robin, N. Ferrand, et al. 2021. The evolution of comparative phylogeography: Putting the geography (and more) into comparative population genomics. *Genome Biology and Evolution* 14:evab176. DOI:10.1093/gbe/evab176

Eernisse, D. J., and A. G. Kluge. 1993. Taxonomic congruence versus total evidence, and amniote phylogeny inferred from fossils, molecules, and morphology. *Molecular Biology and Evolution* 10:1170–95. DOI:10.1093/oxfordjournals.molbev.a040071

Efimova, V. A. 2005. [*The Problems of Taxon Formation in Plants*.] Rostov-na-Donu: Rostov State University Publ. (in Russian).

Efremov, V. A. 2009. [A theory of concept and the conceptual space.] *Izvestia, Herzen University Journal of Humanities & Sciences*, 104:96–106. ISSN:1992-6464. (in Russian).

Ehresmann, A. C., and J.-P. Vanbremeersch. 1987. Hierarchical evolutive systems: A mathematical model for complex systems. *Bulletin of Mathematical Biology* 49:13–50. DOI:10.1016/S0092-8240(87)80033-2

Eichwald, E. 1821. *De Regni Animalis Limitibus […]*. Dorpati: Lit. Joan. Christ. Schünmanni. ISBN:9785873548835 (2011 reprint).

Eigen, M. 1983. Viral quasispecies. *Scientific American* 269:42–9. PMID:8337597.

Eigen, M., J. McCaskill, and P. Schuster. 1988. Molecular quasi-species. *The Journal of Physical Chemistry* 92:6881–91. DOI:10.1021/j100335a010

Einheuser, I. 2005. Varieties of relativism: Indexical, propositional and factual. In *Relative Truth*, eds. M. García-Carpintero, and M. Kölbel, 187–203. Oxford: Oxford University Press. ISBN:9780199234943.

Eldredge, N. 1985. *Unfinished Synthesis: Biological Hierarchies and Modern Evolutionary Thought*. New York (NY): Oxford University Press. ISBN:9780195036336.

Eldredge, N. 2016. The checkered career of hierarchical thinking in evolutionary biology. In *Evolutionary Theory: A Hierarchical Perspective*, eds. N. Eldredge, T. Pievani, E. Serrelli, et al., 1–16. Chicago (IL): The University of Chicago Press. ISBN:9780226426051.

Eldredge, N., and J. Cracraft. 1980. *Phylogenetic Patterns and the Evolutionary Process*. New York (NY): Columbia University Press. ISBN:023103802X.

Eldredge, N., T. Pievani, E. Serrelli, et al. (eds.). 2016. *Evolutionary Theory: A Hierarchical Perspective*. Chicago (IL): The University of Chicago Press. ISBN:9780226426228.

Ellen, R. F. 2008. *The Categorical Impulse: Essays on the Anthropology of Classifying Behavior*. Oxford: Berghahn Books. ISBN:9781845451554.

Ellis, B. 2002. *The Philosophy of Nature. A Guide to the New Essentialism*. Chesham (UK): Acumen Publication. ISBN:9781902683621.

Emrouznejad, A., and W. Ho (eds.). 2018. *Fuzzy Analytic Hierarchy Process*. Boca Raton (FL): CRC Press. ISBN:9781498732468.

Enfield, N. J. 2023. Scale in language. *Cognitive Science* 47:e13341. DOI:10.1111/cogs.13341

Engels, F. 1964. [*Dialectics of Nature*.] Moscow: Political Literature Publ. (in Russian).

Engler, H. C. A. 1881. Über die morphologische Verhältnisse und die geographische Verbreitung der Gattung *Rhus* [...]. *Botanische Jahrbücher fur Systematik, Pflanzengeschichte und Pflanzengeographie* 1:365–427. ISBN:9783750133099.

Epstein, V. M. 2002. [*Philosophy of Systematics. Book 3. Modern Problems of the Theory of Systematics*.] Moscow: KMK Scientific Press. ISBN:5873171173. (in Russian).

Ereshefsky, M. 1994. Some problems with the Linnaean hierarchy. *Philosophy of Science* 61:186–205. DOI:10.1086/289795

Ereshefsky, M. 1997. The evolution of the Linnaean hierarchy. *Biology and Philosophy* 12:493–519. DOI:10.1023/A:1006556627052

Ereshefsky, M. 2001a. Philosophy of biological classification. In *Encyclopedia of Life Sciences*. https://mrw.interscience.wiley.com/emrw/9780470015902/els/article/a0003447/current/abstract

Ereshefsky, M. 2001b. Names, numbers and indentations: A guide to post-Linnaean taxonomy. *Studies in History and Philosophy of Science* 32:361–83. DOI:10.1016/S1369-8486(01)00004-8

Ereshefsky, M. 2002. Linnaean ranks: Vestiges of a bygone era. *Philosophy of Science* 9:305–15. DOI:10.1086/341854

Ereshefsky, M. 2004. *The Poverty of the Linnaean Hierarchy: A Philosophical Study of Biological Taxonomy*, 2d ed. Cambridge (UK): Cambridge University Press. ISBN:9780521038836.

Ereshefsky, M. 2007a. Species, taxonomy, and systematics. In *Philosophy of Biology*, eds. M. Matthen, C. Stephens, D. M. Gabbay, et al., 403–27. Amsterdam: North-Holland. ISBN:9780444515438.

Ereshefsky, M. 2007b. Psychological categories as homologies: Lessons from ethology. *Biology and Philosophy* 22:659–74. DOI:10.1007/s10539-007-9091-9

Ereshefsky, M. 2008. Systematics and taxonomy. In *A Companion to the Philosophy of Biology*, eds. S. Sarkar, and A. Plutynski, 99–118. Oxford: Blackwell Publication. ISBN:9781405125727.

Ereshefsky, M. 2009. Darwin's solution to the species problem. *Synthese* 175:405–25. DOI:10.1007/s11229-009-9538-4
Ereshefsky, M. 2010a. Microbiology and the species problem. *Biology and Philosophy* 25:553–68. DOI:10.1007/s10539-010-9211-9
Ereshefsky, M. 2010b. Homology: Integrating phylogeny and development. *Biological Theory* 4:225–9. DOI:10.1162/biot.2009.4.3.225
Ereshefsky, M. 2011. Mystery of mysteries: Darwin and the species problem. *Cladistcis* 27:167–79. DOI:10.1111/j.1096-0031.2010.00311.x
Ereshefsky, M. 2017. Species. In *Stanford Encyclopedia of Philosophy*, ed. E. N. Zalta. https://plato.stanford.edu/entries/species/
Erismann, C. 2007. The logic of being: Eriugena's dialectical ontology. In *The Many Roots of Medieval Logic: The Aristotelian and the Non-Aristotelian Traditions*, ed. J. Marenbon, 73–88. Leiden: E. J. Brill. ISBN:9789004164871.
Estabrook, G. F. 1972. Cladistic methodology: A discussion of the theoretical basis for the induction of evolutionary history. *Annual Review of Ecology, Evolution and Systematics* 3:427–56. DOI:10.1146/annurev.es.03.110172.002235
Evangeliou, C. 1996. *Aristotle's Categories and Porphyry*. Leiden: E.J. Brill. ISBN: 9789004085381.
Evans, C.L., S. J. Greenhill, J. Watts, et al. 2021. The uses and abuses of tree thinking in cultural evolution. *Philosophical Transactions of the Royal Society B* 376:20200056. DOI:10.1098/rstb.2020.0056
Fábregas-Tejeda, A., and F. Vergara-Silva. 2017. Hierarchy theory of evolution and the extended evolutionary synthesis: Some epistemic bridges, some conceptual rifts. *Evolutionary Biology* 45: 127–39. DOI:10.1007/s11692-017-9438-3
Faegri, K. 1937. Some fundamental problems of taxonomy and phylogenetics. *The Botanical Review* 3:400–23. DOI:10.1007/BF02870489
Farris, J. S. 1969. A successive approximations approach to character weighting. *Systematic Zoology* 18:374–85. DOI:10.2307/2412182
Farris, J. S. 1982. Outgroup and parsimony. *Systematic Zoology* 31:328–34. DOI:10.1093/sysbio/31.3.328
Farris, J. S. 1983. The logical basis of phylogenetic analysis. In *Advances in Cladistics*, Vol. 2, eds. N. I. Platnick, and V. A. Funk, 7–36. New York (NY): Columbia University Press. ISBN:9780231048088
Farris, J. S. 2008. Parsimony and explanatory power. *Cladistics* 24:825–47. DOI:10.1111/j.1096-0031.2008.00214.x
Felsenstein, J. 1982. Numerical methods for inferring evolutionary trees. *The Quarterly Review of Biology* 57:379–404. DOI:10.1086/412935
Felsenstein, J. 1983. Parsimony in systematics: Biological and statistical issues. *Annual Review of Ecology, Evolution and Systematics* 14:313–33. DOI:10.1146/annurev.es.14.110183.001525
Felsenstein, J. 2004. *Inferring Phylogenies*. Sunderland (MA): Sinauer Assoc. ISBN: 9780878931774
Feyerabend, P. 1999. *Knowledge, Science, and Relativism: 1960–1980*. New York (NY): Cambridge University Press. ISBN:9780521641296.
Filippov, A. E. 1984. [Discrete speciation and Willis' law.] *Zhurnal Obshchei Biologii* 45:410–8. ISSN:0044-4596. (in Russian).
Fischer, G. 1811. *Systema Oryctognosiae*. Moscou: l'Universite Imperiale.
Fisher, D. 2008. Stratocladistics: Integrating temporal data and character data in phylogenetic inference. *Annual Review of Ecology Evolution and Systematics* 39:365–85. DOI:10.1146/annurev.ecolsys.38.091206.095752

Fisher, R. A., A. S. Corbet, and C. B. Williams. 1943. The relation between the number of species and the number of individuals in a random sample of an animal population. *Journal of Animal Ecology* 12:42–58. DOI:10.2307/1411

Fisler, M., C. Crémière, P. Darlu, et al. 2020. The treeness of the tree of historical trees of life. *PLoS ONE* 15:e0226567. DOI:10.1371/journal.pone.0226567

Fitch, W. M., and E. Margoliash. 1967. Construction of phylogenetic trees. *Science* 155:279–84. DOI 10.1126/science.155.3760.279

Fitzhugh, K. 2006. The abduction of phylogenetic hypotheses. *Zootaxa* 1145:1–110. DOI:10.11646/zootaxa.1145.1.1

Fitzhugh, K. 2016. Dispelling five myths about hypothesis testing in biological systematics. *Organisms, Diversity, & Evolution* 16:443–65. DOI:10.1007/s13127-016-0274-6

Flesness, N. R. 2003. International Species Information System (ISIS): Over 25 years of compiling global animal data to facilitate collection and population management. *International Zoo Yearbook* 38:53–61. DOI:10.1111/j.1748-1090.2003.tb02064.x

Floridi, L. 2008. The method of levels of abstraction. *Minds & Machines* 18:303–29. DOI:10.1007/s11023-008-9113-7

Floridi, L., and J. W. Sanders. 2004. Levellism and the method of abstraction. *IEG Research Report 22.11.04*, ed. G. M. Greco. Oxford: Oxford University Press. https://philpapers.org/archive/FLOLAT.pdf

Foucault, M. 1970. *The Order of Things. An Archaeology of the Human Sciences*. New York (NY): Pantheon Books. ISBN:9780679753353.

Francis, A., K. T. Huber, and V. Moulton. 2018. Tree-based unrooted phylogenetic networks. *Bulletin of Mathematical Biology* 80:404–16. DOI:10.1007/s11538-017-0381-3

Franz, N. M., R. Peet, and A. S. Weakley. 2008. On the use of taxonomic concepts in support of biodiversity research and taxonomy. In *The New Taxonomy*, ed. Q. D. Wheeler, 63–86. Boca Raton (FL): CRC Press. ISBN:9780849390883.

Franz, N. M., and R. K. Peet 2009. Towards a language for mapping relationships among taxonomic concepts, *Systematics and Biodiversity* 7:5–20, DOI: 10.1017/S147720000800282X

Franz, N. M., and D. Thau. 2010. Biological taxonomy and ontology development: Scope and limitations. *Biodiversity Informatics* 7:45–66. DOI:10.17161/bi.v7i1.3927

French, R. 1994. *Ancient Natural History. Histories of Nature*. London: Routledge. ISBN:9780415115452.

Friedman, M. 1983. *Foundations of Space-Time Theories*. Princeton (NJ): Princeton University Press. ISBN:9780691610429.

Frigg, R., J. Berkovitz, and F. Kronz. 2020. The ergodic hierarchy. In *The Stanford Encyclopedia of Philosophy*, ed. E. N. Zalta. https://plato.stanford.edu/archives/fall2020/entries/ergodic-hierarchy/

Frigg, R., and J. Nguyen. 2016. Scientific representation. In *The Stanford Encyclopedia of Philosophy*, ed. E. N. Zalta. https://plato.stanford.edu/entries/scientific-representation/

Frontier, S. 1985. Diversity and structure in aquatic ecosystems. *Oceanography and Marine Biology: An Annual Review* 23:253–312. https://api.semanticscholar.org/CorpusID:88644378

Fuchs, L. 1542. *De Historia Stirpium Commentarii Insignes Maximis Impensis Et Vigiliis Elaborati [...]*. Basileae [Basel]: Isingriniana.

Futuyama, D. J. 1979. *Evolutionary Biology*. Sunderland: Sinauer Assoc. ISBN:9780878931996.

Galich, A. 1985. *Songs and Poems*. Ann Arbor (MI): Ardis. ISBN:9780882339528.

Galich, A. 2006. [*Poems and Poetries*.] St. Petersburg: Akademicheskii Proect Publication. ISBN:9785733103051. (in Russian).

García-Carpintero, M., and M. Kölbel (eds.). 2008. *Relative Truth*. Oxford: Oxford University Press. ISBN:9780199234943.

Gärdenfors, P. 2000. *Conceptual Spaces*. Cambridge (MA): MIT Press. ISBN:9780262071994.

Gardner, H. 1985. *The Mind's New Science. A History of the Cognitive Revolution*. New York (NY): Basic Books Publ. ISBN:0465046347.

Gardner, M. 1958. *Logic Machines and Diagrams*. New York (NY): McGraw-Hill Book. ISBN:978710804099 (1982 reprint).

Garey, M. R., and D. S. Johnson. 1979. *Computer and Intractability: A Guide to the Theory of NP-Completeness*. San Francisco (CA): W.H. Freeman. ISBN:9780716710455.

Garnett, S. T., L. Christidis, S. Conix, et al. 2020. Principles for creating a single authoritative list of the world's species. *PLoS Bioliogy* 18:e3000736. DOI:10.1371/journal.pbio.3000736

Gascuel, O. 2000. On the optimization principle in phylogenetic analysis and the minimum-evolution criterion. *Molecular Biology and Evolution* 17:401–5. DOI:10.1093/oxfordjournals.molbev.a026319

Gasparski, W. 1994. The concept of hierarchy in general systems theory. *Studies in Human Ecology.* 11:141–8. PMID:7633484. https://pubmed.ncbi.nlm.nih.gov/7633484/

Gauthier, J., R. Estes, and K. de Queiroz. 1988. A phylogenetic analysis of Lepidosauromorpha. In *The phylogenetic Relationships of the Lizard Families*, eds. R. Estes, and G. Pregill, 15–98. Stanford (CA): Stanford University Press. ISBN:9780804714358.

Gay, B., and S. Weaver. 2011. Theory building and paradigms: A primer on the nuances of theory construction. *American International Journal of Contemporary Research* 1:24–32. https://www.aijcrnet.com/journals/Vol_1_No_2_September_2011/4.pdf

Gaydenko, P. P. 2000. [*The history of New European Philosophy in its Connection with Science*.] Moscow: PER SE. ISBN:9785987120743. (in Russian).

Gaydenko, P. P. 2003. [*Scientic Rationality and Philosophical Mind*.] Moscow: Progress-Traditsia. ISBN:5898261427. (in Russian).

Gaydenko, V. P. 2001. [On the basic concepts of the doctrine of Thomas Aquinas.] In [*Thomas Aquinas. Ontology and Theory of Knowledge: Fragments of Essays*], 3–32. Moscow: Insitute of Philosophy. ISBN:520102050X. (in Russian).

Gayon, J. 1998. *Darwinism's Struggle for Survival: Heredity and the Hypothesis of Natural Selection*. Cambridge (UK): Cambridge University Press. ISBN:9780521039673.

Gecow, A. 2025. Two coherent definitions of the life process derived from the half-chaos theory and the (unintentional) purposeful information theory. *BioSystems* 256:105533. DOI:10.1016/j.biosystems.2025.105533

Geeraerts, D. (ed.). 2006. *Cognitive Linguistics: Basic Readings*. Berlin: Mouton de Gruyter. ISBN:9783110190847.

Gelashvili, D. B., V. N. Yakimov, D. I. Iudin, et al. 2010. [Fractal aspects of taxonomic diversity.] *Zhurnal Obshchei Biologii* 71:115–30. ISSN:0044-4596. (in Russian).

Gelashvili, D. B., D. I. Yudin, G. S. Rosenberg, et al. 2013. [*Fractals and Multifractals in Bioecology*.] Nizhny Novgorod: Nizhny Novgorod State University Publ. ISBN:9785913262462. (in Russian).

Geoffroy de Saint-Hilaire, É. 1830. *Principes de Philosophie Zoologique [...]*. Paris: Pichon et Didier. ISBN:9780331083569 (2018 reprint).

Geoffroy, M. and Berendsohn W. G. 2003. The concept problem in taxonomy: Importance, components, approaches. In *MoReTax: Handling Factual Information Linked to Taxonomic Concepts in Biology*, ed. W. G. Berendsohn, 5–14. Bonn: Federal Agency for Nature Conservation. https://www.researchgate.net/publication/285456910_The_concept_problem_in_taxonomy_Importance_components_approaches

Gesner, C. 1542. *Catalogus Plantarum Latinè, Graecè, Germanicè, & Gallicè [...]*. Zürich: Christoph. Froschoverum.

Gesner, C. 1560. *Nomenclator aquatilium animantium. Icones animalium aquatilium in mari [...]*. Zürich: Christoph. Froschoverum. ISBN:9785519152907 (2015 reprint).

Ghilarov, A. M. 1996. What does 'biodiversity' mean: Scientific problem or convenient myth? *Trends in Ecology and Evolution* 11:304–06. DOI:10.1016/0169-5347(96)20016-3

Ghilarov, A. M. 2001. [The relationship of biodiversity with productivity: Science and politics]. *Priroda* 2:20–4. https://vivovoco.astronet.ru/VV/JOURNAL/NATURE/02_01/GILYAROV.PDF (in Russian).

Ghilarov, M. S. 1964. [Modern concepts of homology.] *Uspekhi Sovremennoi Biologii* 57:300–16. ISSN:0042-1324. (in Russian).

Ghiselin M.T. 1969. The principles and concepts of systematic biology. In *Systematic Biology: Proceedings of the International Conference*, 45–66. Washington (DC): The National Academy of Science. ISBN:9780309016926.

Ghiselin, M. T. 1997. *Metaphysics and the Origin of Species*. New York (NY): State University of New York Press. ISBN:9780791434680.

Ghiselin, M. T. 2005. Homology as a relation of correspondence between parts of individuals. *Theory in Biosciences* 124:91–103. DOI:10.1007/BF02814478

Gibbs, R. W. (Ed.). 2008. *The Cambridge Handbook of Metaphor and Thought*. Cambridge (UK): Cambridge University Press. ISBN:9780521841061.

Gilbert, S. F., and J. A. Bolker. 2001. Homologies of process and modular elements of embryonic construction. *Journal of Experimental Zoology* 291:1–2. DOI:10.1002/jez.1

Gildea, S., and F. Zúñiga. 2016. Referential hierarchies: A new look at some historical and typological patterns. *Linguistics* 54:483–529. DOI:10.1515/ling-2016-0007

Gill, T. 1872. Arrangement of the families of fishes. *Smithsonian Miscellaneous Collections* 247:1–95. https://www.biodiversitylibrary.org/page/17584650#page/5/mode/1up

Gilmour, J. S. L. 1940. Taxonomy and philosophy. In *The New Systematics*, ed. J. Huxley, 461–74. Oxford (UK): Oxford University Press. ISBN:9780403017867.

Gilmour, J. S. L. 1961. Taxonomy. In *Contemporary Botanical Thought*, ed. A. M. MacLeod, and L. S. Cobley, 27–45. Chicago (IL): Quadrangle Book. ASIN B0000CL6NB.

Gilson, E. 1955. *History of Christian Philosophy in the Middle Ages*. New York (NY): Sheed and Ward. ISBN:9780394428666.

Gilson, E. 1960. *The Christian Philosophy of Saint Augustine*. New York (NY): Random House. ISBN:9780394419305.

Gilson, E. 1984. *From Aristotle to Darwin and Back Again. A Journey in Final Causality, Species, and Evolution*. Notre Dame (IN): University of Notre Dame Press. ISBN:9780722035177.

Gilson, E. 1991. *The Spirit of Mediaeval Philosophy*. Notre Dame (IN): University of Notre Dame Press. ISBN:9780268017408.

Giribet, G., G. Hormiga, and G. D. Edgecombe. 2016. The meaning of categorical ranks in evolutionary biology. *Organisms Diversity & Evolution* 16:427–30. DOI:10.1007/s13127-016-0263-9

Giseke, P. D. 1792. *Caroli a Linné […] Praelectiones in Ordines Naturales Plantarum*. Hamburg: Benj. Gottl. Hoffmanni. ISBN:9781343182158 (2015 reprint).

Gladyshev, G. P. 2017. Hierarchical thermodynamics: Foundation of extended Darwinism. *Imperial Journal of Interdisciplinary Research* 3:1576–88. ISSN:2454-1362.

Gogarten, J. P. 1994. Which is the most conserved group of proteins? Homology-orthology, paralogy, xenology, and the fusion of independent lineages. *Journal of Molecular Evolution* 39:541– 3. DOI:10.1007/bf00173425

Goldfuss, G. A. 1817. *Über die Entwicklungsstufen des Thieres*. Nürnberg: Leonhard Schrag. ISBN:9781248784426 (2012 reprint).

Golovanov, M. 2004. [Hierarchical data structures in relational BD.] *RSDN Magazine* #1. https://www.opennet.ru/docs/RUS/hierarchy_sql/. (in Russian).

Gontier, N. 2011. Depicting the Tree of Life: The philosophical and historical roots of evolutionary tree diagrams. *Evolution, Education and Outreach* 4:515–38. DOI 10.1007/s12052-011-0355-0

Gontier, N. (ed.). 2015a. *Reticulate Evolution, Lateral Gene Transfer, Hybridization and Infectious Heredity*. Cham (Switzerland): Springer International Publication. ISBN: 9783319163444.

Gontier, N. 2015b. Uniting micro- with macroevolution into an extended synthesis: Reintegrating life's natural history into evolution studies. . In *Macroevolution. Explanation, Interpretation and Evidence*, eds. E. Serrelli, and N. Gontier, 227–278. Cham (Switzerland): Springer International Publishing Switzerland. ISBN:9783319150444.

Gonzáles, G. L. 2015. Homology, an (un)solved problem. *Teorema* 34:211–23. DOI:10.1080/09515089.2014.949004

Göpel, T. 2024. Synthesizing foundations in evolutionary morphology and a plea for conceptual explicitness regarding homology concepts. *Frontiers in Ecology and Evolution* 12:1343346. DOI:10.3389/fevo.2024.1343346

Göpel, T., and S. Richter. 2016. The word is not enough: On morphemes, characters and ontological concepts. *Cladistics* 32:682–90. DOI:10.1111/cla.12145

Gordon, A. D. 1980. On the assessment and comparison of classifications. In *Analyse de Donnees et Informatique*, ed. R. Tomassone, 149–60. Le Chesnay (France): INRIA. ISBN:9782726102329.

Gordon, A. D. 1987. A review of hierarchical classification. *Journal of the Royal Statistical Society, Ser. A (General)* 150:119–37. DOI:10.2307/2981629

Górska, E. 2001. The approach of cognitive grammar to the part-whole relation. *Anglica* 11:85–99. https://www.academia.edu/38176472/_The_approach_of_cognitive_grammar_to_the_part_whole_relation_

Górska, E. 2002. On partonomy and taxonomy. *Studia Anglica Posnaniensia* 39:103–11. http://ifa.amu.edu.pl/sap/files/39/05Gorska.pdf

Gotthelf, A., and J. G. Lennox (eds.). 1987. *Philosophical Issues in Aristotle's Biology*. Cambridge (UK): Cambridge University Press. ISBN:9780521310918.

Gould, S. J. 2002. *The Structure of Evolutionary Theory*. Cambridge (MA): Harvard University Press. ISBN:9780674006133.

Gouvêa, D. Y. 2023. Historicizing the homology problem. *Studies in History and Philosophy of Science* 99:56–66. DOI:10.1016/j.shpsa.2023.01.005

Gower, J. C. 1988. Numerical techniques as an aid to objectivity. In *Prospects in Systematics*, ed. D. L. Hawksworth, 234–51 Oxford: Clarendon Press. ISBN:9780198577072.

Grande, L., and O. Rieppel. 1994. *Interpreting the Hierarchy of Nature: From Systematic Patterns to Evolutionary Process Theories*. San Diego (CA): Academic Press. ISBN:9780122951206.

Grant, T., and A. G. Kluge. 2004. Transformation series as an ideographic character concept. *Cladistics* 20:32–41. DOI:10.1111/j.1096-0031.2004.00003.x

Grant, V. 1963. *The Origin of Adaptations*. New York (NY): Columbia University Press. ISBN:9780231086486.

Greene, E. L. 1909. *Landmarks of Botanical History. A Study of Certain Epochs in the Development of the Science of Botany. Pt 1. Prior to 1562*. Washington (DC): Smithsonian Miscellaneous Collections 54. ISBN:9781343207356.

Greene, E. L. 1910. Certain aspects of the species question. *The American Midland Naturalist* 1:245–63. ISSN:0003-0031.

Greene, R. M. 1984. Structuralism and its heuristic implications. *Learning Disability Quarterly* 7:359–62. DOI:10.2307/1510236

Gregg, J. R. 1950. Taxonomy, language and reality. *American Naturalist* 84:419–35. DOI:10.1086/281639

Gregg, J. R. 1954. *The Language of Taxonomy*. New York (NY): Columbia University Press. ISBN:9780804728058.

Gregg, J. R. 1967. Finite linnaean structures. *The Bulletin of Mathematical Biophysics* 29:191–206. DOI:10.1007/bf02476893

Gregor, S. 2006. The nature of theory in information systems. *Management Information Systems Quarterly* 30:614–42. DOI:10.2307/25148742

Grene, M. 1969. Hierarchy: One word, how many concepts?. In *Hierarchical Structures*, eds. L. L. Whyte, A. G. Wilson, and D. Wilson, 56–8. New York (NY): Elsevier. ISBN:9780444000699.

Grene, M. 1974. Is genus to species as matter to form? Aristotle and taxonomy. *Synthese* 28:51–69. DOI:10.1007/978-94-010-2224-8_6

Grene, M. 1987. Hierarchies in biology. *American Scientist* 75:504–10. https://www.jstor.org/stable/27854792

Greuter, W. 2004. Recent developments in international biological nomenclature. *Turkish Journal of Botany* 28:17–26. ISSN:1300-008X.

Greuter, W., D. L. Hawksworth, J. McNeill, et al. 1998. Draft BioCode (1997): The prospective international rules for the scientific names of organisms. *Taxon* 47:127–50. DOI:10.2307/1224030

Griffing, L. R. 2011. Who invented the dichotomous key? Richard Waller's watercolors of the herbs of Britain. *American Journal of Botany* 98:1911–23. DOI:10.3732/ajb.1100188

Griffiths, A. J. 1981. A numericlature of the yeasts. *Antonie Van Leeuwenhoek* 47:547–63. DOI:10.1007/BF00443241

Griffiths, G. C. D. 1973. Some fundamental problems in biological classification. *Systematic Zoology* 22:338–43. DOI:10.2307/2412942

Griffiths, G. C. D. 1974. On the foundations of biological systematics. *Acta Biotheoretica* 23:85–131. DOI:10.1007/bf01556343

Griffiths, G. C. D. 1976. The future of Linnean nomenclature. *Systematic Zoology* 25:168–73. DOI:10.2307/2412743

Griffiths, P. E. 1996. Darwinism, process structuralism, and natural kinds. *Philosophy of Science* 63:S1–S9. DOI:10.1086/289930

Griffiths, P. E. 1999. Squaring the circle: Natural kinds with historical essences. In *Species: New Interdisciplinary Essays*, ed. R. A. Wilson, 209–28. Cambridge (MA): MIT Press. ISBN:9780262731232.

Griffiths, P. E. 2007. The phenomena of homology. *Biology and Philosophy* 22:643–58. DOI:10.1007/s10539-007-9090-x

Grinin, L. E., A. V. Markov, and A. V. Korotayev. 2009. Aromorphoses in biological and social evolution: Some general rules for biological and social forms of macroevolution. *Social Evolution & History* 8:6–50. ISSN:1681-4363.

Grötschel, M., and J. Lügger. 1999. Scientific information systems and metadata. In *Classification in the Information Age*, eds. W. Gaul, and H. Locarek-Junge, 3–20. Berlin: Springer-Verlag. ISBN:9783540658559.

Grover, N. 2014. A study of various fuzzy clustering algorithms. *International Journal of Engineering Research* 3:177–81.

Gumperz, J. J., and S. C. Levinson. 1996. *Rethinking Linguistic Relativity*. Cambridge (UK): Cambridge University Press. ISBN:9780521448901.

Gunderson, L., and C. S. Holling (eds.). 2001. *Panarchy: Understanding Transformations in Systems of Humans and Nature*. Washington (DC): Island Press. ISBN:9781559638579.

Gunderson, L. H., C. R. Allen, and A. Garmestani (eds.). 2022. *Applied Panarchy: Applications and Diffusion across Disciplines*. Washington (DC): Island Press. ISBN:9781642830903.

Gusev, P. G. 1995. [Classification in pre-literate cultures.] In [*Towards a Theory of Classification*,] ed. S. S. Rozova, 44–59. Novosibirsk: Novosibirsk State University Publ. ISBN:5887420049. (in Russian).

Gutmann, W.F. 1977. Phylogenetic reconstruction: Theory, methodology, and application to chordate evolution. In *Major Patterns in Vertebrate Evolution*, eds. M. K. Hecht, P. C. Goody, and B. M. Hecht, 645–70. New York (NY): Plenum Press. ISBN:9781468488531.

Hacking, I. 1983. *Representing and Intervening. Introductory Topics in the Philosophy of Natural Science*. Cambridge (UK): Cambridge University Press. ISBN:9780521282468.

Hacking, I. 2007. Trees of logic, trees of Porphyry. In *Advancements of Learning. Essays in Honour of Paolo Rossi*, ed. J. L. Heilbron, 221–63. Florence: Leo S. Olschki. ISBN:9788822256553.

Haeckel, E. 1866. *Generelle Morphologie der Organismen. Allgemeine Grundzuge der organischenFormen-Wissenschaft[...]*,Bd.1,2.Berlin:GeorgReimerVerlag.ISBN:97831 10101850 (1988 reprint).

Haeckel, E. 1891. *Anthropogenie Oder Entwickelungsgeschichte des Menschen: Keimes- und Stammesgeschichte*, Bd. 2. Leipzig: Wilhelm Engelmann Verlag. ISBN:9783743442276 (2016 reprint).

Haeckel, E. 1895. *Systematische Phylogenie. Entwurf eines Natürlichen Systems der Organismen auf Grund ihrer Stammesgeschichte, T. 3 [...]*. Berlin: Georg Reimer Verlag. ISBN:9781332454013 (2018 reprint).

Hagedorn, G., G. Rambold, and S. Martellos. 2010. Types of identification keys. In *Tools for Identifying Biodiversity: Progress and Problems*, eds. P. L. Nimis, and L. R. Vignes, 59–64. ISBN:9788883032950.

Hall, B. K. (ed.). 1994a. *Homology, the Hierarchical Basis of Comparative Biology*. San Diego (CA): Academic Press. ISBN:9780123189202.

Hall, B. K. 1994b. Introduction. In *Homology: The Hierarchical Basis of Comparative Morphology*, ed. B. K. Hall, 1–19. London: Academic Press. ISBN:9780123195838.

Hall, B. K. 1996. Baupläne, phylotypic stages, and constraint: Why there are so few types of animals? *Evolutionary Biology* 29:251–61. DOI:10.1007/s12052-012-0424-z

Hall, B. K. 2007. Homology and homoplasy. In *Philosophy of Biology*, eds. M. Matthen, and, C. Stephens, 429–53. Amsterdam: Elsevier. ISBN:9780444515438.

Hall, B. K. 2012. Evolutionary developmental biology (evo–devo): Past, present, and future. *Evolution, Education and Outreach* 5:184–93. DOI:10.1007/s12052-012-0418-x

Hallam, A., and P. B. Wignall. 1997. *Mass Extinctions and Their Aftermath*. Oxford (UK): Oxford Univ. Press. ISBN:9780198549161.

Hamilton, A. 2014. Historical and conceptual perspectives on modern systematics: Groups, ranks, and the phylogenetic turn. In *The Evolution of Phylogenetic Systematics*, ed. A. Hamilton, 89–115. Berkeley (CA): University of California Press. ISBN:9780520276581.

Hammand, E. B. 1918. The tree of knowledge. *The Sewanee Review* 26:168–72.

Hampton, A. J. B., and J. P. Kenney. 2021. *Christian Platonism: A History*. Cambridge (UK): Cambridge University Press. ISBN:9781108491983.

Hansell, R. I. C., and B. Ewing. 1973. The detection and estimation of character weighting in classifications. *Journal of Theoretical Biology* 39:297–314. DOI:10.1016/0022-5193(73)90099-4

Harris, E. S. J., and B. D. Mishler. 2009. The delimitation of phylogenetic characters. *Biological Theory* 4:230–4. DOI:10.1162/biot.2009.4.3.230

Haszprunar, G. 1986. Die Klado-evolutionäre Klassifikation-Versuch einer Synthese. *Zeitschrift für zoologische Systematik und Evolutionsforschung* 24: 89–109. ISSN:0044-3808.

Haszprunar, G. 1991. The types of homology and their significance for evolutionary biology and phylogenetics. *Journal of Evolutionary Biology* 5:13–24. DOI:10.1046/J.1420-9101.1992.5010013.X

Hathaway, R. F. 1969. *Hierarchy and the Definition of Order in the Letters of Pseudo-Dionysius*. Hague: Marlinus Nijhott. ISBN:9789401184687.

Hawking, S., and G. Ellis. 1975. *The Large Scale Structure of Space-Time*. Cambridge (UK): Cambridge University Press. ISBN:9780521099066.

Hawkins, H. L. 1935. The species problem. *Nature*, 136:574–5. DOI:10.1038/136574a0

Hawksworth, D. L., and F. A. Bisby. 1988. Systematics: The keystone of biology. In *Prospects in Systematics*, ed. D. L. Hawksworth, 3–30. Oxford: Clarendon Press. ISBN:9780198577072.

Hawksworth D. L., J. McNeill, P. H. A. Sneath, et al. 1994. Towards a harmonized bionomenclature for life on Earth. [...]. *Bulletin of Zoological Nomenclature* 51:188–216. ISSN:0007-5167.

Hawley, K., and A. Bird. 2011. What are natural kinds? *Philosophical Perspectives* 25:205–21. DOI:10.1111/j.1520-8583.2011.00212.x

Heard, S.B., D. L. Hauser 1995. Key evolutionary innovations and their ecological mechanisms. *Historical Biology* 10:151–73. DOI:10.1080/10292389509380518

Hedden, H. 2010. *The Accidental Taxonomist.* Medford (NJ): Information Today Inc. ISBN:9781573873970.

Heise, H., and M. P. Starr. 1968. Nomenifers: Are they christened or classified? *Systematic Zoology* 17:458–67. DOI:10.1093/sysbio/17.4.458

Heisenberg, W. 2007. *Physics and Philosophy: The Revolution in Modern Science.* New York (NY): Harper Perennial. ISBN:9780061209192.

Heiss, J., D. H. Huson, and M. Steel. 2025. Transformations to simplify phylogenetic networks. *Bulletin of Mathematical Biology* 87:20. DOI:10.1007/s11538-024-01398-7

Heljakka, A., M. Trapp, J. Kannala, et al. 2023. Disentangling model multiplicity in deep learning. *ArXiv:2206.08890v2 [cs.LG].* DOI:10.48550/arXiv.2206.08890

Heller, J. L. 1964. The early history of binomial nomenclature. *Huntia* 1:33–70. www.huntbotanical.org/admin/uploads/06hibd-huntia-1-pp33-70.pdf

Hellerstein, J. M., M. Stonebraker, and J. Hamilton. 2007. *Architecture of a Database System.* Norwell (MA): Now Publishers Inc. ISBN:9781601980786.

Hellström, N. P. 2012. Darwin and the Tree of Life: The roots of the evolutionary tree. *Archives of Natural History* 39:234–52. DOI:10.3366/anh.2012.0092

Hempel, C. G. 1966. *Philosophy of Natural Science.* Englewood Cliffs (NJ): Prentice-Hall. ISBN:9780136638230.

Henle, K., S. G. Potts, W. E. Kunin, et al. (eds.). 2014. *Scaling in Ecology and Biodiversity Conservation.* Sofia: Pensoft Publ. ISBN:9789546427397.

Hennig, W. 1950. *Grundzüge einiger Theorie der phylogenetische Systematik.* Berlin: Deutscher Zentralverlag. ISBN:9783874291880 (1980 reprint).

Hennig, W. 1965. Phylogenetic systematics. *Annual Review of Entomology* 10:97–116. (reprinted in Sober 1994, pp. 257–76).

Hennig, W. 1966. *Phylogenetic Systematics.* Urbana (IL): University Illinois Press. ISBN:9780252068140.

Hennig, W. 1969. *Die Stammesgeschichte der Insekten.* Frankfurt: Verlag Waldemar Kramer. ISBN:9780471278481.

Herbst, Ph. G. 1976. *Alternatives to Hierarchies.* Leiden: H. E. Steofert Kroese. ISBN:9781468469479.

Hester, P.T., and K. M. Adams. 2014. *Systemic Thinking: Fundamentals for Understanding Problems and Messes.* Cham (Switzerland): Springer International Publ. ISBN:9783319076287.

Hey, J. 2001. *Genes, Categories, and Species. The Evolutionary and Cognitive Cause of the Species Problem.* New York (NY): Oxford University Press. ISBN:9780195144772.

Hey, J., R. Waples, M. Arnold, et al. 2003. Understanding and confronting species uncertainty in biology and conservation. *Trends in Ecology and Evolution* 18:597–603. DOI:10.1016/j.tree.2003.08.014

Heyes, C. 2003. Four routes of cognitive evolution. *Psychological Reviews* 110:713–27. DOI:10.1037/0033-295X.110.4.713

Heyes, C. 2012. New thinking: The evolution of human cognition. *Philosophical Transactions of the Royal Society B* 367:2091–6. DOI:10.1098/rstb.2012.0111

Heyes, C. M., and H. C. Plotkin. 1988. Replicators and interactors in cultural evolution. In *What the Philosophy of Biology Is*, ed. M. Ruse, 139–62. Dordrecht: Kluwer Academic Publication. ISBN:9789401070201.

Hilgevoord, J., and J. Uffink. 2016. The uncertainty principle. In *The Stanford Encyclopedia of Philosophy*, eds. E. N. Zalta, and U. Nodelman. https://plato.stanford.edu/archives/spr2024/entries/qt-uncertainty/

Hjørland, B. 2009. Concept theory. *Journal of the American Society for Information Science and Technology* 60:1519–36. DOI:10.1002/asi.21082

Hjørland, B. 2008. Core classification theory: A reply to Szostak. *Journal of Documentation* 64: 333–42. DOI:10.1108/00220410810867560

Hjørland, B. 2016. Knowledge organization. *Knowledge Organization* 43:475–84. https://www.isko.org/cyclo/knowledge_organization

Hjørland, B. 2017. Classification. *Knowledge Organization* 44: 97–128. https://www.isko.org/cyclo/classification

Ho, M. W. 1988. How rational can rational morphology be? A post-Darwinian rational taxonomy based on a structuralism of process. *Theoretical Biology Forum* 81:11–55. ISSN:2283-7175.

Ho, M. W. 1992. Development, rational taxonomy and systematics. *Theoretical Biology Forum* 85:193–211. ISSN:2282-2593.

Ho, S. Y. W. 2014. The changing face of the molecular evolutionary clock. *Trends in Ecology and Evolution* 29:496–503. DOI:10.1016/j.tree.2014.07.004

Hofer-Szabó, G., M. Rédei, and L.E. Szabó. 2013. *The Principle of Common Cause*. Cambridge (UK): Cambridge University Press. ISBN:9781139094344.

Hoffman, A. 1979. Community paleoecology as an epiphenomenal science. *Paleobiology* 5:357–79. DOI:10.1017/S0094837300016894

Hogben, L. 1940. Problems of the origins of species. In *The New Systematics*, ed. J. Huxley, 270–86. London: Oxford University Press. ISBN:9780403017867.

Holland, J. H. 1992. *Adaptation in Natural and Artificial Systems: An Introductory Analysis with Applications to Biology, Control, and Artificial Intelligence*. Cambridge (MA): The MIT Press. ISBN:9780262581110.

Holland, J. H. 2014. *Complexity: A Very Short Introduction*. Oxford: Oxford University Press. ISBN:9780191639548.

Holopainen, T. 1996. *Dialectic and Theology in the Eleventh Century*. Leiden: Brill. ISBN:9789004105775.

Holt, B. G., and K. A. Jønsson. 2014. Reconciling hierarchical taxonomy with molecular phylogenies. *Systematic Biology* 63:1010–7. DOI:10.1093/sysbio/syu061

Honorius Augustodunensis. 1001–1200. n.d. *Clavis Physicae*. The open Internet source: Bibliotheca Gallica (https://gallica.bnf.fr/ark:/12148/btv1b90665224/f1.item). Holding Institution: Bibliothèque nationale de France, Département des Manuscrits.

Hörandl, E. 2006. Paraphyletic versus monophyletic taxa. *Taxon* 55:564–70. DOI:10.2307/25065631

Hörandl, E. 2010. Beyond cladistics: Extending evolutionary classifications into deeper time levels. *Taxon* 59:345–50. DOI:10.1002/tax.592001

Hossfeld, U., and L. Olsson. 2005. The history of the homology concept and the "Phylogenetisches Symposium". *Theory in Biosciences* 124:243–53. DOI:10.1007/BF02814486

Hrbacek, K., and T. Jech. 1999. *Introduction to Set Theory*. New York (NY): M. Dekker Publ. ISBN:9780585243412.

Hubbs, C. L. 1944. Concepts of homology and analogy, *The American Naturalist* 78:289–307. DOI:10.1086/281202

Hübner, K. 1988. *Critique of Scientific Reason*. Chicago (IL): The University of Chicago Press. ISBN:9780226357096.

Huelsenbeck, J. P., B. Rannala, and J. P. Masly. 2000. Accommodating phylogenetic uncertainty in evolutionary studies. *Science* 288:2349–50. DOI:10.1126/science.288.5475.2349

Hug, L. A., B. J. Baker, K. Anantharaman, et al. 2016. A new view of the tree of life. *Nature Microbiology* 11:16048. DOI:10.1038/nmicrobiol.2016.48

Hughes, A. J., and D. M. Lambert. 1984. Functionalism, structuralism, and "ways of seeing". *Journal of Theoretical Biology* 111:787–800. DOI:10.1016/S0022-5193(84)80267-2

Hull, D. L. 1966. Phylogenetic numericlature. *Systematic Zoology* 15:14–7. DOI:10.2307/sysbio/15.1.14
Hull, D. L. 1968. The syntax of numericlature. *Systematic Zoology* 17:472–4. DOI:10.1093/sysbio/17.4.472
Hull, D. L. 1970. Contemporary systematic philosophies. *Annual Review of Ecology, Evolution and Systematics* 1:19–54. DOI:10.1146/annurev.es.01.110170.000315
Hull, D. L. 1978. A matter of individuality. *Philosophy of Science* 45:335–60. DOI:10.1086/288811
Hull, D. L. 1980. Sociobiology: Another new synthesis. In *Sociobiology: Beyond Nature/Nurture?* eds. G. W. Barlow, and I. Silverberg, 77–96. Boulder (CO): Westview Press. ISBN:9780367303297.
Hull, D. L. 1988. *Science as a Process*. Chicago (IL): The University of Chicago Press. ISBN:9780226360508.
Hull, D. L. 1998. Taxonomy. In *Routledge Encyclopedia of Philosophy*. London: Taylor & Francis Group. https://www.rep.routledge.com/articles/thematic/taxonomy/v-1
Hull, D. L. 1999a. The use and abuse of Sir Karl Popper. *Biology and Philosophy* 14:481–504. DOI:10.1023/A:1006554919188
Hull, D. L. 1999b. On the plurality of species: Questioning the party line. In *Species: New Interdisciplinary Essays*, ed. R. A. Wilson, 23–48. Cambridge (MA): MIT Press. ISBN:9780262731232.
Hull, D. L. 2001. The role of theories in biological systematics. *Studies in History and Philosophy of Biological and Biomedical Sciences* 32:221–38. DOI:10.1016/S1369-8486(01)00006-1
Hull, D. L. 2002. Recent philosophy of biology: A review. *Acta Biotheoretica* 50:117–28. DOI:10.1023/a:1016337212804
Humphries, C. J. 1983. Primary data in hybrid analysis. In: *Advances in Cladistics*, Vol. 2, eds. N. I. Platnick, and V. A. Funk, 89–103. New York (NY): Columbia University Press. ISBN:9780231048088.
Humphries, C. J. 2005. Systematics: Historical overview. In *Encyclopedia of Life Sciences*. DOI:10.1038/npg.els.0003337
Hunn, E. 1977. *Tzeltal Folk Zoology: The Classification of Discontinuities in Nature*. New York (NY): Academic Press. ISBN:9780123617507.
Hunn, E. S., and D. H. French. 2000. Alternatives to taxonomic hierarchy: The Sahapin case. In *Ethnobotany: A Reader*, ed. P. E. Minnis, 118–28. Norman (OK): University of Oklahoma Press. ISBN:978-0806131801.
Hurford, J. R. 1987. *Language and Number: The Emergence of a Cognitive System*. Oxford: Basil Blackwell. ISBN:9780631155683.
Huson, D. H., and C. Scornavacca. 2011. A survey of combinatorial methods for phylogenetic networks. *Genome Biology and Evolution* 3:23–35. DOI:10.1093/gbe/evq077
Huth, M., and M. Ryan. 2004. *Logic in Computer Science: Modelling and Reasoning about Systems*, 2nd ed. Cambridge: Cambridge University Press. ISBN: 9780521543101.
Huxley, J. 1942. *Evolution, the Modern Synthesis*. London: Allen and Unwin. ISBN:9780028468006 (1975 edition).
Ilg, U. 2014. *Quasi lignum vitae*. The Tree of Life as an image of mendicant identity. In *The Tree: Symbol, Allegory, and Mnemonic Device in Medieval Art and Thought*, eds. P. Salonius, and W. Andrea, 187–212. Turnhout: Brepols Publ. ISBN:9782503548395.
Ilyin, V. V. 2003. [*Philosophy of Science*.] Moscow: Moscow State University Publ. ISBN:5211048210. (in Russian).
Ingram, A. 2024. Understanding the flatarchy organizational structure: Pros, cons, and examples. *Bonsai*. https://www.hellobonsai.com/blog/flatarchy-organizational-structure
Iordansky, N. N. 1994. [*Macroevolution. A Systemic Theory*.] Moscow: Nauka. ISBN:5020057290. (in Russian).

Iordansky, N. N. 2001. [*Evolution of Life.*] Moscow: Academia. ISBN:9785534096330. (in Russian).
Isaacson, W. 2008. *Einstein: His Life and Universe.* New York (NY): Pocket Books. ISBN:9781847390547.
Ivanter E. V. 2012. [Biodiversity: The story of a misconception.] *Printsipy Ekologii* 4:61–4. https://ecopri.ru/journal/article.php?id=1802. (in Russian).
Ivlev, Yu. V. 2025. [*Logic*,] 4th ed. Moscow: Prospect. ISBN:9785392297719. (in Russian).
Jackson, F. 1982. Epiphenomenal qualia. *The Philosophical Quarterly* 32:127–36. DOI:10.2307/2960077
Jacob, E. 1991. Classification and categorization: Drawing the line. *2nd ASIS SIG/CR Classification Research Workshop*, eds. B. H. Kwasnik, and R. Fidel, 63–80. Silver Spring (MD): American Society for Information Science. DOI:10.7152/acro.v2i1.12548
Jacob, E. 2004. Classification and categorization: A difference that makes a difference. *Library Trends*, 52:515–40. https://www.ideals.illinois.edu/handle/2142/1686
Jaeger, J. 2019. Dynamic structures in evo-devo: From morphogenetic fields to evolving organisms. In *Perspectives on Evolutionary and Developmental Biology*, ed. G. Fusco, 335–55. Padova: Padova University Press. ISBN:9788869381409.
Jahn, T. L. 1961. Man versus machine: A future problem in protozoan taxonomy. *Systematic Zoology* 10:179– 92. DOI:10.2307/2411616
Jamniczky, H. A. 2005. Biological pluralism and homology. *Philosophy of Science* 72:687–98. DOI:10.1086/508108
Jansen, L., and S. Schulz. 2014. Crisp islands in vague seas: Cases of determinate parthood relations in biological objects. In *Mereology and the Sciences: Parts and Wholes in the Contemporary Scientific Context*, eds. C. Calosi, and P. Graziani, 163–88. ISBN:9783319053561.
Jardine, C. J., N. Jardine, and R. Sibson. 1967. The structure and construction of taxonomic hierarchies. *Mathematical Biosciences* 1:173–9. DOI:10.1016/0025-5564(67)90032-6
Jardine, N. 1967. The concept of homology in biology. *British Journal for the Philosophy of Science* 18:125–39. DOI:10.1093/bjps/18.2.125
Jardine, N. 1969. A logical basis for biological classification. *Systematic Zoology* 18:37–52. DOI:10.2307/2412409
Jardine, N., and R. Sibson. 1968. A model for taxonomy. *Mathematical Biosciences* 2:465–82. DOI:10.1016/0025-5564(70)90077-5
Jardine, N., and R. Sibson. 1971. *Mathematical Taxonomy (Probability and Mathematical Statistics).* New York (NY): John Wiley & Sons. ISBN:9780471440505.
Jardine, W. 1866. Report of a Committee "appointed to report on the changes which they may consider desirable to make desirable, if any, in the Rules of Zoological Nomenclature [...]". In *Report of the Thirty-Fifth Meeting of the British Association for the Advancement of Science (1865)*, 25–42. London: John Murray. www.biodiversitylibrary.org/item/93098#page/96/mode/1up
Jech, T. 2003. *Set Theory.* Berlin: Springer-Verlag. ISBN:9783540440857.
Jenner, R. 2024. *Ancestors in Evolutionary Biology. Linear Thinking about Branching Trees.* Cambridge (UK): Cambridge University Press. ISBN:9781107105935.
Jenyns, L. 1833. Some remarks on genera and subgenera, and on the principles on which they should be established. *The Magazine of Natural History* 6:385–90. ISSN:0374-5481.
Jiang, J., A. Rago, and F. Leofante. 2024. Recourse under model multiplicity via argumentative ensembling. *ArXiv:2312.15097v2 [cs.LG].* DOI:10.48550/arXiv.2312.15097
John, E. O. 2009. Instrumentalism in science: Comments and criticisms. *Global Journal of Humanities* 8:1–9. ISSN:1596-6232.
Johnson, D. S. 1908. Aspects of the species question. *The American Naturalist* 42:217–81. ISSN:1537-5323.
Johnson, K. 2007. Natural history as stamp collecting: A brief history. *Archives of Natural History* 34:244–58. DOI:10.3366/anh.2007.34.2.244

Jonas, A. E. G. 1994. The scale politics of spatiality. *Environment and Planning D: Society and Space* 12:257–64. DOI:10.1068/d120257

Jonas, A. E. G. 2006. Pro scale: Further reflections on the 'scale debate' in human geography. *Transactions of the Institute of British Geographers, New Series* 31:399–406. DOI:10.1111/j.1475-5661.2006.00210.x

Jones, K. T. 1998. Scale as epistemology. *Political Geography* 17:25–8. DOI:10.1016/S0962-6298(97)00049-8

Jønsson, K. A., P.-H. Fabre, J. D. Kennedy, et al. 2016. A supermatrix phylogeny of corvoid passerine birds (Aves: Corvides). *Molecular Phylogenetics and Evolution* 94:87–94. DOI:10.1016/j.ympev.2015.08.020

Jorge, G., and Newton L. 2016. Medieval theories of the categories. In *Stanford Encyclopedia of Philosophy*, ed. E. N. Zalta. https://plato.stanford.edu/archives/win2016/entries/medieval-categories/

Jung, J. (1662) 1747. *Ioachimi Iungi lubecensis [...] Opuscula Botanico-Physica.* Coburg: Typus Georgii Ottonis. ISBN:9781173363604 (2011 reprint).

Junkkari, M., and M. Niinimäki. 1999. A path-oriented approach to hierarchical concept structures. *PennState CiteSeerX.* https://webpages.tuni.fi/utacs_history/cs/reports/pdf/A-1998-4.pdf

Jussieu, A. L. 1789 *Genera Plantarum Secundum Ordines Naturales Disposita [...].* Paris: Viduam Herissant et Theophilum Barrois. ISBN:9780332255293 (2018 reprint).

Jussieu, A. L. 1824. *Principes de la Method Naturelle des Végétaux.* Paris: F. G. Levrault. ISBN:9782329434711 (2020 reprint).

Kafanov, A. I. 2005. [*Historical and Methodological aspects of General and Marine Biogeography.*] Vladivostok: Far East State University Publication. ISBN:574441682X. (in Russian).

Kafanov, A. I., and Sukhanov V. V. 1981. [On the relationship between the number and volume of taxa]. *Zhurnal Obshchei Biologii* 42:345–50. ISSN:0044-4596. (in Russian).

Kahana, A., A. MacLeod, H. Mehr, et al. 2024. Constructing the molecular Tree of Life using assembly theory and mass spectrometry. *ArXiv:2408.09305 [q-bio.PE].* DOI:10.48550/arXiv.2408.09305

Kälin, J. A. 1945. Die Homologie als Ausdruck ganzheitlicher Baupläne von Typen. *Bulletin de la Société Fribourgeoise des Sciences Naturelles* 37:135–61. ISSN:0366-3256.

Kant, I. 2009. The metaphysical foundations of natural science. In *Philosophy of Science: An Historical Anthology*, eds. T. McGrew, M. Alspector-Kelly, and F. Allho, 232–7. Oxford (UK): Blackwell Publ. ISBN:9781405175425.

Kaplan, P. 2004. Morphology and disparity through time. In *Encyclopedia of Life Sciences.* John Wiley & Sons. DOI:10.1038/npg.els.0001637

Karamanolis, G. E., and A. D. R. Sheppard. 2007. *Studies on Porphyry.* London: University of London. ISBN:9781905670123.

Kasavin, I. T. 2024. [The notion of taxon: Holism and elementarism.] *Voprosy Filosofii* 11:27–37. DOI:10.21146/0042-8744-2024-11-27-37. (in Russian).

Kaup, J. J. 1844. *Classification der Säugethiere und Vögel.* Darmstadt: C. W. Leske. ISBN:9780428385668 (2018 reprint).

Kearney, M., and O. Rieppel. 2006. Rejecting "the given" in systematics. *Cladistics* 22:369–77. DOI:10.1111/j.1096-0031.2006.00110.x

Keller, R. A., R. N. Boyd, and Q. D. Wheeler. 2003. The illogical basis of phylogenetic nomenclature. *The Botanical Review* 69:93–110. DOI:10.1663/0006-8101(2003)069[0093:TIBOPN]2.0.CO;2

Kendig, C. 2014. Towards a multidimensional metaconception of species. *Ratio* 27:155–72. DOI:10.1111/rati.12037

Kendig, C. (ed.). 2016. *Natural Kinds and Classification in Scientific Practice.* Abingdon: Routledge. ISBN:9781848935402.

Kennedy, J. B., R. Kukla, and T. Paterson. 2005. Scientific names are ambiguous as identifiers for biological taxa: Their context and definition are required for accurate data integration. In *Data Integration in the Life Sciences*, eds. B. Ludäscher, and L. Raschid, 80–95. Berlin: Springer-Verlag. DOI:10.1007/11530084_8

Kenrick, D., D. Becker, J. Butner, et al. 2003. Evolutionary cognitive science: Adding what and why to how the mind works. In *From Mating to Mentality: Evaluating Evolutionary Psychology*, eds. K. Sterelny, and J. Fitness, 13–38. Hove (UK): Psychology Press. ISBN:9780203484708.

Kerner von Marilaun, A. 1895. *The Natural History of Plants, Their Forms, Growth, Reproduction, and Distribution*. New York (NY): H. Holt and Company. DOI:10.5962/bhl.title.54631

Kesić, S. 2016. Systems biology, emergence and antireductionism. *Saudi Journal of Biological Sciences* 23:584–91. DOI:10.1016/j.sjbs.2015.06.015

Kinman, K. E. 1994. *The Kinman System: Toward a Stable Cladisto-Eclectic Classification of Organisms, Living and Extinct, 48 Phyla, 269 Classes, 1,719 Orders*. Hays (KA): K. E. Kinman. ASIN:B0000EHVS4.

Kirschner, M., and J. Gerhart. 1998. Evolvability. *Proceedings of the National Academy of Sciences of the United States of America* 95:8420–7. DOI:10.1073/pnas.95.15.8420

Kitcher, P. 1984. Species. *Phylosophy of Science* 51:308–33. DOI:10.1086/289182

Kitching, I. J., P. L. Forey, C. J. Humphries, et al. 1998. *Cladistics: The Theory and Practice of Parsimony Analysis*. 2nd ed. Oxford: Oxford University Press. ISBN:9780198501381.

Kittay, E. F. 1987. *Metaphor, Its Cognitive Force and Structure*. Oxford: Oxford University Press. ISBN:9780198242468.

Klapisch-Zuber, C. 1991. The genesis of the family tree. In *I Tatti Studies: Essays in the Renaissance*, Vol. 4, 105–29. Florence: Villa i Tatti. ISBN:9788822227911.

Klein, L. S. 1991. [*Archeological Typology*.] Leningrad: Academy of Sciences of USSR. (in Russian).

Kleineberg, M. 2017. Integrative levels. *Knowledge Organization* 44:349–79. https://www.isko.org/cyclo/integrative_levels

Kleisner, K. 2007. The formation of the theory of homology in biological sciences. *Acta Biotheoretica* 55:317– 40. DOI:10.1007/s10441-007-9023-8

Klima, G. 2017. The Medieval problem of universals. In *Stanford Encyclopedia of Philosophy*, ed. E. N. Zalta. https://plato.stanford.edu/entries/universals-medieval/

Klir, G. J. 2001. *Facets of Systems Science*, 2d ed. Dodrecht: Springer Science+Business Media. ISBN:9781461355014.

Klix, F. 1985. *Erwachendes Denken. Eine Entwicklungsgeschichte der menschlichen Intelligenz*. Berlin: Deutscher Verlag der Wissenschaften. ISBN:9783860250846.

Kluge, A. G. 1984. The relevance of parsimony to phylogenetic inference. In *Cladistics: Perspectives on the Reconstruction of Evolutionary History*, ed. T. Duncan, and T. F. Stuessy, 24–38. New York (NY): Columbia University Press. ISBN:0231054300.

Kluge, A. G. 1997a. Sophisticated falsification and research cycles: Consequences for differential character weighting in phylogenetic systematics. *Zoologica Scripta* 26:349–60. DOI:10.1111/j.1463-6409.1997.tb00424.x

Kluge, A. G. 1997b. Testability and the refutation and corroboration of cladistic hypotheses. *Cladistics* 13:81–96. DOI:10.1111/j.1096-0031.1997.tb00242.x

Kluge, A. G. 1998a. Total evidence or taxonomic congruence: Cladistics or consensus classification. *Cladistics* 14:151–8. DOI:10.1111/j.1096-0031.1998.tb00328.x

Kluge, A. G. 2004. On total evidence: For the record. *Cladistics* 20:205–7. DOI:10.1111/j.1096-0031.2004.00020.x

Kluge, A. G. 2009. Explanation and falsification in phylogenetic inference: Exercises in Popperian philosophy. *Acta Biotheoretica* 57:171–86. DOI:10.1007/s10441-009-9070-4

Kluge, N. Yu. 1998b. [*Principles of Systematics of Living Beings.*] St. Petersburg: Saint Petersburg State University Publ. https://www.geokniga.org/books/11874. (in Russian).

Kluge, N. Yu. 1999a. [A system of alternative nomenclature of supraspecific taxa.] *Entomologicheskoye Obozrenie* 78:224–43. ISSN:0367-1445. (in Russian).

Kluge, N. Yu. 1999b. [Linnean and post-Linnean systematics of supraspecific taxa and new principles of nomenclature.] *Russian Ornithological Journal, Express Issue*, 3–21. ISSN:0869- 4362. (in Russian).

Kluge, N. Yu. 2000. [*The Modern Insect Taxonomy. Part I. The Principles of Systematics of Living Organisms and the General System of Insects, with a Classification of Apterygota and Palaeoptera.*] St. Petersburg: Lan. ISBN: 5811402163. (In Russian).

Knapp, S., G. Lamas, E. N. Lughadha, et al. 2004. Stability or stasis in the names of organisms: The evolving codes of nomenclature. *Philosophical Transactions: Biological Sciences* 359:611–22. DOI:10.1098/rstb.2003.1445

Knowles, L. L., and W. P. Maddison. 2002. Statistical phylogeography. *Molecular Ecology* 11:2623–35. DOI:10.1046/j.1365-294X.2002.01410.x

Knox, E. B. 1998. The use of hierarchies as organizational models in systematics. *Biological Journal of the Linnean Society* 63:1–49.

Kober, G. 2008. *Biology Without Species: A Solution to the Species Problem.* Doctoral dissertation. Boston (MA): Boston University. https://philpapers.org/rec/KOBBWS

Koestler, A. 1968. *The Ghost in the Machine.* New York (NY): The Macmillan Co. ISBN:9780090838806.

Kokić, T. 2009. Orbital model of species. In *Symposium: Panel "Darwin Today"* 17:157–72. Sarajevo: Academy of Sciences and Arts of Bosnia and Herzegovina. https://www.academia.edu/561309/Orbital_Model_of_Species

Koren, S. 2020. *The Hierarchy Problem: From the Fundamentals to the Frontiers.* PhD dissertation. Santa Barbara (CA): University of California. https://arxiv.org/pdf/2009.11870

Korzhinsky, S. I. 1893. [Flora of the East of European Russia in its systematic and geographical rations, Pt. 1.] *Izvestia Tomsk State Univerersity* 5: 71–299. (in Russian).

Kosko, B. 1993. *Fuzzy Thinking: The New Science of Fuzzy Logic.* New York (NY): Hyperion. ISBN:9780006547136.

Koslicki, K. 2008. Natural kinds and natural kind terms. *Philosophy Compass* 3/4:789–802. DOI:10.7939/R3GF0NB01

Kövecses, Z. 2010. *Metaphor. A Practical Introduction*, 2d ed. New York (NY): Oxford University Press. ISBN:9780195374940.

Koyré, A. 1973. *Éludes d'histoire de la pensée scientifique.* Paris: Gallimard. ISBN:9782070703357.

Kozo-Polyansky, B. M. 1922. [*Introduction to Phylogenetic Systematics of Higher Plants.*] Voronezh: Priroda i Kultura. (in Russian).

Kozo-Polyansky, B. M. 1949. The styles of representation of the evolutionary system of plants. *Botanichesky Zhurnal* 34:245–52. ISSN:0006-8136.

Kraichak, E., A. Crespo, P. K. Divakar, et al. 2017. A temporal banding approach for consistent taxonomic ranking above the species level. *Scientific Reports* 7:2297. DOI:10.1038/s41598-017-02477-7

Kranke, N. 2024. How phenograms and cladograms became molecular phylogenetic trees. *Journal of the History of Biology* 57:423–43. DOI:10.1007/s10739-024-09782-8

Kraus, O. 2008. The Linnean foundations of zoological and botanical nomenclature. *Zootaxa* 1950:9–20.

Krell, F.-T. 2004. Parataxonomy *vs.* taxonomy in biodiversity studies—pitfalls and applicability of 'morphospecies' sorting. *Biodiversity and Conservation* 13:795–812. DOI:10.1023/B:BIOC.0000011727.53780.63

Kripke, S. 1980. *Naming and Necessity.* Cambridge (MA): Harvard University Press. ISBN:9780674598461.

Kudrin, B. I. 2002. [Mathematics of cenoses: Specific, rank-specific, parameter-ranked hyperbolic *H*-distributions and the laws of Lotka, Zipf, Pareto, Mandelbrot.] *Cenologicheskie Issledovania* 19:357–412. https://www.kudrinbi.ru/public/10524/index.htm. (in Russian).

Kuhn, T. S. 1962. *The Structure of Scientific Revolutions*. Chicago (IL): The University of Chicago Press. ISBN:9780226458083.

Kull, K. 2019. Steps towards the natural meronomy and taxonomy of semiosis: Emotin between index and symbol? *Sign Systems Studies* 47:88–104. DOI: 10.12697/SSS.2019.47.1-2.03

Kumar, S. 2005. Molecular clocks: Four decades of evolution. *Nature Reviews, Genetics* 6:654–62. DOI:10.1038/nrg1659

Kuntner, M., and I. Agnarsson. 2006. Are the Linnean and phylogenetic nomenclatural systems combinable? Recommendations for biological nomenclature. *Systematic Biology* 55:774–84. DOI:10.1080/10635150600981596

Kuprevich, V. F. 1949. [*The Species Problem in Heterotrophic and Autotrophic Plants.*] Leningrad: USSR Academy of Sciences. (in Russian).

Kuprevich, V. F. 1950. [Species as a stage in the evolution of heterotrophic and autotrophic plants.] In [*Problems of Botany*] 1:149–62. Leningrad: USSR Academy of Sciences. (in Russian).

Kupriyanov, A. V. 2005. [*Prehistory of Biological Systematics.*] St. Petersburg: Saint Petersburg European University. ISBN:5943800433. (in Russian).

Kuraev, V. I., and F. V. Lazarev. 1988. [*Precision, Truth, and Growth of Knowledge.*] Moscow: Nauka. ISBN:5020079618. (in Russian).

Kusakin, O. G., and A. L. Drozdov. 1994. [*The Phyleme of the Organic World. 1. Prolegomena to the Construction of the Phyleme.*] St. Petersburg. ISBN:5020258776. (in Russian).

Kuznetsov, A. N. 1999. [The limb structural plans and the evolution of tetrapod running techniques.] *Zoologicheskie Issledovania* 3:3–91. ISSN:1025-532X. (in Russian).

Kwok, R. B. H. 2011. Phylogeny, genealogy and the Linnaean hierarchy: A logical analysis. *Journal of Mathematical Biology* 63:73–108. DOI:10.1007/s00285-010-0364-6

Labrosse, J. J., J. Ganssle, R. Oshana, et al. 2007. *Embedded Software*. London: Newnes Publication. ISBN:9780750685832.

Lam, H. J. 1936. Phylogenetic symbols, past and present. *Acta Biotheoretica* 2:153–94. DOI:10.1007/bf01556309

Lam, H. J. 1957. What is a taxon? *Taxon* 6:213–5. DOI:10.2307/1217460

Lam, H. J. 1959. Taxonomy. General principles and angiosperms. In *Vistas in Botany. A Volume in Honour of The Bicentenary of the Royal Botanic Gardens, Kew*, eds. W. B. Turrill, R. C. Rollins, and G. Taylor, 3–75. London: Pergamon Press. ISBN:9780080098494.

Lamarck, J.-B. 1778. *Flore françoise, ou Description succincte de toutes les plantes qui croiffent naturellement en France [...]*, T. 1. Paris: Impr. Royal. ISBN:9781166064693 (2010 reprint).

Lamarck, J.-B. 1809. *Philosophie zoologique , ou Exposition des considérations relatives à l'histoire naturelle des animaux [...]*. Paris: Muséum d'Histoire Naturelle (Jardin des Plantes).

Lamarck, J.-B. 1963. *Zoological Philosophy. An Exposition with Regard to the Natural History of Animals*. New York (NY): Hafner Publ. ISBN:9780226468105.

Lane, D. 2006. Hierarchy, complexity, society. In *Hierarchy in Natural and Social Sciences*, ed. D. Pumain, 81–119. Dordrecht: Springer-Verlag. ISBN:9781402041266.

Langacker, R. W. 1990. *Concept, Image, and Symbol: The Cognitive Basis of Grammar*. Berlin: Mouton de Gruyter. ISBN:9783110172805.

Langacker, R. W. 2007. Cognitive grammar. In *The Oxford Handbook of Cognitive Linguistics*, eds. D. Geeraerts, and H. Cuyckens, 421–61. Oxford: Oxford University Press. ISBN:9780195143782.

Langacker, R. W. 2009. *Investigations in Cognitive Grammar*. Berlin: Mouton de Gruyter. ISBN:9783110214345.

Langefors, B. 1977. Information systems theory. *Information Systems* 2:207–19. DOI:10.1016/0306-4379(77)90009-6

Lankester, E. R. 1870. On the use of the term homology in modern zoology, and the distinction between homogenetic and homoplastic agreements. *Annals and Magazine of Natural History, Ser. 5* 6:34–43. DOI:10.1080/00222937008696201

Lansing, J. 1995. Genus, species, and vantages. In *Language and the Cognitive Construal of the World*, eds. J. R. Taylor and R. E. MacLaury, 365–75. Berlin: Mouton de Gruyter. ISBN:9783110143010.

LaPorte, J. 2017. Modern essentialism for species and its animadversions. In *The Routledge Handbook of Evolution and Philosophy*, ed. R. Joyce, 182–93. New York (NY): Routledge. ISBN:9780367573072.

LaPorte, J. 2022. Rigid designators. In *The Stanford Encyclopedia of Philosophy*, eds. E. N. Zalta, and U. Nodelman. https://plato.stanford.edu/archives/win2022/entries/rigid-designators/

Larson, J. L. 1968. The species concept of Linnaeus. *Isis* 59:291–9. DOI:10.1086/350398

Larson, J. L. 1971. *Reason and Experience: The Representation of Natural Order in the Work of Carl von Linné*. Berkely (CA): University of California Press. ISBN:9780520018341.

Lassaline, M. E., E. J. Wisniewski, and D. L. Medin. 1992. Basic levels in artificial and natural categories: Are all basic levels created equal? *Advances in Psychology* 93:327–78.

Laszlo, E. 1972a. *Introduction to Systems Philosophy. Toward a New Paradigm of Contemporary Thought*. New York (NY): Harper Torchbooks. ISBN:9780061317620.

Laszlo, E. 1972b. *The Systems View of The World: The Natural Philosophy of the New Developments in the Sciences*. New York (NY): George Braziller. ISBN: 9780807606360.

Latreille, P. A. 1801. *Histoire naturelle, générale et particulière des Crustacés et des Insectes. T. 1*. Paris: F. Dufart. DOI:10.5962/bhl.title.15764

Latreille, P. A. 1806. *Genera Crustaceorum et Insectorum secundum ordinem naturalem […]*, T. 1. Parisiis & Argentorati [Strasbourg]: Amand Koenig. DOI:10.5962/bhl.title.5093

Latreille, P. A. 1810. *Considérations générales sur l'ordre naturel des animaux […]*. Paris, F. Schoell. DOI:10.5962/bhl.title.34917

Laubheimer, P. 2018. Polyhierarchies improve findability for ambiguous IA categories. *NN/g* (an Internet source). https://www.nngroup.com/articles/polyhierarchy/

Laubichler, M. D. 2000. Homology in development and the development of the homology concept. *American Zoologist* 40:777–88. DOI:10.1093/icb/40.5.777

Laubichler, M. D., and J. Maienschein (eds.). 2008. *Evolving Pathways: Key Themes in Evolutionary Developmental Biology*. New York (NY): Cambridge University Press. ISBN:9780521880244.

Laurin, M. 2007. The advantages of phylogenetic nomenclature over Linnean nomenclature. In *Animal Names*, eds. A. Minelli, G. Ortalli, and G. Sanga, 67–98. Venice: Istituto Veneto di Scienze, Lettere ed Arti. ISBN:8888143386.

Laurin, M. 2010. The subjective nature of Linnaean categories and its impact in evolutionary biology and biodiversity studies. *Contributions to Zoology* 79:131–46. DOI:10.1163/18759866-07904001

Laurin, M. 2024. *The Advent of PhyloCode. The Continuing Evolution of Biological Nomenclature*. Boca Raton (FL): CRC Press. ISBN:9780367552886.

Le Quesne, W. J. 1982. Compatibility analysis and its applications. *Zoological Journal of the Linnean Society* 74:267–75. DOI:10.1111/j.1096-3642.1982.tb01151.x

Lebedev, V. L., S. A. Kosushkin, and A. A. Bannikova. 2024. [The genetic species concept as applied to multilocus nuclear data.] *Zhurnal Obshchei Biologii* 85: 269–83. DOI:10.31857/S0044459624040014. (in Russian).

Leclerc, B. 1985. La comparaison des hierarchies: Indices et metriques. *Mathématiques et Sciences Humaines* 92:5–40. https://www.numdam.org/article/MSH_1985__92__5_0.pdf

Lee, M. S. Y. 1998. Phylogenetic uncertainty, molecular sequences, and the definition of taxon names. *Systematic Biology* 47:719–26. DOI:10.1080/106351598260707

Lee, M. S.Y., and A. Skinner. 2008. Hierarchy and clade definitions in phylogenetic taxonomy. *Organisms Diversity & Evolution* 8:17–20. DOI:10.1016/j.ode.2006.08.002

Lefebvre, H. 1991. *The Production of Space*. Hoboken (NJ): Wiley-Blackwell. ISBN:9780631181774.

Lefebvre, H. 2003. *The Urban Revolution*. Minneapolis (MN): University of Minnesota Press. ISBN:9780816641604.

Lennox, J. G. 1980. Aristotle on genera, species, and "the more and the less". *Journal of the History of Biology* 13:321–46. DOI:10.1007/BF00125746

Lennox, J. G. 2021. Aristotle's biology. In *Stanford Encyclopedia of Philosophy*, ed. E. N. Zalta. https://plato.stanford.edu/archives/fall2021/entries/aristotle-biology/

Lennox, J. G., and R. Bolton (eds.). 2010. *Being, Nature, and Life in Aristotle. Essays in Honor of Allan Gotthelf.* Cambridge (UK): Cambridge University Press. ISBN:9780521768443.

Lenormand, T., D. Roze, and F. Rousset. 2009. Stochasticity in evolution. *Trends in Ecology and Evolution* 24:157–65. DOI:10.1016/j.tree.2008.09.014

Lesch, J. E. 1990. Systematics and the geometrical spirit. In *The Quantifying Spirit in the 18th Century*, eds. T. Frangsmyr, J. L. Heilbron, and R. E. Rider, 73–111. Berkeley (CA): University of California Press. ISBN:9780585071985.

Levich, A. P. (ed.). 1996. *On the Way to Understanding the Time Phenomenon*, Pt. 1. Moscow: Moscow State University Publ. ISBN:9789810213602.

Levich, A. P. (ed.). 2009. *Towards Understanding the Phenomenon of Time: Constructions of Time in Natural Science*, Pt. 3. Moscow: Progress–Traditsia. ISBN:5898262970.

Levin, S. A. 1992. The problem of pattern and scale in ecology: The Robert H MacArthur Award Lecture. *Ecology* 73: 1943–67. DOI:10.2307/1941447

Levins, R. 1966. The strategy of model building in population biology. *American Scientist* 54:421–31.

Levins, R. 1973. The limits of complexity. In *Hierarchy Theory*, ed. H. Pattee, 109–27. New York (NY): George Braziller. ISBN:9780807606742.

Lévi-Strauss, C. 1966. *The Savage Mind*. London: Weidenfeld & Nicolson. ISBN:9 780226474847.

Lewens, T. 2009. Evo–devo and "typological thinking": An exculpation. *Journal of Experimental Zoology* 312B:789–96. DOI:10.1002/jez.b.21292

Li, W. 2002. Zipf's Law everywhere. *Glottometrics* 5:14–21. https://citeseerx.ist.psu.edu/document?repid=rep1&type=pdf&doi=15a4931ebba2372075f1cdddc58f3623fac36951

Lima, M. 2017. *The Book of Circles. Visualizing Spheres of Knowledge*. New York (NY): Princeton Architectural Press. ISBN:9781616895280.

Lin, H. 2024. Bayesian epistemology. In *The Stanford Encyclopedia of Philosophy*, eds. E. N. Zalta, and U. Nodelman. https://plato.stanford.edu/archives/sum2024/entries/epistemology-bayesian/

Lindley, J. 1835. *A Key to Structural, Physiological, and Systematics Botany*. London: Longman. ISBN:9781145520752 (2010 reprint).

Lindley, J. 1836. *The Vegetable Kingdom: or The Structure, Classification, and Uses of Plants,* [...]. London: Bradbury & Evans. ISBN:9780265543450 (2018 reprint).

Lines, J. L., and T. R. Mertens. 1970. *Principles of Biosystematics*. Chicago (IL): Educational Methods. ISBN:9781682862650.

Linnaeus, C. 1735. *Systema naturae, sive Regna tria naturae Systematice Proposita Per Classes, Ordines, Genera, & Species*. Leiden: Haak.

Linnaeus, C. 1736. *Fundamenta Botanica: Quae Majorum Operum Prodromi Instar Theoriam Scientiae Botanice*. Amsterdam: Salomonem Schouten. ISBN:9789060460641 (2010 reprint).

Linnaeus, C. 1737. *Genera Plantarum: Eorumque Characteres Naturales Secundum Numerum, Figuram, Situm [...]*. Lugduni Batavorum: Conradum Wishoff.

Linnaeus, C. 1738. *Classes Plantarum, Seu Systemata Plantarum [...]*, Pars II. Lugduni Batavorum: Conradum Wishoff.

Linnaeus, C. 1749. *Amoenitas Academiae, seu, Dissertationes Variae Physicae Medicae, Botanicae [...]*. Holmiae et Lipsiae: apud Godofredum Kiesewetter.

Linnaeus, C. 1751. *Philosophia Botanica in Qua Explicantur Fundamenta Botanica [...]*. Stockholm: Godofr. Kiesewetter. ISBN:9781104629731 (2009 reprint).

Linnaeus, C. 1762. *Fundamentum fructificationis [...]*. Upsala: J. M. Graberg.

Linnaeus, C. 1767. *Systema naturae, Vol. 1, Part II. Ed. duodecima reformata*. Holmiae: Laurentii Salvii. ISBN:9780243506941 (2016 reprint).

Lipictz, A. 1993. The local and the global: Regional individuality or intcrregionalism? *Transactions of the Institute of British Geographers, New Series* 18: 8–18. DOI:10.2307/623066

Little, F. J. 1964. The need for a uniform system of biological numericlature. *Systematic Zoology* 13:191–4. DOI:10.2307/sysbio/13.1-4.191

Lloyd, A. C. 1962. Genus, species and ordered series in Aristotle. *Phronesis* 7:67–90. DOI:10.1163/156852862x00043

Lobanov, A. L. 1972. [Logical analysis and classification of the existing forms of diagnostic keys.] *Entomologicheskoie Obozrenie*, 51:668–81. ISSN:0367-1445. (in Russian).

Loreau, M. 2010. Linking biodiversity and ecosystems: Towards a unifying ecological theory. *Philosophical Transactions of the Royal Society, B Biological Sciences* 365:49–60. DOI:10.1098/rstb.2009.0155

Löther, R. 1972. *Die Beherrschung der Mannigfaltigkeit: Philosophische Grundlagen der Taxonomie*. Jena: Gustav Fischer. https://www.max-stirner-archiv-leipzig.de/dokumente/Loether-Mannigfaltigkeit.pdf

Lotsy, J. P. 1916. *Evolution by Means of Hybridization*. Hague: M. Nijhoff. ISBN:9789401183819.

Lotsy, J. P. 1931. On the species of the taxonomist in its relation to evolution. *Genetica* 13:1–16. DOI:10.1007/BF01725037

Love, A. C. 2007. Functional homology and homology of function: Biological concepts and philosophical consequences. *Biology and Philosophy* 22:691–708. DOI:10.1007/s10539-007-9093-7

Lovejoy, A. O. 1936. *The Great Chain of Being: A Study of the History of an Idea*. Cambridge (MA): Harvard University Press. ISBN:9780674361539 (1976 reprint).

Løvtrup, S. 1977. Phylogenetics: Some comments on cladistic theory and method. In *Major Patterns of Vertebrate Evolution*, ed. M. K. Hecht, P. C. Goody, and B. M. Hecht, 805–22. New York (NY): Plenum Press. ISBN:9780306356148.

Løvtrup, S. 1984. Ontogeny and phylogeny. In *Beyond Neo-Darwinism*. London: Academic Press P. 159–190.

Løvtrup, S. 1989. On divergent and progressive evolution. *Alternatve Life-History Styles of Animals*, ed. M. N. Bruton, 55–69. Dordrecht: Kluwer Academic Press. ISBN:9789061936626.

Lubischev, A. A. 1923. [On the form of the natural system of organisms.] *Izvestia Biological Research Institute, Perm' State University* 2:99–110. (in Russian).

Lubischev, A. A. 1969. Philosophical aspects of taxonomy. *Annual Review of Entomology* 14:19–38. DOI:10.1146/annurev.en.14.010169.000315

Lubischev, A. A. 1972. [To logic of systematics]. In [*Problems of Evolution*], Vol. 2. Novosibirsk: Nauka. (in Russian).

Lücking, R. 2019. Stop the abuse of time! Strict temporal banding is not the future of rank-based classifications in fungi (including lichens) and other organisms. *Critical Reviews in Plant Sciences* 38:199–253. DOI:10.1080/07352689.2019.1650517

Ludwig, D., and S. Ruphy. 2024. Scientific pluralism. In *The Stanford Encyclopedia of Philosophy*, eds. E. N. Zalta, and U. Nodelman. https://plato.stanford.edu/archives/fall2024/entries/scientific-pluralism/

Lull, R. 1515. *Arbor Scientiae, venerabilis et caelitus [...]*. Lugdun: Opera Gilberti de Villiers.

Lunkevich, V. V. 1960. [*From Heraclitus to Darwin. Essays on the History of Biology*,] 2nd ed., Vol. 2. Moscow: Uchpedgiz. (in Russian).

Luo, J., D. E. Whitney, C. Y. Baldwin, et al. (eds.). 2009. *Measuring and Understanding Hierarchy as an Architectural Element in Industry Sectors*. Chicago (IL): Industry Studies Assoc. https://www.hbs.edu/ris/Publication%20Files/09-144.pdf

Luschei, E. C. 1962. *The Logical System of Lesniewski*. Amsterdam: North Holland Publication. ISBN:9780720422160.

Lyubarsky, G. Yu. 1991. [Objectivation of the category of taxonomic rank.] *Zhurnal Obshchei Biologii* 52:613–6. ISSN:0044-4596. (in Russian).

Lyubarsky, G. Yu. 1996. [*Archetype, Style and Rank in Biological Systematics*.] Moscow: KMK Scientific Press. ISBN:5873170053. (in Russian).

Lyubarsky, G. Yu. 2018. [*Origins of Hierarchy: The History of Taxonomic Rank*.] Moscow: KMK Scientific Press. ISBN:9785950082962. (in Russian).

Maat, J. 2004. *Philosophical Languages in the Seventeenth Century: Dalgarno, Wilkins, Leibniz*. Dordrecht: Kluwer. ISBN:9789401037716.

Mabee, P. M., M. Ashburner, Q. Cronk, et al. 2007. Phenotype ontologies: The bridge between genomics and evolution. *Trends in Ecology and Evolution* 22:345–50. DOI:10.1016/j.tree.2007.03.013

MacArthur, R. H., and E. O. Wilson. 1967. *The Theory of Island Biogeography*. Princeton (NJ): Princeton University Press. ISBN:9780691080505.

MacLeay, W. S. 1819. *Horae Entomologicae, or Essay on the Annulose Animals*, Vol. 1(Pt. 1), 2. London: S. Bagster. DOI:10.5962/bhl.title.48636

Maddison, D. R., D. L. Swofford, and W. P. Maddison. 1997. NEXUS: An extensible file format for systematic information. *Systematic Biology* 46:590–621. DOI:10.1093/sysbio/46.4.590

Madsen, M. W. 2016. Cognitive metaphor theory and the metaphysics of immediacy. *Cognitive Science* 40:881–908. DOI:10.1111/cogs.12320

Magnol, P. 1689. *Prodromus Historiae Generalis Plantarum in Quo Familiae Plantarum [...]*. Monspelij: Gabrielis & Honorati Pech. ASIN:B07QYNGG9W (2019 reprint).

Mahner, M. 1993. What is a species? *Journal for General Philosophy of Science* 24:103–26. DOI:10.1007/BF00769517

Mahner, M., and M. Bunge. 1997. *Foundations of Biophilosophy*. Frankfurt: Springer-Verlag. ISBN:9783662033685.

Mai, J.E. 2009. The boundaries of classification. In *20th Annual ASIS SIG/CR Workshop: Bridging Worlds, Connecting People: Classification Transcending Boundaries*. https://digital.lib.washington.edu/ojs/index.php/acro/issue/view/874

Mai, J.-E. 2004. Classification in context: Relativity, reality, and representation. *Knowledge Organization* 31:39–48. DOI:10.5771/0943-7444-2004-1-39

Majid, A., N. J. Enfield, and M. van Staden. 2006. Parts of the body: Cross-linguistic categorisation (Special Issue). *Language Sciences* 28:137–360. DOI:10.1016/S0388-0001(06)00005-2

Makovelsky, A. O. 2004. [*History of Logic*.] Moscow: Kuchkovo Pole. ISBN:5860900813. (in Russian).

Malaterre, C., and F. Merlin. 2015. The (in)determinism of biological evolution: Where does the stochastic character of evolutionary theory come from?. In *Handbook of Evolutionary Thinking in the Sciences*, eds. T. Heams, P. Huneman, G. Lecointre, et al., 349–67. Dordrecht: Springer-Verlag. DOI:10.1007/978-94-017-9014-7_17

Mallet, J. 1995. A species definition for the modern synthesis. *Trends in Ecology and Evolution* 10:294–99. DOI:10.1016/0169-5347(95)90031-4
Mandelbrot, B. 1966. Information theory and psycholinguistics: A theory of word frequencies. In *Readings in Mathematical Social Sciences*, eds. P. F. Lazarsfield, and N. W. Henry, 350–68. Cambridge: The MIT Press.
Mandelbrot, B. B. 1982. *The Fractal Geometry of Nature*. San Francisco (CA): W. H. Freeman. ISBN:9780716711865.
Manson, S. M. 2008. Does scale exist? An epistemological scale continuum for complex human: Eenvironment systems. *Geoforum* 39:776–88. DOI:10.1016/j.geoforum.2006.09.010
Marenbon, J. 1988. *Early Medieval Philosophy (480–1150): An Introduction*, 2nd ed. London: Routledge. ISBN:9780415000703.
Margolis, E., and S. Laurence. 2011. Concepts. In *The Stanford Encyclopedia of Philosophy*, ed. E. N. Zalta. https://plato.stanford.edu/entries/concepts/
Margulis, L., and, K. V. Schwartz (eds.). 1997. *Five Kingdoms: An Illustrated Guide to the Phyla of Life on Earth*. San Francisco (CA): W.H. Freeman & Co. ISBN:0613923383.
Markman, A. B., and E. J. Wisniewski 1997. Similar and different: The differentiation of basic-level categories. *Journal of Experimental Psychology* 23: 54–70. DOI:10.1037/0278-7393.23.1.54
Markov, A. V., and E. B. Neimark. 1998. [*Quantitative Patterns of Macroevolution. An Attempt of Applying a Systemic Approach to the Analysis of the Development of Supraspecific Taxa*.] Moscow: Paleontological Institute of RAS. (in Russian).
Marsh, F. L. 1944. *Evolution, Creation, and Science*. Washington (DC): Review & Herald Publication. https://pdf.amazingdiscoveries.org/eBooks/EVOLUTION_CREATION_AND_SCIENCE.pdf
Marshall, D. C., K. B. R. Hill, M. Moulds, et al. 2016. Inflation of molecular clock rates and dates: Molecular phylogenetics, biogeography, and diversification of a global cicada radiation from Australasia (Hemiptera: Cicadidae: Cicadettini). *Systematic Biology* 65:16–34. DOI:10.1093/sysbio/syv069
Marston, S. A., J. P. Jones, and K. Woodward. 2005. Human geography without scale. *Transactions of the Institute of British Geographers, New Series* 30:416–32. DOI: 10.1111/j.1475-5661.2005.00180.x
Martin, J., D. Blackburn, and E. O. Wiley. 2010. Are node-based and stem-based clades equivalent? Insights from graph theory. *PLoS Current* 2:RRN1196. DOI:10.1371/currents.RRN1196
Martynov, A. V. 2012. Ontogenetic systematics: The synthesis of taxonomy, phylogenetics, and evolutionary developmental biology. *Paleontological Journal* 46:833–64. DOI:10.1134/S0031030112080072
Marume, S. B. M., and A. S. Chikasha. 2016. The concept hierarchy in organisational theory and practice. *International Journal of Engineering Science Invention* 5:55–8. ISSN:2319–6734.
Masalova, S. 2013. Flexible rationality as a cognitive model. *International Journal of Cognitive Research in Science, Engineering and Education* 1:19–24. ISSN:2334-847X.
Maslin, T. P. 1952. Morphological criteria of phyletic relationships. *Systematic Zoology* 1:49–70. ISSN:0039-7989.
Massey, D. 1999. *Power-Geometries and the Politics of Space-Time*. Heidelberg: University of Heidelberg. ISBN:3885705028.
Matthen, M. 2007. Defining vision: What homology thinking contributes. *Biology and Philosophy* 22:675–89. DOI:10.1007/s10539-007-9088-4
Matthewson, J., and M. Weisberg. 2009. The structure of tradeoffs in model building. *Synthese* 170:169–90. DOI:10.1007/s11229-008-9366-y
Maturana, H., and F. J. Varela. 1980. *Autopoiesis and Cognition: The Realization of the Living*. Dodrecht: D. Reidel Publication. ISBN:9789027710161.

Maurer, A. 1982. *Medieval Philosophy: An Introduction*, 2nd ed. Toronto: Pontifical Institute of Mediaeval Studies. ISBN:9780888447043.

Mavrodiev, E. V. 2002. [Once again about 'Gregg's paradox' and its decision.] *Zhurnal Obshchei Biologii* 63:236–8. ISSN:0044-4596. (in Russian).

Mayden, R. L. 1997. A hierarchy of species concepts: The denouement in the saga of the species problem. In *Species. The Units of Biodiversity*, eds. M. F. Claridge, A. H. Dawah, and M. R. Wilson, 381–424. London: Chapman & Hall. ISBN:9780412631207.

Mayden, R. L. 1999. Consilience and a hierarchy of species concepts: Advances toward closure on the species puzzle. *Journal of Nematology* 31:95–116. PMCID:PMC2620363.

Mayden, R. L. 2002. On biological species, species concepts and individuation in the natural world. *Fish and Fisheries* 3:171–96. DOI:10.1046/j.1467-2979.2002.00086.x

Mayden, R. L., and R. M. Wood. 1995. Systematics, species concepts, and the evolutionarily significant unit in biodiversity and conservation biology. In *Evolution and the Aquatic Ecosystem: Defining Unique Units in Population Conservation*, ed. J. L. Nielson, 58–113. Bethesda: American Fisheries Soc. ISBN: 9780913235942.

Mayr, E. 1942. *Systematics and the Origin of Species, from the Viewpoint of Zoologist*. New York (NY): Columbia University Press. ISBN:9780674862500 (1999 reprint).

Mayr, E. 1963. *Animal Species and Evolution*. Cambridge (MA): Belknap Press. ISBN:9780674037502.

Mayr, E. 1965. Classification and phylogeny. *American Zoologist* 5:165–74. DOI:10.1093/icb/5.1.165

Mayr, E. 1968. Theory of biological classification. *Nature* 220:545–8. DOI:10.1038/220545a0

Mayr, E. 1969. *Principles of Systematic Zoology*. New York (NY): McGraw Hill. ISBN:9780070411432.

Mayr, E. 1976. *Evolution and the Diversity of Life: Selected Essays*. Cambridge (MA): Belknap Press. ISBN:9780674271043.

Mayr, E. 1978. Origin and history of some terms in systematic and evolutionary biology. *Systematic Zoology* 27:83–8. DOI:10.2307/2412818

Mayr, E. 1982. *The Growth of Biological Thought: Diversity, Evolution, and Inheritance*. Cambridge (MA): Belknap Press. ISBN:9780674364462.

Mayr, E. 1987. The ontological status of species: Scientific progress and philosophical terminology. *Biology and Philosophy* 2:145–66. DOI:10.1007/BF00057959

Mayr, E. 1988. *Toward a New Philosophy of Biology*. New York (NY): Cambridge University Press. ISBN:9780674896666.

Mayr, E., and W. J. Bock. 2002. Classifications and other ordering systems. *Journal of Zoological Systematics and Evolutionary Research* 40:169–94. DOI:10.1046/j.1439-0469.2002.00211.x

Mazzocchi, F. 2018. Knowledge Organization System (KOS): An introductory critical account. *Knowledge Organization* 45: 54–78. https://www.isko.org/cyclo/kos

McCabe, T. 2008. Studying species definitions for mutual nonexclusiveness. *Zootaxa* 1939: 1–9. DOI:10.11646/zootaxa.1939.1.1

McCabe, T. 2021. *Descent and Logic in Biosystematics*. Juneau (AK): Perseverant Publication. ISBN:9780578758480.

McClure, E. F. 1975. Ethno-anatomy: The structure of the domain. *Anthropological Linguistics* 17:78–88. https://www.jstor.org/stable/30027558

McCord Adams, M. 1989. *William Ockham, Pt. 2*. Notre Dame (IN): University of Notre Dame Press. ISBN:9780268019457.

McFadden, J. 2001. *Quantum Evolution: The New Science of Life*. New York (NY): W.W. Norton & Co. ISBN:9780393050417.

McGhee, G. R. 1999. *Theoretical Morphology: The Concept and its Applications*. New York (NY): Columbia University Press. ISBN:9780231106177.

McGrade, A. S. (ed.). 2003 *The Cambridge Companion to Medieval Philosophy*. Cambridge (UK): Cambridge University Press. ISBN:9780521000635.

McGuire, M. 2007. Some problems for an ontology of complexity. In *Worldviews, Science, and Us. Philosophy and Complexity*, eds. C. Gershenson, D. Aerts, and B. Edmonds, 181–202. Singapore: World Scientific Publ. ISBN:9789812705488.

McIlwaine, I. C. 2009. Universal Decimal Classification (UDC). In *Encyclopedia of Library and Information Sciences*, 3d ed., Vol. 2, 5432–39. ISBN:9780849397127.

McKelvey, B. 1982. *Organizational Systematics. Taxonomy, Evolution, Classification*. Berkeley (CA): University of California Press. ISBN:9780520042254.

McKenna, M. C., and S. K. Bell. 1997. *Classification of Mammals Above the Species Level*. New York (NY): Columbia University Press. ISBN:9780231110136.

McNeill, J. 1982. Phylogenetic reconstruction and phenetic taxonomy. *Zoological Journal of the Linnean Society* 74:337–44. DOI:10.1111/j.1096-3642.1982.tb01156.x

McShea, D. W. 2016. Hierarchy: The source of teleology in evolution. In *Evolutionary Theory: A Hierarchical Perspective*, eds. N. Eldredge, T. Pievani, E. Serrelli, et al., 86–102. Chicago (IL): The University of Chicago Press. ISBN:9780226426051.

Meakin, P. 1998. *Fractals, Scaling and Growth far from Equilibrium*. Cambridge (UK): Cambridge University Press. ISBN:139780521189811.

Medin, D. L., and S. Atran. 2004. The native mind: Biological categorization and reasoning in development and across cultures. *Psychological Review* 111:960–83. DOI:10.1037/0033-295X.111.4.960

Medin, D. L., E. B. Lynch, S. Atran, et al. 1996. The basic level and privilege in relation to goals, theories, and similarity. *Proceedings of the Third International Conference on Multistrategy Learning*, 68–80. Palo Alto (CA): AAAI Press. ISBN:9781577350101.

Mednikov, B. M. 1974. [On the reality of the higher systematic categories of vertebrates.] *Zhurnal Obshchei Biologii* 35: 659–65. ISSN:0044-4596. (in Russian).

Meier, R., and S. Richter. 1992. Suggestion for a more precise usage of proper names of taxa. Ambiguities related to the stem lineage concept. *Zeitschrift fur Zoologische Systematik und Evolutionsforschung* 30:81–8. DOI:10.1111/j.1439-0469.1992.tb00159.x

Mella, P. 2006. Organizations and orgonizations. The holonic view of organizations. *International Journal of Knowledge, Culture and Change Management in Organisations* 5:111–23. ISSN:1447-9575.

Mella, P. 2009. *The Holonic Revolution: Holons, Holarchies and Holonic Networks*. Pavia: Pavia University Press. ISBN:9788896764008.

Merkulov, I. P. (ed.). 1996. [*Evolutionary Epistemology: Problems and Perspectives*.] Moscow: ROSSPEN. ISBN:520101884X. (in Russian).

Mermin, N. D. 1989. What's wrong with this pillow? *Physics Today* 42:9–11. DOI:10.1063/1.2810963

Mervis, C.B., and E. Rosch. 1981. Categorization of natural objects. *Annual Review of Psychology* 32:89–115. DOI:10.1146/annurev.ps.32.020181.000513

Mesarovic, M. D., and D. Macko. 1969. Foundations for a scientific theory of hierarchical systems. In *Hierarchical Structures*, eds. L. L. Whyte, A. G. Wilson, and D. Wilson, 29–50. New York (NY): American Elsevier Publication. ISBN:9780444000699.

Mesarovic, M. D., D. Macko, and Y. Takahara. 1970. *Theory of Hierarchical, Multilevel, Systems*. New York (NY): Academic Press. ISBN:9780124915503.

Messaris, P. 1994. *Visual "Literacy": Image, Mind, and Reality*. Boulder (CO): Westview Press. ISBN:9780813316673.

Meyen, S. V. 1977. [Taxonomy and meronomy.] In [*Methodological Issues in Geological Sciences*], 25–33. Kïev: Naukova Dumka. (in Russian).

Meyen, S. V. 1978. [Basic aspects of the organismal typology.] *Zhurnal Obshchei Biologii* 39:495–508. ISSN:0044-4596. (in Russian).

Meyen, S. V. 1984. [Principles of historical reconstructions in biology.] In [*Systemity and Evolution*,] ed. Yu. A. Shreyder, 7–32. Moscow: Nauka. (in Russian).

Meyen, S. V. 1989. [*An Introduction to the Theory of Stratigraphy.*] Moscow: Nauka. ISBN:5020046337. (in Russian).

Meyen, S. V., and Yu. A. Shreyder. 1976. [Methodological issues of the theory of classification.] *Voprosy Filosofii*, 12:67–79. ISSN:0042-8744. (in Russian).

Meyer A. 1926. *Logik der Morphologie im Rahmen einer Logik der gesamten Biologie*. Berlin: Springer-Verlag. DOI:10.1007/978-3-642-50733-5

Michener, C. D. 1957. Some bases for higher categories in classification. *Systematic Zoology* 6:160–73. DOI:10.2307/sysbio/6.4.160

Michener, C. D. 1963. Some future developments in taxonomy. *Systematic Zoology* 12: 151–72. DOI: 10.2307/2411757

Mickevich, M. F. 1978. Taxonomic congruence. *Systematic Biology* 27:143–58. DOI:10.2307/2412969

Mickevich, M. F. 1982. Transformation series analysis. *Systematic Zoology* 31:461–78. DOI:10.1093/sysbio/31.4.461

Migne, J.-P. 1847. *Patrologiae Cursus Completus [...]*, T. 68. Paris: Ven. apud. Editorem. ISBN:9789004237049 (2010 reprint).

Mikhailov, A. I. (ed.). 1983. [*Research on Non-Classical Logic and Formal Systems.*] Moscow: Nauka. (in Russian).

Mikhailov, K. E. 2003. [Typological comprehension of "biological species" and the ways to stabilize an "around-species" taxonomy in birds.] *Ornithology* 30:9–24. (in Russian). ISSN:0474-7313.

Mill, J. S. 1882. *A System of Logic, Ratiocinative and Inductive [...]*, 8th ed. New York (NY): Harper & Brothers. ISBN:9780530088556 (2019 reprint).

Miller, C. R., and M. Hartzog. 2020. "Tree thinking": The rhetoric of tree diagrams in biological thought. *An Interdisciplinary Journal of Rhetorical Analysis and Invention* 15:1–61. DOI:10.13008/2151-2957.1290

Miller, J. G. 1978. *Living Systems*. New York (NY): McGraw-Hill. ISBN:9780070420151.

Miller, W. 2016. The species problem: Concepts, conflicts, and patterns preserved in the fossil records. In *Species and Speciation in the Fossil Record*, eds. W. D. Allmon, and M. M. Yacobucci, 28–58. Chicago (IL): The University of Chicago Press. ISBN:9780226377445.

Mindell, D. P., and A. Meyer. 2001. Homology evolving. *Trends in Ecology and Evolution* 16:434–40. DOI:10.1016/S0169-5347(01)02206-6

Minelli, A. 1993. *Biological Systematics. The State of the Art*. London: Chapman & Hall. ISBN:9780412626203.

Minelli, A. 1995. The changing paradigms of biological systematics: New challenges to the principles and practice of biological nomenclature. *Bulletin of Zoological Nomenclature* 52:303–9. www.biodiversitylibrary.org/item/44798#page/333/mode/1up

Minelli, A. 1998. Molecules, developmental modules and phenotypes: A combinatorial approach to homology. *Molecular Phylogenetics and Evolution* 9:340–7. DOI:10.1006/mpev.1997.0490

Minelli, A. 2000. The ranks and the names of species and higher taxa, or A dangerous inertia of the language of natural history. In *Cultures and Institutions of Natural History. Essays in the History and Philosophy of Science*, eds. M. T. Ghiselin, and A. E. Leviton, 339–51. San Francisco (CA): California Academy of Sciences. ISBN:9781333633899.

Minelli, A. 2003. *The Development of Animal Form: Ontogeny, Morphology, and Evolution.* Cambridge (MA): Cambridge University Press. ISBN:9780511072413.

Minelli, A. 2015. Biological systematics in the evo–devo era. *European Journal of Taxonomy* 125:1–23. DOI:10.5852/ejt.2015.125

Minelli, A. 2016. Tracing homologies in an ever-changing world. *Rivista di Estetica* 62:40–55. DOI:10.4000/estetica.1174

Minelli, A. 2017. Evolvability and its evolvability. In *Challenging the Modern Synthesis: Adaptation, Development, and Inheritance*, eds. P. Huneman, and D. Walsh, 211–38. Oxford: Oxford University Press. ISBN:9780199377176.

Minelli, A. 2019. The galaxy of the non-Linnaean nomenclature. *History and Philosophy of the Life Sciences* 41:31. DOI:10.1007/s40656-019-0271-0

Minelli, A. 2020. Taxonomy needs pluralism, but a controlled and manageable one. *Megataxa* 1:9–18.

Minelli, A. 2022. The species before and after Linnaeus: Tension between disciplinary nomadism and conservative nomenclature. In *Species Problems and Beyond. Contemporary Issues in Philosophy and Practice*, eds. J. S. Wilkins, F. E. Zachos, and I. Ya. Pavlinov, 191–226. Boca Raton (FL): CRC Press. ISBN:9780367855604.

Minelli, A. 2023a. A refreshed approach to homology: Prioritizing epistemology over metaphysics. *Journal of Morphology* 284:e21533. DOI:10.1002/jmor.21533

Minelli, A. 2023b. Linear listing order and hierarchical classification: History, conflict, and use. *European Journal of Taxonomy* 908:1–26. DOI:10.5852/ejt.2023.908.2331

Minelli, A., and G. Fusco. 2013. Homology. In *The Philosophy of Biology: A Companion for Educators*, ed. K. Kampourakis, 289–322. Dordrecht: Springer Science + Business Media. DOI:10.1007/978-94-007-6537-5_15

Minelli, A., G. Fusco, and S. Sartori. 1991. Self-similarity in biological classifications. *BioSystems* 26:89–97. DOI:10.1016/0303-2647(91)90040-R

Minelli, A., and T. Pradeu (eds.). 2014. *Towards a Theory of Development*. London: Oxford University Press. ISBN:9780199671434.

Minto, B. 2008. *The Pyramid Principle: Logic in Writing and Thinking*. London: Pearson Education. ISBN:9780273710516.

Mirkin, B. M. 1985. [*Theoretical Foundations of Contemporary Phytocaenology*.] Moscow: Nauka. (in Russian). https://lib.agu.site/books/116/170/

Mirkin, B. M. 1989. Plant taxonomy and syntaxonomy: A comparative analysis. *Vegetatio* 82: 35–40. https://www.jstor.org/stable/20038459

Mishler, B. D. 1990. Phylogenetic analogies in the conceptual development of science. *Proceedings of the Biennial Meeting of the Philosophy of Science Association (1990)*, Vol. 2: 225–35. Chicago (IL): The University of Chicago Press. https://doi.org/10.1086/psaprocbienmeetp.1990.2.193070

Mishler, B. D. 2009. Three centuries of paradigm changes in biological classification: Is the end in sight? *Taxon* 58:61–7. DOI:10.1002/tax.581009

Mishler, B. D. 2021. *What, if Anything, are Species?* Boca Raton (FL): CRC Press. ISBN:9781498714549.

Mishler, B. D. 2022. Ecology, evolution, and systematics in a post-species world. In *Species Problems and Beyond. Contemporary Issues in Philosophy and Practice*, eds. J. S. Wilkins, F. E. Zachos, and I. Ya. Pavlinov, 177–90. Boca Raton (FL): CRC Press. ISBN:9780367855604.

Mivart, G. J. 1870. On the use of the term "homology". *Annals and Magazine of Natural History, Ser.* 4, 6:112–27. ISSN:0374-5481.

Mones, E., L. Vicsek, and T. Vicsek. 2012. Hierarchy measure for complex networks. *PLoS ONE* 7:e33799. DOI:10.1371/journal.pone.0033799

Monsch, K. A. 2006. The PhyloCode, or alternative nomenclature: Why it is not beneficial to palaeontology, either. *Acta Palaeontologica Polonica* 51:521–4. DOI:10.1080/02724634.2018.1528450

Morgan, G. J., and W. B. Pitts, 2008. Evolution without species: The case of mosaic bacteriophages. *The British Journal for the Philosophy of Science* 5:745–65. DOI:10.1093/bjps/axn038

Morin, E. 2007. Restricted complexity, general complexity. In *Worldviews, Science, and Us. Philosophy and Complexity*, eds. C. Gershenson, D. Aerts, and B. Edmonds, 5–29. Singapore: World Scientific Publ. ISBN:9789812705488.

Morison, R. 1672 *Plantarum Umbelliferarum Distributio Nova, Per Tabulas Cognationes Et Affinitatis Ex Libro Naturae*. Oxford: Theatro Sheldoniano. ISBN:9781240831456 (2011 reprint).

Morville, P., and L. Rosenfeld. 2007. *Information Architecture for the World Wide Web*, 3rd ed. Sebastopol (CA): O'Reilly & Associates. ISBN:9780596527341.

Morville, P., L. Rosenfeld, and J. Arango. 2015. *Information Architecture*, 4th ed. Sebastopol (CA): O'Reilly Media. ISBN:9781491911686.

Moshkov, M. 2019. *Comparative Analysis of Deterministic and Nondeterministic Decision Trees*. Cham (Switzerland): Springer Nature Switzerland. ISBN:9783030417277.

Moss, W. W., and J. A. Hendrickson. 1973. Numerical taxonomy. *Annual Review of Entomology* 18:227–58. ISSN:0066-4170.

Motroshilova, N. V. (ed.). 2000. [*History of Philosophy. West-Russia-East. The First Book. Philosophy of Antiquity and the Middle Ages*], 3rd ed. Moscow: Greek-Latin Cabinet. ISBN:9785829113759. (in Russian).

Müller, G. B. 2003. Homology: The evolution of morphological organization. In *Origination of Organismal Form: Beyond the Gene in Developmental and Evolutionary Biology*, ed. G. B. Muller, and S. A. Newman, 52–69. Cambridge (MA): The MIT Press. ISBN:9780262134194.

Mungall, C. J., G. V. Gkoutos, C. L. Smith, et al. 2010. Integrating phenotype ontologies across multiple species. *Genome Biology* 11:R2. DOI:10.1186/gb-2010-11-1-r2

Muntermann, J., R. C. Nickerson, and U. Varshney. 2015. Towards the development of a taxonomic theory. *21st Americas Conference on Information Systems, Puerto Rico*, Vol. 5. Red Hook (NY): Curran Associates, 3536–51. ISBN:9781510814936.

Muro, E. M., F. J. Ballesteros, B. Luque, et al. 2025. The emergence of eukaryotes as an evolutionary algorithmic phase transition. *Proceedings of the National Academy of Sciences* 122: art. e2422968122. DOI: 10.1073/pnas.2422968122

Murphy, G. L. 2002. *The Big Book of Concepts*. Cambridge (MA): The MIT Press. ISBN:9780262632997.

Murray, I. A. 1784. *Caroli a Linne equitis Systema Vegitabilium secundum classes ordines genera species [...]*. Gottingae: Jo. Christ. Dieterich.

Naef, A. 1919. Idealistische Morphologie und Phylogenetik (zur Methodik der systematischen Morphologie). *Zeitschrift für Vererbungslehre* 22:279– 82. DOI:10.1007/BF01882544

Naef, A. 1931. Allgemeine Morphologie. I. Die Gestalt als Begriff und Idee. In *Handbuch der vergleichenden Anatomie der Wirbeltiere*, Vol. 1, ed. L. Bolk, E. Göppert, and E. Kallius, 77– 118. Berlin: Urban und Schwarzenberg. ISBN:9783743449770 (2016 reprint).

Nalimov, V. V. 1979. [*A Probabilistic Model of Language. On Relation Between Natural and Artificial Languages*.] Moscow: Nauka. ISBN:9785458489232 (2024 reprint). (in Russian).

Naomi, S.-I. 2011. On the integrated frameworks of species concepts: Mayden's hierarchy of species concepts and de Queiroz's unified concept of species. *Journal of Zoological Systematics and Evolutionary Research* 49:177–84. DOI:10.1111/j.1439-0469.2011.00618.x

Naomi, S.-I. 2014. Proposal of an integrated framework of biological taxonomy: A phylogenetic taxonomy, with the method of using names with standard endings in clade nomenclature. *Bionomina* 7:1–44. DOI:10.11646/bionomina.7.1.1

Nascimento, F. F., M. D. Reis, and Z. Yang. 2017. A biologist's guide to Bayesian phylogenetic analysis. *Nature Ecology and Evolution* 1:1446–54. DOI:10.1038/s41559-017-0280-x

Nathan, M. J. 2017. Pluralism is the answer! What is the question? *Philosophy and Theory in Biology* 11:1–14. DOI:10.3998/ptpbio.16039257.0011.015

Nathan, M. J., and J. Cracraft. 2020. The nature of species in evolution. In *The Theory of Evolution: Principles, Concepts, and Assumptions*, eds. S. M. Scheiner and D. P. Mindell, 102–22. Chicago (IL): The University of Chicago Press. ISBN:9780226671161.

Nawar, T. 2021. Augustine on active perception, awareness, and representation. *Phronesis* 66:84–110. DOI:10.1163/15685284-bja10035

Nazarov, V. I. 1991. [*The Doctrine of Macroevolution: on the Way to a New Synthesis.*] Moscow: Nauka. ISBN:5020040959. (in Russian).

Nebel, M., C. Pfabel, A. Stock, et al. 2011. Delimiting operatonal taxonomic units for assessing ciliate environmental diversity using small-subunit rRNA gene sequences. *Environmental Microbiology Reports* 3:154–8. DOI:10.1111/j.1758-2229.2010.00200.x

Needham, J. G. 1911. The law that inheres in nomenclature. *Science, New Series*, 33:813–16. DOI:10.1126/science.33.856.813

Neff, N. A. 1986. A rational basis for a priory character weighting. *Systematic Zoology* 35:110–23. DOI:10.2307/2413295

Nei, M., and S. Kumar. 2000. *Molecular Evolution and Phylogenetics.* Oxford: Oxford University Press. ISBN:9780195135855

Nelson, G. 1978. Ontogeny, phylogeny, paleontology, and the biogenetic law. *Systematic Zoology* 27:324–45. DOI:10.2307/2412883

Nelson, G. 1979. Cladistic analysis and synthesis: Principles and definitions, with a historical note on Adanson's Familles des Plantes (1763–1764). *Systematic Zoology* 28:1–21. DOI:10.2307/2412995

Nelson, G. 1994. Homology and systematics. In *Homology: The Hierarchical Basis of Comparative Morphology*, ed. B. K. Hall, 101–49. New York (NY): Academic Press. ISBN:9780123195838.

Nelson, G., and N. Platnick. 1981. *Systematics and Biogeography: Cladistics and Vicariance.* New York (NY): Columbia University Press. ISBN:9780231045742.

Nelson, G. J. 1974. Classification as an expression of phylogenetic relationships. *Systematic Zoology* 22:344–59.

Neretina, S. S. 1994. [*Word and Text in Medieval Culture.*] Moscow: Gnosis. ISBN:5733304650. (in Russian).

Neretina, S. S. 2013. [Is Christian Platonism possible?]. *Electronic Philosophical Magazine Vox/Voice* 15:1–39. https://vox-journal.org. (in Russian).

Newman, M. 2005. Power laws, Pareto distributions and Zipf's law. *Contemporary Physics* 46:323–51. DOI:10.1080/00107510500052444

Nicolson, D. H. 1991. A history of botanical nomenclature. *Annals of the Missouri Botanical Garden* 78:33–56. DOI:10.2307/2399589

Nielsen, C., and P. Martinez. 2003. Patterns of gene expression: homology or homocracy? *Development Genes and Evolution* 213:149–54. DOI:10.1007/s00427-003-0301-4

Nitecki, M. H. (ed.). 1990. *Evolutionary Innovations.* Chicago (IL): The University of Chicago Press. ISBN:9780226586953.

Nixon, K. C., and J. M. Carpenter. 2000. On the other "Phylogenetic systematics". *Cladistics* 16:298–318.

Nonaka, S., K. Majima, Y. Kamitani, et al. 2021. Brain hierarchy score: Which deep neural networks are hierarchically brain-like? *iScience* 24:103013. DOI:10.1016/j.isci.2021.103013

Norenzayan, A., and S. J. Heine 2005. Psychological universals: What are they and how can we know? *Psychological Bulletin* 131:763–84. DOI:10.1037/0033-2909.131.5.763

Novick, A. 2018. The fine structure of 'homology'. *Biology and Philosophy* 33:6. DOI:10.1007/s10539-018-9617-3

Novick, L. R., K. Catley, and E. Schreiber. 2012. Understanding evolutionary history: An introduction to tree thinking. *Semantic Scholar.* https://api.semanticscholar.org/CorpusID:49357188

Nowak, M. A. 1992. What is a quasispecies. *Trends in Ecology and Evolution* 7:118–21. DOI:10.1016/0169-5347(92)90145-2

Nylander, J. A., F. Ronquist, J. P. Huelsenbeck, et al. 2004. Bayesian phylogenetic analysis of combined data. *Systematic Biology* 53:47–67. DOI:10.1080/10635150490264699

O'Hara, R. J. 1988. Homage to Clio, or toward an historical philosophy of evolutionary biology. *Systematic Zoology* 37:142–55. DOI:10.2307/2992272

O'Hara, R. J. 1991. Representations of the natural system in the nineteenth century. *Biology and Philosophy* 6:255–74. DOI:10.1007/BF02426840

O'Hara, R. J. 1996. Trees of history in systematics and philology. In *New Perspectives on the History of Life: Essays on Systematic Biology as Historical Narrative*, eds. M. C. C. de Pinna, and M. Ghiselin, 81–8. Milan: Museo Civico di Storia Naturale di Milano. ISBN:9780520260856.

O'Hara, R. J. 1997. Population thinking and tree thinking in systematics. *Zoologica Scripta* 26:323–9. DOI:10.1111/j.1463-6409.1997.tb00422.x

O'Meara, J. J. O. (ed.). 1987. *Eriugena. Periphyseon (the Division of Nature).* Montréal: Bellarmin. ISBN:9781610976305.

O'Neill, R.V., D. L. DeAngelis, J. B. Waide, et al. 1986. *A Hierarchical Concept of Ecosystems.* Princeton (NJ): Princeton University Press. ISBN:9780691084374.

O'Neill, R. V., A. R. Johnson, and A. W. King. 1989. A hierarchical framework for the analysis of scale. *Landscape Ecology* 3:193–205. DOI:10.1007/BF00131538

Ochman, H., E. Lerat, and V. Daubin. 2005. Examining bacterial species under the specter of gene transfer and exchange. *Proceedings of the National Academy of Sciences of the United States of America* 102 (Suppl. 1):6595–9. DOI:10.1073/pnas.0502035102

Ochoa, C. 2024. Deconstructing the extended evolutionary synthesis: Do we need a new theory of evolution? In *Unity and Disunity in Evolutionary Biology. Deconstructing Darwinism*, eds. R. G. Delisle, M. Esposito, and D. Ceccarelli, 403–39. Cham (Switzerland): Springer Nature Switzerland. ISBN:9783031426285.

Ochoa, C. 2025. *Hierarchical Evolutionary-Developmental Theory. Towards a New Theory of Evolution.* Cham (Switzerland): Springer Nature Switzerland. ISBN:9783031887222.

Odum, E. P. 1971. *Fundamentals of Ecology*, 3d ed. Philadelphia (PA): W. B. Saunders. ISBN:0721669417.

Ogiela, L., and M. R. Ogiela. 2012. *Advances in Cognitive Information Systems.* Berlin: Springer-Verlag. ISBN:9783642252457.

Ogilby, W. 1838. Further observations on "Rules for Nomenclature". *The Magazine of Natural History N.S.* 2:275–84. www.biodiversitylibrary.org/item/19511#page/293/mode/1up

Oken, L. 1809. *Grundzeichnung des natürlichen Systems der Erze.* Jena: Friedrich Frommann.

Oken, L. 1847. *Elements of Physiophilosophy.* London: Royal Society. ISBN:9781150213281 (2012 reprint).

Oliver, I., and A. J. Beattie. 1996. Invertebrate morphospecies as surrogates for species: A case study. *Conservation Biology* 10:99–109. DOI:10.1046/j.1523-1739.1996.10010099.x

O'Malley, M. A., and J. Dupré. 2005. Fundamental issues in systems biology. *BioEssays* 12:1270–6. DOI:10.1002/bies.20323

O'Meara, D. J. 1987. The Chain of Being in the light of recent work on Platonic hierarchy. In *Jacob's Ladder and the Tree of Life. Concepts of Hierarchy and the Great Chain of Being*, eds. M. L. Kuntz, and P. G. Kuntz, 15–30. New York (NY): Peter Lang. ISBN:0820402338.

Omland, K. E., L. G. Cook, and M. D. Crisp. 2008. Tree thinking for all biology: The problem with reading phylogenies as ladders of progress. *BioEssays* 30:854–67. DOI:10.1002/bies.20794

Ong, W. J. 1958. *Ramus, Method, and the Decay of Dialogue. From the Art of Discourse to the Art of Reason.* Cambridge (MA): Harvard University Press. ISBN:9780226629766.

Oren, A. 2004. Prokaryote diversity and taxonomy: Current status and future challenges. *Philosophical Transactions of the Royal Society B* 359: 623–38. DOI:10.1111/j.1574-6976.1998.tb00379.x

Orlov, A. I. 2006. [*Non-Numeric Statistics.*] Moscow: Bauman Tech University Publ. ISBN:9785703832776.

Orlov, Yu. K. 1976. [Generalized Zipf–Mandelbrot law and frequency structures of information units of different levels.] In [*Computational Linguistics*], ed. E. K. Guseva, 179–202. Moscow: Nauka. (in Russian).

Osborn, H. F. 1936. *Proboscidea. A Monograph of the Discovery, Evolution, Migration and Extinction of the Mastodonts and Elephants of the World.* New York (NY): The American Museum Press. ISBN:9789333450881 (2018 reprint).

Oseledchik, M., M. Ivleva, and V. Ivlev. 2017. Knowledge as a non-equilibrium dynamic system. *Advances in Social Science, Education and Humanities Research* 124:1–5.

Oskolski, A. 2007. [The taxon as an ontological problem.] *Archive of the Zoological Museum of Moscow State University* 48: 213–60. ISSN:0134-8647. (in Russian).

Oskolski, A. 2011. The taxon as an ontological problem. *Biosemiotics* 4:201–22. DOI:10.1007/BF00128837

Otsuka, J. 2017. The causal homology concept. *Philosophy of Science* 84:1128–39. DOI:10.1086/694038

Otsuka, J. 2019. Species as models. *Philosophy of Science* 86:1075–86. DOI:10.1086/705519

Owen, R. 1843. *Lectures on the Comparative Anatomy and Physiology of the Invertebrate Animals.* London: Longman. ISBN:9781290925495 (2016 reprint).

Owen, R. 1848. *On the Archetype and Homologies of the Vertebrate Skeleton.* London: J. van Voorst. ISBN:9781230235370 (2013 reprint).

Paasch, J. T. 2014. Scotus and Ockham on universals and individuation. In *Debates in Medieval Philosophy. Essential Readings and Contemporary Responses*, ed. J. Hause, 371–94. London: Routledge. ISBN:9780415505420.

Pace, N. R. 2009. Mapping the Tree of Life: Progress and prospects. *Microbiology and Molecular Biology Reviews* 73:565–76. DOI:10.1128/MMBR.00033-09

Padial, J. M., A. Miralles, I. De la Riva, et al. 2010. The integrative future of taxonomy. *Frontiers in Zoology* 7 (16):1–14. DOI:10.1186/1742-9994-7-16

Padt, F., P. Opdam, N. Polman, et al. (eds.). 2014. *Scale-Sensitive Governance of the Environment.* Hoboken (NJ): Wiley-Blackwell. ISBN:9781118567159.

Page, R. D. 2006. Taxonomic names, metadata, and the semantic web. *Biodiversity Informatics* 3:1–15. DOI:10.17161/BI.V3I0.25

Page, R. D. M. 2004. Phyloinformatics: towards a phylogenetic database. In *Data Mining in Bioinformatics*, eds. J. T. L. Wang, J. Z. Mohammed, H. T. T. Toivone, et al., 219–41. London: Springer. ISBN:9781852336714.

Page, R. D. M., and E. C. Holmes. 1998. *Molecular Evolution: A Phylogenetic Approach.* London: Blackwell Science. ISBN:9780865428898.

Panaccio, C. 2004. *Ockham on Concepts.* Aldershot: Ashgate Publishing. ISBN: 9780754632283.

Panchen, A. L. 1992. *Classification, Evolution, and the Nature of Biology.* New York (NY): Cambridge University Press. ISBN:9780521315784.

Panchen, A. L. 1994. Richard Owen and the concept of homology. In *Homology: The Hierarchical Basis of Comparative Morphology*, ed. B. K. Hall, 21–62. London: Academic Press. ISBN:9780123195838.

Panchen, A. L. 1999. Homology: History of a concept. In *Homology*, eds. G. R. Bock, and G. Cardew, 5–23 Chichester: John Wiley & Sons. ISBN:9780470515662.

Pankhurst, R. J. (ed.). 1975. *Biological Identification with Computers: Proceedings of a Meeting Held at King's College […].* London: Academic Press. ISBN:0125448503.

Parrochia, D. 2016. Classification. In *The Internet Encyclopedia of Philosophy*, Eds. J. Fieser, and B. Dowden. Martin (TN): University of Tennessee at Martin. https://iep.utm.edu/classification-in-science/

Parrochia, D., and P. Neuville. 2013. *Towards a General Theory of Classifications.* Basel: Birkhäuser/Springer. ISBN:9783034806084.

Pattee, C. C. 1970. The problem off biological hierarchy. In *Organization Stability and Process*, ed. C. H. Waddington, 117–36. Chicago (IL): Routledge. ISBN:9780202363837 (2010 edition).

Pattee, H. 1969. Physical conditions for primitive functional hierarchies. In: *Hierarchical Structures*, eds. L. L. Whyte, A. G. Wilson, and D. Wilson, 161–71. New York (NY): American Elsevier. ISBN:9780444000699.

Pattee, H. H. (ed.). 1973a. *Hierarchy Theory: The Challenge of Complex Systems.* New York (NY): George Braziller Publ. ISBN:9780807606742.

Pattee, H. H. 1973b. Unsolved problems and potential applications of hierarchy theory. In *Hierarchy Theory: The Challenge of Complex Systems*, ed. H. H. Pattee, 129–56. New York (NY): George Braziller Publication. ISBN:9780807606742.

Pattee, H. H. 1978. The complementarity principle in biological and social structures. *Journal of Social and Biological Structures* 1:191–200. DOI:10.1016/S0140-1750(78)80007-4

Patten, B. C. 1975 Ecosystem linearization: An evolutionary design problem. *The American Naturalist* 109:529–39. DOI:10.1086/283024

Patterson, C. 1977. The contribution of paleontology to teleostean phylogeny. In *Major Patterns of Vertebrate Evolution*, ed. M. K. Hecht, P. C. Goody, and B. M. Hecht, 579–644. New York (NY): Plenum Press. ISBN:9780306356148.

Patterson, C. 1982. Morphological characters and homology. In *Problems of Phylogenetic Reconstruction*, eds. K. A. Joysey, and A. E. Friday, 21–74. London: Academic Press. ISBN:9780123912503.

Patterson, C. 1983. How does ontogeny differ from phylogeny? In *Development and Evolution*, eds. B. C. Goodwin, N. Holder, and C. C. Wylie, 1–31. Cambridge (UK): Cambridge University Press. ISBN:9780521249492.

Patterson, C. 1988. Homology in classical and molecular biology. *Molecular Biology and Evolution.* 5:603–25. DOI:10.1093/oxfordjournals.molbev.a040523

Patterson, D. J., J. Cooper, P. M. Kirk, et al. 2010. Names are key to the big new biology. *Trends in Ecology and Evolution* 25:686–91. DOI:10.1016/j.tree.2010.09.004

Patterson, D. J., D. Remsen, W. A. Marino, et al. 2006. Taxonomic indexing: Extending the role of taxonomy. *Systematic Biology* 55:367–73.

Pauly, G. B., D. Hillis, and D. Cannatella. 2009. Taxonomic freedom and the role of Official Lists of Species Names. *Herpetologica* 65:115–28. DOI:10.1655/08-031R1.1

Pavlinov, I. Ya. 1987. [The uncertainty principle in cladistic systematics.] *Zhurnal Obshchei Biologii* 48:589–601. ISSN:0044-4596. (in Russian).

Pavlinov, I.Ya. 1990a. [*Cladistic Analysis (Methodological Problems).*] Moscow: Moscow State University Publication. ISBN:5211009185. (in Russian).

Pavlinov, I. Ya. 1990b. [The outgoup concept in cladistics.] *Zhurnal Obshchei Biologii* 51:304–15. ISSN:0044-4596. (in Russian).

Pavlinov, I. Ya. 1992a. [An evolutionary cone metaphor and the problem of reliability of phylogenetic reconstructions.] *Zhurnal Obshchei Biologii* 53:293–306. ISSN:0044-4596. (in Russian).

Pavlinov, I.Ya. 1992b. [If there is the biological species, or what is the "harm" of taxonomy.] *Zurnal Obshchei Biologii* 53:757–67. ISSN:0044-4596. (in Russian).

Pavlinov, I. Ya. 1998. On the problems of axiomatic foundation of evolutionary cladistics. In *Program and Abstracts of the 17th Meeting of The Willly Hennig Society "Hennig XVII"*, 67. São Paulo.

Pavlinov, I. Ya. 2003. [It is normal for taxonomic approaches to be diverse.] *Zhurnal Obshchei Biologii* 64: 275–91. ISSN:0044-4596. (in Russian).

Pavlinov, I. Ya. 2005. [*Introduction to Contemporary Phylogenetics.*] Moscow: KMK Scientific Press. ISBN:9785041154813 (in Russian).

Pavlinov, I. Ya. 2007a. [Etudes on metaphysics of contemporary systematics.] *Archives of the Zoological Museum of Moscow State University* 48: 123–82. ISSN:0134-8647 (in Russian).

Pavlinov, I. Ya. 2007b. [Phylogenetic thinking in contemporary biology.] *Zhurnal Obshchei Biologii* 68:19–34. ISSN:0044-4596. (in Russian).

Pavlinov, I. Ya. 2009. [The species problem: Another look.] In [*Species and Speciation: An Analysis of New Ideas and Trends*], eds. A. F. Alimov and S. D. Stepanyanz, 250–71. St. Petersburg: Zoological Institute RAN. ISBN:9785873175895. (in Russian).

Pavlinov, I. Ya. 2010. [Comments on biomorphics (ecomorphological systematics).] *Zhurnal Obshchei Biologii* 71:187–92. ISSN:0044-4596. (in Russian).

Pavlinov, I. Ya. 2011a. [How it is possible to develop a taxonomic theory.] *Zoologicheskie Issledovania* 10:45–100. ISSN:1025-532X. (in Russian).

Pavlinov, I. Ya. 2011b. Morphological disparity: An attempt to widen and to formalize the concept. In *Research in Biodiversity: Models and Applications*, ed. I. Ya. Pavlinov, 341–64. Rijeka (Croatia): InTech Open Access Publ. ISBN:9789535144175.

Pavlinov, I. Ya. 2011c. Concepts of rational taxonomy in biology. *Biology Bulletin Reviews* 1: 60–78. DOI:10.1134/S2079086411030078

Pavlinov, I. Ya. 2012. The contemporary concepts of homology in biology: A theoretical review. *Biology Bulletin Reviews* 2:36–54. DOI:10.1134/S2079086412010057

Pavlinov, I. Ya. 2013. The species problem: Why again? In *The species Problem: Ongoing Issues*, ed. I. Ya. Pavlinov, 3–37. Rijeka: InTech Open Access Publ. ISBN:9789535109570.

Pavlinov, I. Ya. 2015. [*Nomenclature in Systematics. History, Theory, Practice.*] Moscow: KMK Scientific Press. ISBN:9785990715745. (in Russian).

Pavlinov, I. Ya. 2018. [*Foundations of Biological Systematics: Theory and History.*] Moscow: KMK Scientific Press. ISBN:9785604074992. (in Russian).

Pavlinov, I. Ya. 2020. Multiplicity of research programs in the biological systematics: A case for scientific pluralism. *Philosophies* 5:7. DOI:10.3390/philosophies5020007

Pavlinov, I. Ya. 2021a. *Biological Systematics. History and Theory*. Boca Raton (FL): CRC Press. ISBN:9780367654450.

Pavlinov, I. Ya. 2021b. *Horses, Tapirs, Rhinoceroses, and Other Perissodactyles: Evolution and Diversity*. Moscow: Phiton XXI Publication. ISBN:9785906811981.

Pavlinov, I. Ya. 2022a. *Taxonomic Nomenclature. What's in a Name: Theory and History*. Boca Raton (FL): CRC Press. ISBN:9781032015071.

Pavlinov, I. Ya. 2022b. The species problem from a conceptualist's standpoint. In *Species Problems and Beyond. Contemporary Issues in Philosophy and Practice*, eds. J. S. Wilkins, F. E. Zachos, and I. Ya. Pavlinov, 245–78. Boca Raton (FL): CRC Press. ISBN:9781032221472.

Pavlinov, I. Ya. 2023. *The Species Problem. A Conceptual History*. Boca Raton (FL): CRC Press. ISBN:9781032386812.

Pavlinov, I. Ya. 2024. An outline of a conceptual framework for contemporary biological systematics. *Biology Bulletin* 51(Suppl. 1):S1–S22. DOI:10.1134/S1062359024609625

Pavlinov, I. Ya., Y. G. Puzachenko, A. Y. Puzachenko, et al. 1995. To zipf or not to zipf, or why are there so few scientists supposing there are no genera in the nature at all? *Zhurnal Obshchei Biologii* 56:152–8. ISSN:0044-4596.

Payne, R. W., and D. A. Preece 1980. Identification keys and diagnostic tables: A review. *Journal of the Royal Statistical Society, Ser. A* 143:253–92. DOI:10.2307/2982129

Pecarič, D., and J. Pecarič. (eds.). 2019. *Inequalities and Zipf–Mandelbrot Law*. Zagreb: University of Zagreb. ISBN:9789531976701.

Peirce, C. S. 1920. Tree of Porphyry. In *Dictionary of Philosophy and Psychology*, New ed., Vol. 2, ed. J. M. Baldwin, 713–4. New York (NY): Macmillan. ISBN:9780844610481 (1960 edition).

Pellegrin, P. 1986. *Aristotle's Classification of Animals. Biology and the Conceptual Unity of Aristotelian Corpus*. Berkeley (CA): University California Press. ISBN:9780520055025.

Pellegrin, P.1987. Logical difference and biological difference: The unity of Aristotle's thought. In *Philosophical Issues in Aristotle's Biology*, eds. A. Gotthelf and J. Lennox, 313–38. New York (NY): Cambridge University Press. ISBN:9780521310918.

Pellegrin, P. 1990. Taxonomie, moriologie, division. In *Biologie, Logique et Métaphysique chez Aristote*, eds. D. Deverreux and P. Pellegrin, 37–48. Paris: CNRS. ISBN:9782222044031.

Penn, M. J., N. Scheidwasser, J. Penn, et al. 2023. Leaping through tree space: Continuous phylogenetic inference for rooted and unrooted trees. *Genome Biology and Evolution* 15 (12). DOI:10.1093/gbe/evad213

Pesenko, Yu. A. 1989. [Methodological analysis of systematics. I. Formulation of the problem, and principal taxonomic schools.] *Trudy Zoologicheskogo Instituta AN USSR*, 206:8–119. ISSN:0206-0477 (in Russian).

Pesenko, Yu. A. 1991. [Methodological analysis of systematics. II. Phylogenetic reconstructions as scientific hypotheses.] *Trudy Zoologicheskogo Institute AN USSR*, 234:61–155. ISSN:0206-0477. (in Russian).

Petit-Thouars, L.-M. 1822. *Histoire particulière des plantes orchidées recuellies [...]*. Paris: Arthus Bertrand, Treuttel & Wurtz. www.biodiversitylibrary.org/item/9881#page/3/mode/1up

Petrusiewicz, K. 1959. Teoria ewolucji Darwina jest teoria ecologiczna. *Ekologia Polska* 5:297–302. ISSN:0420-9036.

Philpot, J. H. 1897. *The Sacred Tree, or the Tree in Religion and Myth*. London: Macmillan & Co. ISBN:9780766129399 (2003 reprint).

Piaget, J. 1970. *Structuralism*. New York (NY): Basic Books Publ. ISBN:9780465082384.

Piaget, J. 1971. *Genetic Epistemology*. New York (NY): W.W. Norton. ISBN:9780393005967.

Pichler, F. 2000. On the construction of A. Koestler's holarchical networks. http://www.cast.uni-linz.ac.at/Department/Publications/Pubs1998/holons.doc

Pietruszczak, A. 2005. Pieces of mereology. *Logic and Logical Philosophy* 14:211–34. DOI:10.12775/LLP.2005.014

Pietsch, T.W. 2012. *Trees of Life. A Visual History of Evolution*. Baltimore (MA): John Hopkins University Press. ISBN:9781421411859.

Pik, A., I. Oliver, and A. J. Beattie. 1999. Taxonomic sufficiency in ecological studies of terrestrial invertebrates: A case study investigating ant community structure within revegetated agricultural land. *Australian Journal of Ecology* 24: 555–62. DOI:10.1046/j.1442-9993.1999.01003.x

Platnick, N. I. 1979. Philosophy and the transformation of cladistics. *Systematic Zoology* 28: 537–46. DOI:10.2307/2412566

Platnick, N. I., and E. S. Gaffney. 1977. Systematics: A Popperian perspective. *Systematic Zoology* 26:360–5. DOI:10.1093/sysbio/26.3.360

Platt, J. 1969. Theorems on boundaries in hierarchical systems. In: *Hierarchical Structures*, eds, L. L. Whyte, A. G. Wilson, and D. Wilson, 201–13. New York (NY): American Elsevier. ISBN:9780444000699.

Pleijel, F., and G. W. Rouse. 2000. Least-inclusive taxonomic unit: A new taxonomic concept for biology. *Proceedings of the Royal Society, Ser. B* 267:627–30. DOI:10.1098/rspb.2000.1048

Podani, J. 2013. Tree thinking, time and topology: Comments on the interpretation of tree diagrams in evolutionary/phylogenetic systematics *Cladistics* 29: 315–27. DOI:10.1111/j.1096-0031.2012.00423.x

Podlazov, A. V. 2009. [Zipf's Law and models of competitive growth.] In [*Nonlinearity in Contemorary Natural Science*,] ed. G. G. Malinetsky, 229–56. Moscow: Editorial URSS. ISBN:9785382009841.

Pogue, M. G., and M. F. Mickevich. 1990. Character definitions and character state delineation: The bête noire of phylogenetic inference. *Cladistics* 6:319–61. DOI:10.1111/j.1096-0031.1990.tb00549.x

Pokrovsky, M. P. 2014. [*Introduction to the Classiology*.] Yekaterinburg: Institute of Geology & Geochemistry. ISBN:9785943321085. (in Russian).

Polanyi, M. 1968. Life's irreducible structure. *Science* 160:1308–12. DOI:10.1126/science.160.3834.1308

Polanyi, M. 1998. *Personal Knowledge: Towards a Post-Critical Philosophy*, 2d ed. London: Taylor & Francis Group. ISBN:9780226232621 (2005 reprint).

Polyansky, V. I. 1959. [On the integrity of the species and its evolution.] *Bulleten Moskovskogo Obshestva Ispytatelej Prirody, Biology*. 64:5–13. ISSN:0027-1403. (in Russian).

Popov, I. Yu. 2008. [*Periodical Systems and a Periodical Law in Biology*.] Moscow: KMK Scientific Press. ISBN:9785873175055 (in Russian).

Popper, K. R. 1972. *Objective Knowledge: An Evolutionary Approach*. Oxford: Oxford University Press. ISBN:9780198750246.

Porges, K., I. G. Stewart, U. Hoßfelda, et al. 2019. From idea to law: Theory, concept and terminological formation in Ernst Haeckel's works. *Russian Journal of Developmental Biology* 50:290–302. DOI:10.1134/S1062360419060079

Porphyry. 2003. *Introduction*. Oxford: Clarendon Press. ISBN:9780199246144.

Posey, D. A. 1984. Hierarchy and utility in a folk biological taxonomic system: Patterns in classification of arthropods by the Kayapo indians of Brazil. *Journal of Ethnobiology* 4:123–39. https://ethnobiology.org/sites/default/files/pdfs/JoE/4-2/Posey1984.pdf

Potochnik, A., and B. McGill. 2012. The limitations of hierarchical organization. *Philosophy of Science* 79:120–40. DOI:10.1086/663237

Pozdnyakov, A. A. 2005. [The significance of the Willis' rule for taxonomy.] *Zhurnal Obshchei Biologii* 66:326–35. ISSN:0044-4596. (in Russian).

Pozdnyakov, A. A. 2007. [The ontological status of taxa from the traditional point of view.] In [*The Linnaean Miscellenea*,] ed. I. Ya. Pavlinov, 261–304. Moscow: Moscow State University Publication. ISSN:0134-8647. (in Russian).

Pozdnyakov, A. A. 2014. [The structure of biological diversity.] *Philosofia Nauki* 3:67–92. ISSN:1560-7488. (in Russian).

Pozdnyakov, A. A. 2015. [Typological ideas of S.V. Meyen: Origins, problems and prospects.] *Lethaea Rossica* 2:85–9. ISSN:2224-0446. (in Russian).

Pratt, V. 1984. The essence of Aristotle's zoology. *Phronesis* 29:267–78. DOI:10.1163/156852884x00049

Preobrazhensky, B. V. 2010. [*Metaphysics and Memetamorphoses of Natural Science*.] Vladivostok: Emakho. (in Russian).

Prigogine, I. 1981. *From Being to Becoming. Time and Complexity in the Physical Sciences*. San Francisco (CA): W. H. Freeman & Co. ISBN:9780716711087.

Prigogine, I. 1997. *The End of Certainty: Time, Chaos, and the New Laws of Nature*. New York (NY): Free Press. ISBN:9780684837055.

Prigogine, I., and I. Stengers. 1984. *Order out of Chaos. Man's new Dialogue with Nature*. London: Heinemann. ISBN:9780553340822.

Pringle, C. R. 1991. The 20th meeting of the executive committee of the ICTV. Virus species, higher taxa, a universal database and other matters. *Archives of Virology* 119:303–4. DOI:10.1007/BF01310680

Pullan, M. R., M. F. Watson, J. Kennedy, et al. 2000. The Prometheus taxonomic model: A practical approach to representing multiple classifications. *Taxon* 49:55–75. DOI:10.2307/1223932

Pulteney, R. 1805. *General View of the Writings of Linnaeus*. London: R. Taylor. https://archive.org/details/b28523179/mode/2up

Pumain, D. (ed.). 2006. *Hierarchy in Natural and Social Sciences*. Dordrecht: Springer-Verlag. ISBN:9781402041266.
Punter, D. 2007. *Metaphor*. New York (NY): Routledge. ISBN:9780415281652.
Puzachenko, Yu. G. 2016. [Rank distributions in ecology and nonextensive statistical mechanics.] In *Aspects of Biodiversity*, eds. I. Ya. Pavlinov, M. V. Kalyakin, and A. V. Sysoev, 42–71. ISBN:9785990841666.
Quine, W. V. 1969. *Ontological Relativity and Other Essays*. New York (NY): Columbia University Press. ISBN:9780231083577.
Quine, W. V. 1994. Natural kinds. In *Grue The New Riddle of Induction*, ed. D. Stalker, 42–56. La Salle (IL): Open Court. ISBN:9780812692181.
Rabel, G. 1940. A decimal system for organisms. *Discovery NS* 3:16–23. ISSN:2149-6307.
Radnitzky, G., and W. W. Bartley (eds.). 1993. *Evolutionary Epistemology, Rationality, and the Sociology of Knowledge*. La Salle (IL): Open Court Publication. ISBN:9780812690392.
Ragan, M. A. 2009. Trees and networks before and after Darwin. *Biology Direct* 4:43. DOI:10.1186/1745-6150-4-43
Raguenaud, C., J. Kennedy, and P. J. Barclay, 1999. Database support for taxonomy. *Prometheus technical report #1* (Napier University). https://www.napier.ac.uk/research-and-innovation/research-search/outputs/database-support-for-taxonomy
Ramus, P. 1547. *Dialecticae Institutiones [...]*. Basiliae: Sebastianum Henricpetri. ISBN:9781104869687 (2009 reprint).
Rangel, T. F., R. K. Colwell, G. R. Graves, et al. 2015. Phylogenetic uncertainty revisited: Implications for ecological analyses. *Evolution* 69:1301–12. DOI:10.1111/evo.12644
Rapoport, A. 1986. *General System Theory: Essential Concepts & Applications*. Cambridge (MA): Abacus Press. ISBN:0856261726.
Rasnitsyn, A. P. 1982. Proposal to regulate the names of taxa above the family group. *Bulletin of Zoological Nomenclature* 39:200–7. www.biodiversitylibrary.org/item/44481#page/234/mode/1up
Rasnitsyn, A. P. 1992. [Principles of phylogenetics and systematics.] *Zhurnal Obshchei Biologii* 53:172–85. ISSN:0044-4596. (in Russian).
Rasnitsyn, A. P. 1996. Conceptual issues in phylogeny, taxonomy, and nomenclature. *Contributions to Zoology* 66:3–41. DOI:10.1163/26660644-06601001
Rasnitsyn, A. P. 2002. [Evolutionary process and systematic methodology.] *Trudy Russkogo Entomologicheskogo Obshchestva* 73:5–108. ISSN:1605-7678. (in Russian).
Rasnitsyn, A. P. 2020. [The philosophy of evolutionary biology.] *Zhurnal Obshchei Biologii* 81:54–80. DOI:10.31857/S0044459620010078. ISSN:0044-4596. (in Russian).
Raup, D. M., and J. J. Sepkoski, 1984. Periodicity of extinction in the geologic past. *Proceedings of the National Academy of Science* 81:801–5. DOI:10.1073/pnas.81.3.801
Ravasz, E., and A.-L. Barabási 2003. Hierarchical organization in complex networks. *Physical Review E* 67:0205130. DOI:10.1103/PhysRevE.67.026112
Ray, J. 1686. *Historia Plantarum: Species Hactenus Editas Aliasque Insuper Multas Noviter Inventas & Descriptas Complectens [...]*. London: Mariae Clark. ISBN:9781373926371 (2016 reprint).
Ray, J. 1696. *De Variis Plantarum Methodus Dissertation brevis [...]*. Londoni: S. Smith & B. Walford.
Ray, J. 1714. *The Wisdom of God Manifested in the Works of the Creation [...]*, 6th ed. London: William Innys.
Ray, J. 1733 *Methodus Plantarum Emendate et Aucta [...]*. Londoni: Christianum Andream Myntsing.
Raykawski, W. 2018. Layers and levels: What a column of water tells us about human cognition. *Cognitive Semantics* 4:I04–34. DOI:10.1163/23526416-00401005
Reboiras, F. D., P. Villalba, and V. P. Walter (hrg.). 2002. *Arbor Scientiae. Der Baum des Wissens von Ramon Lull*. Turnhout (Belgium): Brepols Publication. ISBN:9782503512150.

Reck, E., and G. Schiemer. 2025. Structuralism in the philosophy of mathematics. In *The Stanford Encyclopedia of Philosophy*, eds. E. N. Zalta, and U. Nodelman. https://plato.stanford.edu/archives/sum2025/entries/structuralism-mathematics/

Reichenbach, H. 1958. *The Philosophy of Space and Time*. Mineola (NY): Dover Publication. ISBN:9780486604435.

Reif, W.-E. 2003. Problematic issues of cladistics: 1. Ancestor recognition and phylogenetic classification. *Neues Jahrbuch fur Geologie und Palaontologie: Abhandlungen* 230:97–143. ISSN:0077-7749.

Remane, A. 1956. *Grundlagen des Naturlichen Systems, der Vergleichenden Anatomie und der Phylogenetik. Theoretische Morphologie und Systematik*. 2, Aufl. Leipzig: Akad. Verlag. ISBN:9783874290296.

Rensch, B. 1960. *Evolution Above the Species Level*. New York (NY): Columbia University Press. ISBN:9780231022965.

Resnick, I. M. (ed.). 2012. *A Companion to Albert the Great. Theology, Philosophy, and the Sciences*. Leiden: E. J. Brill. ISBN:9789004234079.

Reynolds, P. D. 2016. *A Primer in Theory Construction*. Abingdon: Routledge. ISBN:9780205501281.

Richards, R.A. 2010. *The Species Problem: A Philosophical Analysis*. Cambridge: Cambridge University Press. ISBN:9780521196833.

Richards, R. A. 2013. The species problem: A conceptual problem? In *The species Problem: Ongoing Issues*, ed. I. Ya. Pavlinov, 41–63. Rijeka: InTech Open Access Publ. ISBN:9789535109570.

Richards, R. A. 2016. *Biological Classification: A Philosophical Introduction*. Cambridge (UK): Cambridge University Press.

Richardson, M. K., A. Minelli, M. Coates, et al. 1998. Phylotypic stage theory. *Trends in Ecology and Evolution* 13:158. DOI:10.1016/s0169-5347(98)01340-8

Richter S. 2005. Homologies in phylogenetic analyses—concept and tests. *Theory in Biosciences* 124:105–20. DOI:10.1007/BF02814479

Ricklefs, R. E., and D. Schluter (eds.). 1993. *Species Diversity in Ecological Communities: Historical and Geographical Perspectives*. Chicago (IL): The University of Chicago Press. ISBN:9780226718224.

Riddle, B. R., and D. J. Hafner. 1999. Species as units of analysis in ecology and biogeography: Time to take the blinders off. *Global Ecology and Biogeography* 8:433–41. DOI:10.1046/j.1365-2699.1999.00170.x

Ride, W. D. 1988. Towards a unified system of biological nomenclature. In *Prospects in Systematics*, ed. D. L. Hawksworth, 332–53. Oxford: Clarendon Press. ISBN:9780198577072.

Ridley, M. 1986. *Evolution and Classification. The Reformation of Cladism*. London: Longman. ISBN:0582444977.

Riedl, R. 1978. *Order in Living Organisms*. Chichester: John Wiley & Sons. p. 313 ISBN: 9780608176543.

Riegner, M. F. 2013. Ancestor of the new archetypal biology: Goethe's dynamic typology as a model for contemporary evolutionary developmental biology. *Studies in History and Philosophy of Biological and Biomedical Sciences* 44:735–44. DOI:10.1016/j.shpsc.2013.05.019

Rieppel, O. 1988. *Fundamentals of Comparative Biology*. Basel: Birkhauser Verlag. ISBN:9783764319564.

Rieppel, O. 2003. Popper and systematics. *Systematic Biology* 52:271–80. DOI:10.1080/10635150390192

Rieppel, O. 2005. The philosophy of total evidence and its relevance for phylogenetic inference. *Papéis Avulsos de Zoologia* 45:77–89. DOI:10.1590/S0031-10492005000800001

Rieppel, O. 2006a. The taxonomic hierarchy. *The Systematist* 26:5–9. ISSN:1744-5701.

Rieppel, O. 2006b. On concept formation in systematics. *Cladistics* 22:474–92. DOI:10.1111/j.1096-0031.2006.00114.x

Rieppel, O. 2006c. The *PhyloCode*: A critical discussion of its theoretical foundation. *Cladistics* 22:186–97. DOI:10.1111/j.1096-0031.2006.00097.x

Rieppel, O. 2007. The nature of parsimony and instrumentalism in systematics. *Journal of Zoological Systematics and Evolutionary Research* 45:177–83. DOI:10.1111/j.1439-0469.2007.00426.x

Rieppel, O. 2008. Origins, taxa, names and meanings. *Cladistics* 24:598–610. DOI:10.1111/j.1096-0031.2007.00195.x

Rieppel, O. 2009a. Hennig's enkaptic system. *Cladistics* 25:311–17. DOI:10.1111/j.1096-0031.2009.00260.x

Rieppel, O. 2009b. Total evidence in phylogenetic systematics. *Biology and Philosophy* 24:607–22. DOI:10.1111/j.1096-0031.1991.tb00045.x

Rieppel, O. 2010a. Monophyly and the two hierarchies. In *Beyond Cladistics: The Branching of a Paradigm*, eds. D. M. Williams, and S. Knapp, 147–167. Berkeley: University of California Press. 219–239. ISBN:9780520267725.

Rieppel, O. 2010b. The series, the network, and the tree: Changing metaphors of order in nature. *Biology and Philosophy* 25:475–96. DOI:10.1186/1471-2148-5-33

Rieppel, O. 2016. *Phylogenetic Systematics: Haeckel to Hennig*. Boca Raton (FL): CRC Press. ISNB:9781498754897.

Rieppel, O., M. Rieppel, and L. Rieppel. 2006. Logic in systematics. *Journal of Zoological Systematics and Evolutionary Research* 44:186–92. DOI:10.1111/j.1439-0469.2006.00370.x

Rinard, R. G. 1981. The problem of the organic individual: Ernst Haeckel and the development of the biogenetic law. *Journal of the History of Biology* 14:249–75. DOI:10.1007/BF00141094

Rineau, V., and S. Prin. 2024. Defining phylogenetic relationships: A complete and minimal set of combinatorial properties for triplets. *BioRxiv Preprints*. DOI:10.1101/2024.03.19.585714

Rivas, L. R. 1965. A Proposed code system for storage and retrieval of information in systematic zoology. *Systematic Zoology* 14:131–2. DOI:10.2307/2411736

Robson, G. C. 1928. *The Species Problem: An Introduction to the Study of Evolutionary Divergence in Natural Populations*. Edinburgh: Oliver & Boyd. https://babel.hathitrust.org/cgi/pt?id=mdp.39015005795557&seq=1

Rodendorf, B. B. 1977. [On the rationalization of the names of high-rank taxa in zoology.] *Paleontological Journal* 2:14–22. ISSN:0031-031X (in Russian).

Rohlf, F. J., D. H. Colless, and G. Hart. 1983. Taxonomic congruence: A reanalysis. In *Numerical Taxonomy*, vol 1, ed. J. Felsenstein, 82–6. Berlin: Springer-Verlag. DOI:10.1007/978-3-642-69024-2_10

Rohlf, F. J., and R. R. Sokal. 1981. Comparing numerical taxonomic studies. *Systematic Zoology* 30:459–90. DOI:10.2307/2413054

Rollins, R. C. 1965. On the bases of biological classification. *Taxon* 14:1–6. DOI:10.2307/1216700

Romanes, G. J. 1895. *Darwin, and after Darwin: An Exposition of the Darwinian Theory and a Discussion of the post-Darwinian Questions*, Vol. 2. […]. London: Longmans & Green. ISBN:9780559316449.

Romme, A. G. L. 2021. Ladders of authority, status, responsibility and ideology: Toward a typology of hierarchy in social systems. *Systems* 9:1–20. DOI:10.3390/systems9010020

Rorem, P. 1993. *Pseudo-Dionysius: A Commentary on the Texts and an Introduction to Their Influence*. New York (NY): Oxford University Press. ISBN:9780195076646.

Rosalinde, A. 2010. *Lines of Thought: Diagrammatic Representation and the Scientific Texts of the Arts Faculty, 1200–1500*. Doctoral thesis at Faculty of the Humanities, Leiden University. https://scholarlypublications.universiteitleiden.nl/handle/1887/16029

Rosch, E. 1973. Natural categories. *Cognitive Psychology* 4:328–50. DOI:10.1016/0010-0285(73)90017-0

Rosch, E. 1975. Family resemblances: Studies in the internal structure of categories. *Cognitive Psychology* 7:573–605. DOI:10.1016/0010-0285(75)90024-9

Rosch, E. 1977. Human categorization. In *Advances in Cross-Cultural Psychology*, Vol. 1, ed. N. Warren, 1–49. New York (NY): Academic Press. ISBN:9780126092011.

Rosch, E. 1983. Prototype classification and logical classification: The two systems. In *New Trends in Conceptual Representation*, ed. E. K. Scholnick, 73–86. New York (NY): Psychology Press. ISBN:9780203767658.

Rosch, E. 1988. Principles of categorization. In *Readings in Cognitive Science*, eds. A. Collins, and E. E. Smith, 312–22. DOI:10.1016/b978-1-4832-1446-7.50028-5

Rosch, E., Mervis C.B., Gray W.D. et al. 1976. Basic objects in natural categories. *Cognitive Psychology* 8:382–439. DOI:10.1016/0010-0285(76)90013-X

Rosen, R. 1983. Role of similarity principles in data extrapolation. *American Journal of Physiology* 244:R591–9. DOI:10.1152/ajpregu.1983.244.5.R591

Rössler, O. E. 1983. The chaotic hierarchy. *Zeitschrift für Naturforschung A* 38:788–801. DOI:10.1515/zna-1983-0714

Roth, V. L. 1988. The biological basis of homology. In *Ontogeny and Systematics*, ed. C. J. Humphries, 1–26. New York (NY): Columbia University Press. ISBN:9780231063708.

Roth, V. L. 1991. Homology and hierarchies: Problems solved and unresolved. *Journal of Evolutionary Biology* 4:167–94. DOI:10.1046/j.1420-9101.1991.4020167.x

Roux, J., and M. Robinson-Rechavi. 2010. An ontology to clarify homology-related concepts. *Trends in Genetics* 26:99–102. DOI:10.1016/j.tig.2009.12.012

Rozov, M. A. 1995. [Classification and theory as knowledge systems.] In [*Towards a Theory of Classification*], ed. S. S. Rozova, 81–127. Novosibirsk: Novosibirsk University Publication. ISBN:5887420049. (in Russian).

Rozova, S. S. 1986. [*Classification Problem in Modern Science*.] Moscow: Nauka. (in Russian).

Rozova, S. S. 2014. [*Theory of Social Relay in Epistemology and Philosophy of Science*.] Saarbrucken: Lambert Academic Publication. ISBN:978365918226. (in Russian).

Rüling J. P. 1766 *Commentatio Botanica de Ordinibus Naturalibus Plantarum [...]*. Goettingae: Lit. Fried. and. Rosenbush.

Ruse, M. 1979. Falsifiability, consilience, and systematics. *Systematic Zoology* 28:530–6. DOI:10.2307/sysbio/28.4.530

Ruse, M. 2009. *Monad to Man: The Concept of Progress in Evolutionary Biology*, 2d ed. Cambridge (MA): Harvard University Press. ISBN:9780674582200.

Saaty, T. L., and L. Vargas. 2022. *Models, Methods, Concepts & Applications of the The Analytic Hierarchy Process*, 2d ed. New York (NY): Springer Science+Business Media. ISBN:9781461435976.

Saaty, T. L., and L. G. Vargas. 2012. *Models, Methods, Concepts & Applications of the Analytic Hierarchy Process*, 2d ed. New York (NY): Springer Science+Business Media. ISBN:9781461435969.

Sachs, J. 1906. *History of Botany, 1530–1860*. Oxford: Clarendon Press. ISBN:9781408603901 (2007 reprint).

Saether, O. A. 1983. The canalized evolutionary potential: Inconsistencies in phylogenetic reasoning. *Systematic Zoology* 32:343–59. DOI:10.2307/2413162

Salam, A., and J. C. Taylor. 1990. *Unification of Fundamental Forces*. Cambridge (UK): Cambridge University Press. ISBN:9780521371407.

Salthe, S. N. 1985. *Evolving Hierarchical Systems: Their Structure and Representation*. New York (NY): Columbia University Press. ISBN:9780231060165.

Salthe, S. N. 2006. Summary of the principles of hierarchy theory. https://api.semanticscholar.org/CorpusID:15597110

Salthe, S. N. 2012. Hierarchical structures. *Axiomathes* 22:355–83. DOI:10.1007/s10516-012-9185-0

Salthe, S. N. 2015. Toward a Natural Philosophy of Macroevolution. In *Macroevolution. Explanation, Interpretation and Evidence*, eds. E. Serrelli, and N. Gontier, 163–82. Cham (Switzerland): Springer International Publishing Switzerland. ISBN:9783319150444.

Salvador, A. (ed.). 2013. *International Stratigraphic Guide: A Guide to Stratigraphic Classification, Terminology, and Procedure*, 2d ed. Boulder (CO): Geological Society of America. ISBN:0813774012.

Samofalova, I. A. 2013. [*Contemporasy Issues in the Soil Classifications.*] Perm: Perm Agriculture Academy Publication. ISBN:9785942791544. (in Russian).

Sanderson, M. J., A. Purvis, and C. Henze. 1998. Phylogenetic supertrees: Assembling the trees of life. *Trends in Ecology and Evolution* 13:105–9. DOI:10.1016/S0169-5347(97)01242-1

Santos, G. 2024. Emergence, downward causation, and interlevel integrative explanations. In *New Mechanism: Explanation, Emergence and Reduction*, eds. J. L. Cordovil, G. Santos, and D. Vecchi, 235–65. ISBN:9783031469169.

Santos, G.C. 2015. Upward and downward causation from a relational-horizontal ontological perspective. *Axiomathes* 25:23–40. DOI:10.1007/s10516-014-9251-x

Schafer, A. R. D. 2003. The "kinds" of Genesis 1: What is the meaning of mîn? *Journal of the Adventist Theological Society* 14:86–100. https://digitalcommons.andrews.edu/jats/vol14/iss1/5

Schertl, H.-P., S. J. Mills, and W. V. Maresch (eds.). 2019. *A Compendium of IMA-Approved Mineral Nomenclature.* Stuttgart: E. Schweizerbart'sche Verlagsbuchhandlung. ISBN:9783510654284.

Schindel, D. E., and S. E. Miller. 2010. Provisional nomenclature: The onramp to taxonomic names. In *Systema Naturae 250—The Linnaean Ark*, ed. A. Polaszek, 109–15. Boca Raton (FL): CRC Press. ISBN:9781420095012.

Schmalhausen, I. I. 1939. [*The Routes and Regularities in the Evolutionary Process.*] Moscow/Leningrad: AN SSSR. (in Russian).

Schmalhausen, I. I. 1968. [*Cybernetic Issues in Biology.*] Novosibirsk: Nauka (Siberian branch). ISBN:9785458468817. (in Russian).

Schoch, R. M. 1986. *Phylogeny Reconstruction in Paleontology.* New York (NY): Van Nostrand Reinhold. ISBN:9780442279677.

Schreyder, Yu. A. 1996. [Rank distributions as a system property.] *Tsenologicheskie Issledovania* 1–2:33–42. https://www.kudrinbi.ru/public/441/index.htm. (in Russian).

Schreyder, Yu. A., and A. A. Sharov. 1982. [*Systems and Models.*] Moscow: Radio & Svyaz Publ. ISBN:9785971089667 (2021 reprint). (in Russian).

Schroeder. M. 1991. *Fractals, Chaos, Power Laws: Minutes From an Infinite Paradise.* New York (NY): W.H. Freeman and Co. ISBN:9780486472041.

Schuh, R. T. 2000. *Biological Systematics. Principles and Applications.* Ithaca: Cornell University Press. ISBN:9780801447990.

Schuh, R. T. 2003. The Linnaean system and its 250- year persistence. *The Botanical Review* 69:59–78. DOI:10.1663/0006-8101(2003)069[0059:TLSAIY]2.0.CO;2

Schulz, S., H. Stenzhorn, and M. Boeber. 2008. The ontology of biological taxa. *Bioinformatics* 24, i313–i321. DOI:10.1093/bioinformatics/btn158

Schwartz, S. S. 1980. [*Ecological Patterns of Evolution.*] Moscow: Nauka. (in Russian).

Scopoli, G. A. 1777. *Introductio ad historiam naturalem sistens genera lapideum, plantarum et animalinm [...].* Pragae: Wolfgangum Gerle. ISBN:9781120300997 (2009 reprint).

Scotland, R. W. 2010. Deep homology: A view from systematics. *Bioessays* 32:438–49. DOI:10.1002/bies.200900175

Scott, A. D., and D. A. Baum. 2016. Phylogenetic tree. In *Encyclopedia of Evolutionary Biology*, ed. R. M. Kliman, 270–76. Oxford: Elsevier. DOI:10.1016/b978-0-12-800049-6.00203-1

Scott, A. D., D. M. King, S. W. Ordway, et al. 2022. Phase transitions in evolutionary dynamics. *Chaos* 32:122101. DOI:10.1063/5.0124274

Semenov-Tyan-Shansky, A. 1910. [Taxonomic boundaries of species and its subdivisions: An experience of precise categorization of lower taxonomic units.] *Notes of the Imperial Academy of Sciences, Physical and Mathematical Department* 25:1–29. (in Russian).

Semple, C., and M. Steel. 2003. *Phylogenetics*. Oxford: Oxford University Press. ISBN:9780198509424.

Sen, D. 2014. The uncertainty relations in quantum mechanics. *Current Science* 107: 203–18. DOI:10.13140/2.1.5183.0406

Sepkoski, J. J. 1996. Patterns of Phanerozoic extinction: A perspective from global data bases. In *Global Events and Event Stratigraphy in the Phanerozoic*, ed. O.H. Walliser, 35–51. Berlin—Heidelberg: Springer. ISBN:9783642796364.

Sereno, P. C. 2007. Logical basis for morphological characters in phylogenetics. *Cladistics* 23:565–87. DOI:10.1111/j.1096-0031.2007.00161.x

Sewertzoff A. N. 1929. Directions of evolution. *Acta Zoologica* 10:59–141. DOI:10.1111/j.1463-6395.1929.tb00695.x

Shanahan, T. 2011. Phylogenetic inertia and Darwin's higher law. *Studies in History and Philosophy of Science C* 42:60–8. DOI:10.1016/j.shpsc.2010.11.013

Shapiro, L., and E. Sober. 2007. Epiphenomenalism. The Dos and Don'ts. In *Thinking About Causes: From Greek Philosophy to Modern Physics*, eds. G. Wolter, and P. Machamer, 235–64. Pittsburgh: University of Pittsburgh Press. ISBN:9780822943099

Shapiro, S., and T. K. Kissel. 2024. Classical logic. In *The Stanford Encyclopedia of Philosophy*, eds. E. N. Zalta, and U. Nodelman. https://plato.stanford.edu/archives/spr2024/entries/logic-classical/

Sharkey, M. J. 1989. A hypothesis-independent method of character weighting for cladistic analysis. *Cladistics* 5: 63–86. DOI:10.1111/j.1096-0031.1989.tb00483.x

Shatalkin, A. I. 1983. [Methodological aspects of the application of mathematical methods in systematics.] In [*Theory and Methodology of Biological Classifications*] 46–55. Moscow: Moscow Society of Naturalists. (in Russian).

Shatalkin, A. I. 1988. [*Biological Systematics*.] Moscow: Moscow State University Publication. ISBN:9785211001459. (in Russian).

Shatalkin, A. I. 1990. [Similarity and homology.] *Zhurnal Obshchei Biologii* 51:841–9. ISSN:0044-4596. (in Russian).

Shatalkin, A. I. 1993. [Aristotle and systematics: On the foundations of typology.] *Zhurnal Obshchei Biologii* 54:243–52. ISSN:0044-4596. (in Russian).

Shatalkin, A. I. 1994. [Typological approach to systematics.] *Zhurnal Obshchei Biologii* 55:661–72. ISSN:0044-4596 (in Russian).

Shatalkin, A. I. 1995. [Hierarchies in systematics: A set theory model.] *Zhurnal Obshchei Biologii* 56:277–90. ISSN:0044-4596 (in Russian).

Shatalkin, A. I. 2004. [The highest level of division in the classification of organisms. 1. Prokaryotes and eukaryotes.] *Zhurnal Obshchei Biologii* 65:19–38. ISSN:0044-4596. (in Russian).

Shatalkin, A. I. 2012. [*Taxonomy. Grounds, Principles, and Rules*.] Moscow: KMK Scientific Press. ISBN:9785873178476. (in Russian).

Shifman, I. S. 2007. [*The Old Testament and its World*.] St. Petersburg: Saint Petersburg State University Publ. (in Russian). ISBN:9785000000000.

Shipunov, A. B. 1999. [*Basics of a Theory of Systematics*.] Moscow: Open Lyceum VZMS. (in Russian).

Shishkin, M. A. 2006. Development and lessons of evolutionism. *Russian Journal of Developmental Biology* 37:146–62. DOI:10.1134/S1062360406030039

Shishov, L. L., V. D. Tonkonogov, I. I. Lebedeva, et al. 2004. [*Classification and Diagnostics of the Russian Soils.*] Smolensk: The Ecumene Publication. ISBN:5935200449. (in Russian).

Shubin, N., C. Tabin, and S. Carroll. 2009. Deep homology and the origins of evolutionary novelty. *Nature* 457:818–23. DOI:10.1038/nature07891

Shull, G. H. 1923. The species concept from the point of view of a geneticist. *American Journal of Botany* 5:221–8. ISSN:0002-9122.

Shuman, A. N. 2001. [*Philosophical Logic: Origins and Evolution.*] Minsk: EkonomPress. ISBN:9856479266. (in Russian).

Sibley, C. G., and J. E. Ahlquist. 1990. *Phylogeny and Classification of Birds: A Study in Molecular Evolution.* New Haven (CT): Yale University Press. ISBN:9780300040852.

Siddell, S. G., P. J. Walker, E. J. Lefkowitz, et al. 2020. Binomial nomenclature for virus species: A consultation. *Archives of Virology* 165:519– 25. DOI:10.1007/s00705-019-04477-6

Sierpinski, W. 1965. *Cardinal and Ordinal Numbers.* Warsaw: PWN. ISBN:9780900318023.

Sigwart, J. D. 2022. Taxonomic hierarchies as a tool for coping with the complexity of biodiversity. In *Species Problems and Beyond. Contemporary Issues in Philosophy and Practice*, eds. J. S. Wilkins, F. E. Zachos, and I. Ya. Pavlinov, 227–42. Boca Raton (FL): CRC Press. ISBN:9781032221472.

Simmons, H. 2011. *An Introduction to Category Theory.* Cambridge (UK): Cambridge University Press. ISBN:9780521283045.

Simon, H. A. 1962. The architecture of complexity. *Proceedings of the American Philosophical Society* 106:467–82. DOI:10.1007/978-1-4899-0718-9_31

Simon, H. A. 1973. The organization of complex systems. In *Hierarchy Theory: The Challenge of Complex Systems*, ed. H. H. Pattee, 1–28. New York (NY): George Braziller Publication. ISBN:9780807606742.

Simon, H. A. 1976. How complex are complex systems? In *Proceedings of the Biennial Meeting of the Philosophy of Science Association*, Vol. 2, 507–522. Cincinnati (OH): Philosophy of Science Association. ISBN:9780917586033.

Simon, H. A. 1991. The architecture of complexity. In *Facets of Systems Science*, ed. G.J. Klir, 15–44. Boston (MA): Springer-Verlag. DOI:10.1007/978-1-4899-0718-9_31

Simpson, G. G. 1940. Types in modern taxonomy. *American Journal of Science* 238:413–31. DOI:10.2475/ajs.238.6.413

Simpson, G. G. 1944. *Tempo and Mode in Evolution.* New York (NY): Columbia University Press. ISBN:9780231058476.

Simpson, G. G. 1945. The principles of classification and a classification of mammals. *Bulletin of the American Museum of Natural History* 85:1–350. ISSN:0003-0090.

Simpson, G. G. 1951a. The species concept. *Evolution* 5:285–98. DOI:10.1111/j.1558-5646.1951.tb02788.x

Simpson, G. G. 1951b. *Horses: The Story of the Horse Family in the Modern World and through Sixty Million Years of History.* Oxford: Oxford University Press. ISBN:9780758168276.

Simpson, G. G. 1953. *The Major Features of Evolution.* New York (NY): Columbia University Press. ISBN:9780231018210.

Simpson, G. G. 1961. *Principles of Animal Taxonomy.* New York (NY): Columbia University Press. ISBN:9780231024273.

Siu, R. G. H., and E. T. Reese. 1955. Proposal for a system of biological nomenclature, with special reference to microorganisms. *Farlowia: A Journal of Crypogamic Botany* 4:399–407. www.biodiversitylibrary.org/item/33577#page/425/mode/1up

Skarlato, O. A., and Y. I. Starobogatov. 1974. [Phylogenetics and principles of elaboration of the natural system.] In *Theoretical Issues in Systematics and Phylogeny of Animals*, 30–46. Leningrad: Nauka. (in Russian).

Sklar, A. 1964. On category overlapping in taxonomy. In *Form and Strategy in Science*, eds. J. R. Gregg, and F. T. C. Harris, 395–401. Dordrecht: D. Reidel. ISBN:9789401036030.

Skvortsov, A. K. 1967. [The main stages in the development of the species concept.] *Bulleten Moskovskogo Obshestva Ispytatelei Prirody, Biology*. 72:11–27. ISSN:0027-1403. (in Russian).

Slack, J. M. W., P. W. H. Holland, and C. F. Graham 1993. The zootype and the phylotypic stage. *Nature* 361:490–2. DOI:10.1038/361490a0

Slater, M. H. 2015. Natural kindness. *The British Journal for the Philosophy of Science* 66:375–411. DOI:10.1093/bjps/axt033

Slaughter, M. 1982. *Universal Languages and Scientific Taxonomy in the Seventeenth Century*. Cambridge (UK): Cambridge University Press. ISBN:9780521135443.

Sloan, P. R. 1970. *The History of the Concept of the Biological Species in the Seventeenth and Eighteenth Centuries, and the Origin of the Species Problem*. Doctoral dissertation. San Diego (CA): University of California. https://www.proquest.com/openview/2459127cf68b964cb6df1d210a1d4664/1?cbl=18750&diss=y&pq-origsite=gscholar

Sloan, P. R. 1972. John Locke, John Ray, and the problem of the natural system. *Journal of the History of Biology* 5:1–53. DOI:10.1007/BF02113485

Sloan, P. R. 1979. Buffon, German biology, and the historical interpretation of biological species. *British Journal of History of Science* 12:109–53. DOI:10.1017/S0007087400017027

Sloan, P. R. 1986. From logical universals to historical individuals: Buffon's idea of biological species. In *Histoire du Concept D'espèce dans les Sciences de la Vie*, eds. J. Roger and J. L. Fisher, 101–40. Paris: Fondation Singer-Polignac. ISBN:9782900927199.

Sloan, P. R. 2009. Originating species: Darwin on the species problem. In *The Cambridge companion to the "Origin of Species"*, eds. M. Ruse and R. Richards, 67–86. Cambridge: Cambridge University Press. ISBN: 9780521691291.

Sluys, R. 1991. Species concepts, process analysis, and the hierarchy of nature. *Experientia* 47:1162–70. DOI:10.1007/BF01918380

Sluys, R. 1996. The notion of homology in current comparative biology. *Journal of Zoological Systematics and Evolutionary Research* 34:145–52. DOI:10.1111/j.1439-0469.1996.tb00820.x

Smaling, A. 1992. Varieties of methodological intersubjectivity: The relations with qualitative and quantitative research and with objectivity. *Quality and Quantity* 26:169–80. DOI:10.1007/BF02273552

Smirnov, E. S. 1923. [On the structure of systematic categories.] *Russky Zoologichesky Zhurnal* 3:358–89. ISSN:0044-5134. (in Russian).

Smirnov, E. S. 1969. [*Taxonomic Analysis*.] Moscow: Moscow State University Publ. (in Russian).

Smith, A. B. 1994. *Systematics and the Fossil Record: Documenting Evolutionary Patterns*. Oxford: Blackwell Science. ISBN:9780632036424.

Smith, B. 2004. Beyond concepts: Ontology as reality representation. In *Proceedings of FOIS 2004. International Conference on Formal Ontology and Information Systems, Turin, 4–6 November 2004*, eds. A. Varzi, and L. Vieu. https://api.semanticscholar.org/CorpusID:2132367

Smith, E. E., and D. L. Medin, 1981. *Categories and Concepts*. Cambridge (MA): Harvard University Press. ISBN:9780674102750.

Smith, H. M. 1962. The hierarchy of nomenclatural status of generic and specific names in zoological taxonomy. *Systematic Zoology* 11:139–42. DOI:10.2307/2411877

Smith, N. 1987. Dangers of the empirical turn: Comments on the CURS initiative. *Antipode* 19 59–68. DOI:10.1111/j.1467-8330.1987.tb00149.x

Sneath, P. 2000. Reticulate evolution in bacteria and other organisms: How can we study it? *Journal of Classification* 17:159–63. DOI:10.1007/s003570000014

Sneath, P. H. A., and R. R. Sokal. 1973. *Numercial Taxonomy. The Principles and Methods of Numerical Classification*. San Francisco: W.H. Freeman & Co. ISBN:9780716706977.

Sneppen, K., and P. Minnhagen. 2004. Hierarchy and anti-hierarchy in real and scale free networks. *Physical Review Letters* 92:178702. https://researchprofiles.ku.dk/en/publications/hierarchy-and-anti-hierarchy-in-real-and-scale-free-networks

Sober, E. 1984. Sets, species, and evolution. *Philosophy of Science* 51:334–41. DOI:10.1016/j.sjbs.2017.04.013

Sober E. 1988. *Reconstructing the Past. Parsimony, Evolution, and Inference*. Cambridge (MA): The MIT Press. ISBN:9780262192736.

Sober, E. (ed.). 1994. *Conceptual Issues in Evolutionary Biology*. Cambridge (MA): MIT Press. ISBN:9780262691628.

Sokal, R. R., and F. J. Rohlf. 2009. *Introduction to Biostatistics*, 2d ed. Mineola (NY): Dover Publ. ISBN:9780486469614.

Sokal, R. R., and R. H. A. Sneath. 1963. *Principles of Numerical Taxonomy*. San Francisco (CA): W.H. Freeman & Co. ISBN:9780716706212.

Sokolov, B. S. (ed.). 1980. [*Stratigraphic Classification. Materials for the Problem.*] Leningrad: Nauka. (in Russian).

Solomatin, N. M., and M. F. Menyaev. 1982. [*Information Semantic Systems.*] Moscow: Znanie. ISBN:5060005437. (in Russian).

Soltis, P. S., D. E. Soltis, and J. J. Doyle (eds.). 1992. *Molecular Systematics of Plants*. Dordrecht: Science+Business Media. ISBN:9780412022418.

Soltis, P. S., D. E. Soltis, and J. J. Doyle (eds.). 1998. *Molecular Systematics of Plants, II. DNA Sequencing*. New York (NY): Science+Business Media. ISBN:9780412111310.

Sonnhammer, E. L. L., and E. V. Koonin. 2002. Orthology, paralogy and proposed classification for paralog subtypes. *Trends in Genetics*. 18:619–20. DOI:10.1016/s0168-9525(02)02793-2

Sowa, J. F. 1976. Conceptual graphs for a data base interface. *IBM Journal of Research and Development* 20:336–57. DOI:10.1147/rd.204.0336

Spade, P. V. 1994. *Five texts on the Mediaeval Problem of Universals: Porphyry, Boethius, Abelard, Duns Scotus, Ockham*. Indianapolis (IN): Hackett Publishing. ISBN:9780872202504

Spencer, Q. 2016. Genuine kinds and scientific reality. In *Natural Kinds and Classification in Scientific Practice*, ed. C. Kendig, 184–99. Abingdon: Routledge. ISBN:9781138344839.

Spiridonov, A., and N. Eldredge. 2024. The Bretskyan hierarchy, multiscale allopatry, and geobiomes—on the nature of evolutionary things. *Paleobiology* 50: 194–213. DOI: 10.1017/pab.2023.37

Sprenger, J., and S. Hartmann. 2019. *Bayesian Philosophy of Science*. Oxford: Oxford University Press. ISBN:9780199672110.

Spruit, L. 1994. *Species Intelligibilis. From Perception to Knowledge, Vol. 1. Classical Roots and Medieval Discussions*. Leiden: E. J. Brill. ISBN:9789004098831.

Stafleu, F. A. 1971. *Linnaeus and Linnaeans*. Utrecht: A. Oosthoek. ISBN:9789060460641.

Staley, J. T. 2006. The bacterial species dilemma and the genomic-phylogenetic species concept. *Philosophical Transactions of the Royal Society of London Ser. B* 361:1899–909. DOI:10.1098/rstb.2006.1914

Stamos, D. N. 1996. Was Darwin really a species nominalist? *Journal of the History of Biology* 29:127–44. DOI:10.1007/BF00129698

Stamos, D. N. 2001. Quantum indeterminism and evolutionary biology. *Philosophy of Science* 68:164–84. DOI:10.1086/392871

Stamos, D. N. 2003. *The Species Problem. Biological Species, Ontology, and the Metaphysics of Biology*. Oxford: Lexington Books. ISBN:9780739107782.

Stamos, D. N. 2007. Popper, laws, and the exclusion of biology from genuine science. *Acta Biotheoretica* 55:357–75. DOI:10.1007/s10441-007-9025-6

Stanford, P. K. 2016. Instrumentalism: Global, local, and epistemic. In *The Oxford Handbook of the Philosophy of Science*, ed. P. Humphreys, 318–36. Oxford: Oxford University Press. ISBN:9780199368815.

Stark, Y.S. 1966. [The prohibition principle in the systematics of microorganisms.] *Izvestia Akademii Nauk of the USSR, Biology Series* 5:686–93. ISSN:0002-3329. (in Russian).

Starobogatov, I. Y. 1989. [Natural system, artificial systems and some principles of phylogenetic and systematic research.] *Trudy Zoologicheskogo Instituta AN USSR* 206:191–222. ISSN:0206-0477. (in Russian).

Starobogatov, I.Y. 1991. Problems in the nomenclature of higher taxonomic categories. *Bulletin of Zoological Nomenclature* 48:6–18. www.biodiversitylibrary.org/item/44489#page/22/mode/1up

Stekolnikov, A. A. 2003. [The problem of truth in biological systematics.] *Zhurnal Obshchei Biologii* 64:357–68. ISSN:0044-4596. (in Russian).

Stekolnikov, A. A. 2007. [The truth of classifications in systematics.] *Archive of the Zoological Museum of Moscow State University* 48: 101–22. ISSN:0134-8647. (in Russian).

Stenzel, H. B. 1950. Proposed uniform endings for names of higher categories in zoological systematics. *Science* 112:94. DOI:10.1126/science.112.2899.94

Stepin, V. S. 2005. *Theoretical Knowledge*. Heidelberg: Springer-Verlag. ISBN:9781402030468.

Stevens, P. 1984. Metaphors and typology in the development of botanical systematics 1690–1960, or the art of putting new wine in old bottles. *Taxon* 33:169–211. DOI:10.2307/1221161

Stevens, P. F. 1983. Augustin Augier's "Arbre botanique" (1801), a remarkable early botanical representation of the Natural System. *Taxon* 32:203–11. DOI:10.2307/1221972

Stevens, P. F. 1994. *The Development of Biological Systematics. Antoine-Laurent de Jussieu, Nature, and the Natural System.* New York (NY): Columbia University Press. ISBN:9780231064408.

Stevens, P. F. 2002. Why do we name organisms? Some reminders from the past. *Taxon* 51:11–26. DOI:10.2307/1554959

Stoljar, D. 2024. Physicalism. In *The Stanford Encyclopedia of Philosophy*, eds. E. N. Zalta, and U. Nodelman. https://plato.stanford.edu/archives/spr2024/entries/physicalism/

Storr, G. C. 1780 *Prodromus Methodi Mammalium [...]*. Tubingiae: Reissianis.

Strange, S. K. 1987. Plotinus, Porphyry, and the neoplatonic interpretation of the 'Categories'. In *Philosophie, Wissenschaften, Technik*, ed. W. Haase, 955–74. Berlin: De Gruyter. ISBN:9783110103922.

Strickland, H. E. 1841. On the true method of discovering the natural system in zoology and botany. *Annals and Magazine of Natural History, Ser.* 5, 6:184–94. DOI:10.1080/03745484009443283

Strickland, H. E. 1846. On the structural relations of organized beings. *Philosophical Magazine and Journal of Science* 28:354–64. DOI:10.1080/14786444608645434

Striedter, G. F. 1998. Stepping into the same river twice: Homologues as recurring attractors in epigenetic landscapes. *Brain Behavior and Evolution* 52:218–31. DOI:10.1159/000006565

Striedter, G. F., and R. G. Northcutt. 1991. Biological hierarchies and the concept of homology. *Brain Behavior and Evolution.* 38:177–89. DOI:10.1159/000114387

Stuessy, T. F. 1996. Classification: More than just branching patterns of evolution. *Aliso* 15: 113–24. DOI:10.5642/aliso.19961502.06

Stuessy, T. F. 2000. Taxon names are not defined. *Taxon* 49:231–3. DOI:10.2307/1223709

Stuessy, T. F. 2009. *Plant taxonomy. The Systematic Evaluation of Comparative Data*, 2nd ed. New York (NY): Columbia University Press. ISBN:9780231147125.

Stuessy, T. F., and E. Hörandl. 2014. Evolutionary systematics and paraphyly: Introduction. *Annals of the Missouri Botanical Garden* 100:2–5. DOI:10.3417/2012083

Stuessy, T. F., and C. König. 2008. Patrocladistic classification. *Taxon* 57:594–601. https://www.jstor.org/stable/25066026

Stumme, G. 1999. Hierarchies of conceptual scales. In *Proceedings of the 11th Workshop on Knowledge Acquisition, Modeling and Management*, Vol. 2, eds. T. B. Gaines, R. Kremer, and M. Musen. Banff: 78–95. Banff (AB): Banff Centre Press. ISBN:9783540660446.

Subbotin, A. L. 2001. [*Classification.*] Moscow: Institute of Philosophy RAN. ISBN:5201020461. (in Russian).

Suchecki, K., and J. Hołyst. 2022. Hierarchy depth in directed networks. *Entropy* 24:252. DOI:10.3390/e24020252

Suppes, P. 1978. The plurality of science. In *Proceedings of the Biennial Meeting of the PhilosophyofScienceAssociation*(1978):3–16.DOI:10.1086/psaprocbienmeetp.1978.2.192459

Suppes P. 1992. Axiomatic methods in science. In *Nature, Cognition, and System II. Current Systems-Scientific Research on Natural and Cognitive Systems. Vol. 2: On Complementarity and Beyond*, ed. M. E. Carvallo, 205–232. Dordrecht: Springer-Verlag. ISBN 9789401052344.

Suzuki, D. and H. Saigo. 2025. A category-theoretic interpretation of the homology concept in biology. *PhilSci Archive*. https://philsci-archive.pitt.edu/25010/

Suzuki, D. G., and S. Tanaka. 2017. A phenomenological and dynamic view of homology: Homologs as persistently reproducible modules. *Biological Theory* 12:169–80. DOI:10.1007/s13752-017-0265-7

Svensson, M. E. 2004. Homology and homocracy revisited: gene expression patterns and hypotheses of homology. *Development Genes and Evolution* 214:418–21. DOI:10.1007/s00427-004-0416-2

Sviridov, A. V. 1994. [*Types of Diagnostic Keys and Their Application.*] Moscow: Zoological Museum of MSU. ISBN:5211032624. (in Russian).

Swainson, W. 1837. *On the Natural History and Classification of Birds*, Vol. 2. London: Longman. ISBN:9781402180811 (2001 reprint).

Swofford, D., G. J. Olsen, P. J. Waddell, et al. 1996. Phylogenetic inference. In *Molecular Systematics*, 2nd ed., eds. D. M. Hillis, C. Moritz, and B. K. Mable, 407–514. Sunderland: Sinauer Association. ISBN:9780878932825.

Swoyer, C. 2006. Conceptualism. In *Universals, Concepts, and Qualities: New Essays on the Meaning of Predicates*, eds. E. S. Trawson, and A. Chakrabarti, 127–54. Routledge: CRC Press. ISBN:9780754650324.

Sylvester-Bradley, P. C. 1952. *The Classification and Coordination Of Infraspecific Categories*. London: System Association.

Szalay, F., and W. Bock. 1991. Evolutionary theory and systematics: Relationships between process and patterns. *Zeitschrift für Zoologische Systematik und Evolutions-forschung* 29:1–39. DOI:10.1111/j.1439-0469.1991.tb00442.x

Szostak, R. 2020. The basic concepts classification (BCC). *Knowledge Organization* 47:231–43. https://www.isko.org/cyclo/bcc

Szucsich, N. U., and C. S. Wirkner. 2007. Homology: A synthetic concept of evolutionary robustness of patterns. *Zoologica Scripta* 36:281–9. DOI:10.1111/j.1463-6409.2007.00275.x

Takhtadjan, A. L. 1966. [*The System and Phylogeny of Flowering Plants.*] Moscow–Leningrad: Nauka. (in Russian).

Talmy, L. 2000. *Toward a Cognitive Semantics. Vol. 1. Concept Structuring Systems*. Cambridge (MA): MIT Press. ISBN:9780262700962.

Tarasov, S. V., and V. V. Burakov. 2013. [Relational modeling methods for the hierarchical data structures.] *Informatsionno-Upravlyaushchie Sistemy* 6:58–66. ISSN:1684-8853. (in Russian).

Tassy, P. 2011. Trees before and after Darwin. *Journal of Zoological Systematics and Evolutionary Research* 49:89–101. DOI:10.1111/j.1439-0469.2010.00585.x

Tatarinov, L. P. 1984. Cladistic analysis and phylogenetics. *Paleontologichesky Zhurnal* 3:3–16. ISSN:0031-0301.

Taylor, J. R., and R. E. MacLaury (eds.). 1995. *Language and the Cognitive Construal of the World*. Berlin: Mouton de Gruyter. ISBN:9783110143010.

Tchaikovsky, Yu. V. 1990. [*Elements of Evolutionary Diatropics*.] Moscow: Nauka. ISBN:5020080861. (in Russian).

Tëmkin, I., and N. Eldredge. 2015. Networks and hierarchies: Approaching complexity in evolutionary theory. In *Macroevolution. Explanation, Interpretation and Evidence*, eds. E. Serrelli, and N. Gontier, 183–226. Cham (Switzerland): Springer International Publishing Switzerland. ISBN:9783319150444.

Tennis, J. T. 2015. Foundational, first-order, and second-order classification theory. *Knowledge Organization* 42:244–9. DOI:10.5771/0943-7444-2015-4-244

Tetenyi, P. 2013. Homology of biosynthetic routes: The base in chemotaxonomy. In *Chemistry in Botanical Classification: Medicine and Natural Sciences*, ed. G. Bendz, and J. Santesson, 67–80. New York (NY): Academic Press. ISBN:978-0123942494.

Thagard, P. 2023. Cognitive science. In *The Stanford Encyclopedia of Philosophy*, eds. E. N. Zalta, and U. Nodelman. https://plato.stanford.edu/archives/win2023/entries/cognitive-science/

The Holy. 2025. *The Holy Bible, Berean Standard Bible*. Pittsburgh (PA): Bible Hub.

Theissen, G. 2005. Birth, life and death of developmental control genes: New challenges for the homology concept. *Theory in Biosciences*. 124:199–12. DOI:10.1007/BF02814484

Theophrastus. 1916. *Enquiry Into Plants*, Vol. ss1: Books 1–5. Cambridge (MA): Harvard University Press. ISBN:9780674990883.

Theophrastus. 2010. *On First Principles (Known as his Metaphysics) […]*. Leiden: E. J. Brill. ISBN:9789004179035.

Thompson, W. R. 1952. The philosophical foundations of systematics. *Canadian Entomologist* 84:1–16. DOI:10.4039/Ent841-1

Thorp, W. H. 1940. Ecology and the future of systematics. In *The New Systematics*, ed. J. Huxley, 341–64. London: Oxford University Press. ISBN:9780403017867.

Tilling, S. 1984. Keys to biological identification: Their role and construction. *Journal of Biological Education* 18:293–304. DOI:10.1080/00219266.1984.9654660

Timiryazev, K. A. 1904. [An essay on Darwin's theory.] In [*Theory of Development*,] ed. V. A. Fausek, 49–143. St. Petersburg: Brockhaus & Efron. (in Russian).

Torday, J. S. 2016. Life is simple: Biologic complexity is an epiphenomenon. *Biology*, 5:17. DOI:10.3390/biology5020017

Toulmin, S. E. 1972. *Human Understanding: The Collective Use and Evolution of Concepts*. Princeton (NJ): Princeton University Press. ISBN:978691071855.

Tremblay, B. 2008. Albertus Magnus on the subject of Aristotle's categories. In *Medieval Commentaries on Aristotle's Categories*, ed. L. A. Newton, 73–98. Leiden: E. J. Brill. ISBN:9789047442073.

Treueman, A. E. 1924. The species-concept in palaeontology. *Geological Magazine* 61:355–60. doi:10.1017/S001675680008660X

Troll, W. 1951. Biomorphologie und Biosystematik als typologische Wissenschaften. *Studium Generale* 4:376–89. ISSN:0039-4149.

Trusina, A., S. Maslov, P. Minnhagen, et al. 2004. Hierarchy measures in complex networks. *Physical Review Letters* 92: 178702/1–4. DOI:10.1103/PhysRevLett.92.178702

Tunnicliffe, M., and G. Hunter. 2022. Random sampling of the Zipf–Mandelbrot distribution as a representation of vocabulary growth. *Physica A: Statistical Mechanics and its Applications* 608: 128259. DOI:10.1016/j.physa.2022.128259

Turbin, A. F., and Pratsevityi N. F. 1992. [*Fractal Sets, Functions, Distributions*.] Kïev: Naukova Dumka. ISBN:5922102664. (in Russian).

Turesson, G. 1922. The species and the varieties as ecological units. *Hereditas* 3:100–13.

Turner, D. 2011. *Paleontology. A Philosophical Introduction*. Cambridge (UK): Cambridge University Press. ISBN:9780521116374.

Turrill, W. B. 1925. Species. *The Journal of Botany British and Foreign* 63:359–66. ISSN:0368-1513.

Turrill, W. B. 1938. The expansion of taxonomy with special reference to spermatophyta. *Biological Reviews* 13:342–73. DOI:10.1111/j.1469-185X.1938.tb00522.x

Tversky, A. 1977. Features of similarity. *Psychological Review* 84:327–52. DOI 10.1037/0033-295X.84.4.327

Tversky, B. 1989. Parts, partonomies, and taxonomies. *Developmental Psychology* 25:983–95. DOI:10.1037/0012-1649.25.6.983

Tversky, B. 2013. Where partonomies and taxonomies meet. In *Meanings and Prototypes. Studies in linguistic categorization*, 2d ed., S. L. Tsohatzidis, 334–44. Abingdon: Routledge. ISBN:9780415036122.

Tyler, M. J. 1991. Biological nomenclature, classification and the ethnozoological specieme. Man and a half: Essays in Pacific anthropology and ethnobiology in honour of Ralph Bulmer. *The Journal of the Polynesian Society, Memoirs (Add.)* 48:164–7. ISBN:9780908940011.

Upchurch, P., and C. A. Hunn. 2002. "Time": The neglected dimension in cladistic biogeography? *Geobios* 35: 277–86. DOI:10.1016/S0016-6995(02)00065-7

Urbanowicz, R. J., and W. N. Browne. 2017. *Introduction to Learning Classifier Systems*. Berlin: Springer-Verlag. ISBN: 9783662550069.

Urmantsev, Yu. A. 1988. [General system theory: state, application, and prospects of the development.] In [*System. Symmetry. Harmony*,] ed. V. S. Tyukhtin, 38–124. Moscow: Mysl. ISBN: 9785244001907. (in Russian).

Ursul, A. D. 2010. [*The Nature of Information: A Philosophical Essay*,] 2d ed. Chelyabinsk: Chelyabinsk State Academy of Culture and Arts. ISBN:9785948392486. (in Russian).

Uryson, E. V. 2003. [*Research Problems of the Linguistic World Picture. Analogy in Semantics*.] Moscow: YaSK. ISBN:5944571233. (in Russian).

Usov, S. A. 1867. [*Taxonomic Units and Groups*.] Moscow: University Typogr. (in Russian).

Valentin, R., and S. Prin. 2024. Defining phylogenetic relationships: A complete and minimal set of combinatorial properties for triplets. *License CC BY-NC-ND 4.0*. DOI:10.1101/2024.03.19.585714

Valentine, D. H. 1949. The units of experimental taxonomy. *Acta Biotheoretica* 9:75–88. DOI:10.1007/BF01556761

Valentine D. H., and A. Löve. 1958. Taxonomic and biosystematic categories. *Brittonia* 10:153–66. DOI:10.2307/2804945

Valentine, J. W., and C. L. May. 1996. Hierarchies in biology and paleontology. *Paleobiology* 22:23–33. DOI:10.1017/S0094837300015992

Vamosi, J. C., and Vamosi S. M. 2011. Factors influencing diversification in angiosperms: At the crossroads of intrinsic and extrinsic traits. *American Journal of Botany* 98:460–71. DOI:10.3732/ajb.1000311

van Fraassen, B. 1970. *An Introduction to the Philosophy of Space and Time*. New York (NY): Random House. ISBN:9780231061322.

van Fraassen, B. C. 2008. *Scientific Representation: Paradoxes of Perspective*. Oxford (UK): Oxford University Press. ISBN:97809278220.

Van Valen, L. 1964. An analysis of some taxonomic concepts. In *Form and Strategy in Science. Studies Dedicated to Joseph Henry Woodger on the Occasion of his Seventieth Birthday*, eds. J. R. Gregg, and F. T. C. Harris, 402–15. Dordrecht: D. Reidel Publication. ISBN:9789401036054.

Van Valen, L. 1971. Adaptive zones and the orders of mammals. *Evolution* 25:420–8. DOI:10.1111/j.1558-5646.1971.tb01898.x

Van Valen, L. M. 1973. Are categories in different phyla comparable? *Taxon*, 22:333–73. DOI:10.2307/1219322

Van Valen, L. M. 1982. Homology and causes. *Journal of Morphology* 173:305–12. DOI:10.1002/jmor.1051730307

Vasilakis, D. A. 2019. On the meaning of hierarchy in Dionysius the Areopagite. In *Platonism and Christian Thought in Late Antiquity*, eds. P. G. Pavlos, L. F. Janby, E. K. Emilsson, et al., 181–200. Abington (UK): Routledge. ISBN:9781032092003

Vasiliev, N. A. 1989. [*Imaginary Logic.*] Moscow: Nauka. ISBN:5020079464. (in Russian).

Vasilieva, L. N. 1989. [Typological school of taxonomy.] In [*Methodological problems of biology and ecology*], eds. O. I. Belogurov, and V. Yu. Vereshchagin, 26–43. Vladivostok: Far-East State University Publ. (in Russian). ISBN:574440161X.

Vasilieva, L. N. 1992. [*Platonism in Systematics.*] Vladivostok: Far-East Branch Acad. Sci. Publ. (in Russian).

Vasilieva, L. N. 1998. [Hierarchical model of evolution.] *Zhurnal Obshchei Biologii* 59:5–23. ISSN 0044-4596. (in Russian).

Vasilieva, L. N. 1999. *Systematics in Mycology.* Stuttgart: Schweizerbartische Verlagsbuchhandlung. ISBN:9783443590802.

Vasilieva, L. N. 2001. [Classification of organisms and structuralism in biology.] Zhurnal Obshchei Biologii 62:371–85. ISSN:0044-4596. (in Russian).

Vasilieva, L. N. 2007. [Linnaean hierarchy and "extensional thinking".] *Archives of the Zoological Museum of Moscow State University* 48:183–212. ISSN:0134-8647 (in Russian).

Vasilyeva, L. N., and S. L. Stephenson. 2008. The Linnaean hierarchy and 'extensional thinking'. *The Open Evolution Journal* 2:55–65. DOI:10.2174/1874404400802010055

Vasyukov, V. L. 2005. [*Quantum Logic.*] Moscow: PerSe. ISBN:5929201420. (in Russian).

Vdovina, G. V. 2009. [*The Language of the Non-obvious. Teachings on Signs in Scholasticism of the 17th Century.*] Moscow: St. Thomas Institute of Philosophy, Theology, and History. ISBN:9785942420499. (in Russian).

Velichkovsky, B. M. 2006. [*Cognitive Science. Foundations of Psychology of Knowing.*] Moscow: Academia Publ. ISBN:5769529830 (in Russian).

Vences, M., J. M. Guayasamin, A. Miralles, et al. 2013. To name or not to name: Criteria to promote economy of change in Linnaean classification schemes. *Zootaxa* 3636: 201–44. DOI:10.11646/zootaxa.3636.2.1

Verboon A. R. 2014. Medieval tree of Porphyry: An organic structure of logic. In *The Tree: Symbol, Allegory, and Mnemonic Device in Medieval Art and Thought*, eds. P. Salonius, and W. Andrea, 95–116. Turnhout: Brepols Publ. ISBN:9782503548395.

Verdier, N. 2006. Hierarchy: A short history of a word in western thought. In *Hierarchy in Natural and Social Sciences*, ed. D. Pumain, 13–38. Dordrecht: Springer-Verlag. ISBN:9781402041266.

Vilenkin, N. Ya. 1964. *Successive Approximation.* London: Pergamon Book Press. ISBN: 9780080106496.

Villegas, C., and V. Triviño. 2023. Typology and organismal dispositions in evo–devo: A metaphysical approach. *Revista de Estudios de la Ciencia y la Tecnología* 12:79–102. DOI:10.14201/art202312179102

Vityaev, E. E., and B. C. Kostin. 1992. [Natural classification as a law of nature.] In [*Intellectual Support for Activities in the Complex Subject Areas*], 4:107–15. Novosibirsk: Siberian Branch Acad. Sci. Publ. (in Russian).

Vogt, L. 2002. Testing and weighting characters. *Organisms Diversity & Evolution* 2:319–33. DOI:10.1078/1439-6092-00051

Vogt, L. 2009. The future role of bio-ontologies for developing a general data standard in biology: Chance and challenge for zoo-morphology. *Zoomorphology* 128:201–17. DOI:10.1007/s00435-008-0081-5

Vogt, L., T. Bartolomaeus, and G. Giribet. 2010. The linguistic problem of morphology: Structure versus homology and the standardization of morphological data. *Cladistics* 26:301–325. DOI:10.1111/j.1096-0031.2009.00286.x

Vogt, L., M. Nickel, R. A. Jenner, et al. 2013. The need for data standards in zoomorphology. *Journal of Morphology* 274:793–808. DOI:10.1002/jmor.20138

Voigt, W. 1973. *Homologie und Typus in der Biologie. Weltanschaulich-philiosophische und erkenntnistheoretisch-methodologische Probleme*. Jena: Gustav Fischer.

von Baer, K. E. 1828. *Über die Entwicklungsgeschichte der Thiere. Beobachtungen und Reflexion, Vol. 1*. Königsberg: Gebrudern Borntrager. ISBN:3487108712 (1999 reprint).

von Goethe, J. W. 1957. [*Selected Issues on Natural History*.] Moscow: Academy of Sciences PublEication. (in Russian).

von Hofsten, N. 1958. Linnaeus's conception of Nature. *Årsbok /Kungliga Vetenskaps-societetens i Upsala* 1957:65–105. ISSN:0348-7849f.

von Wettstein, R. 1898. *Grundzüge der geographisch-morphologischen Methode der Pflanzensystematik*. Jena: G. Fischer. ISBN:9780266643951 (2018 reprint).

von Winterfeldt, D., and E. Ward. 1986. *Decision Analysis and Behavioral Research*. Cambridge (UK): Cambridge University Press. ISBN:0521273048X.

Voronin, Yu. A. 1985. [*Classification Theory and its Applications*.] Novosibirsk: Nauka. ISBN:9785001397205. (in Russian).

Voss, E. G. 1952. The history of keys and phylogenetic trees in systematic biology. *The Michigan Botanist* 51:1–25. ISSN:2576-8727.

Voyshvillo, E. K. 1989. [*Concept as a Form of Thinking: Logical and Epistemological Analysis*.] Moscow: Moscow State University Publ. ISBN:5211002059. (in Russian).

Waddington, C. H. 1962. *New Pattern in Genetics and Development*. New York (NY): Columbia University Press. ISBN:9781125278345.

Wägele, J.-W. 2005. *Foundations of Phylogenetic Systematics*. Munich: Dr. Friedrich Pfeil Verlag. ISBN:9783899370560.

Wagner, A. 2011. *The Origins of Evolutionary Innovations: a Theory of Transformative Change in Living Systems*. Oxford: Oxford University Press. ISBN:9780199692606.

Wagner, G. P. 1989. The biological homology concept. *Annual Review of Ecology, Evolution, and Systematics* 20:51–69. DOI:10.1146/annurev.es.20.110189.000411

Wagner, G. P. 1994. Homology and the mechanisms of development. In *Homology, the Hierarchical Basis of Comparative Biology*, ed. B. K. Hall, 273–99. San Diego: Academic Press. ISBN:9780123195838.

Wagner, G. P. 2014. *Homology, Genes, and Evolutionary Innovation*. Princeton (NJ): Princeton University Press. ISBN:9780691156460.

Wagner, G. P., and P. F. Stadler. 2003. Quasi-independence, homology and the unity of type: A topological theory of characters. *Journal of Theoretical Biology* 220:505–27. DOI:10.1006/jtbi.2003.3150

Wahl, R. 1991. *Cartesian rationalism. Colby Quarterly* 27:100–5. https://digitalcommons.colby.edu/cq/vol27/iss2/7

Wäldchen, J., H. C. Wittich, M. Rzanny, et al. 2022. Towards more effective identification keys: A study of people identifying plant species characters. *People and Nature* 4:1603–15. DOI:10.1002/pan3.10405

Wallace, W. A. 1999. Introduction. In *Albertus Magnus. On Animals. A Medieval Summa Zoologica*, Vol. I, 1–42. Baltimore (MD): Johns Hopkins University Press. ISBN: 9780814254516.

Ward, D. F., and M. C. Stanley. 2004. The value of RTUs and parataxonomy versus taxonomic species. *New Zealand Entomology* 27:3–9. DOI:10.1080/00779962.2004.9722118

Wartofsky, M. W. 1979. *Models: Representation and Scientific Understanding*. Boston (MA): Springer Science + Business Media. ISBN:9789027707369.

Watrous, J. E., and Q. D. Wheeler. 1981. The out-group comparison method of character analysis. *Systematic Zoology* 30:1–11. DOI:10.1093/sysbio/30.1.1

Weber, M. 1949. *On the Methodology of Social Sciences*. Glencoe (IL): The Free Press. ISBN:9780029343609 (1997 reprint).

Webster, G., and B. Goodwin. 1996. *Form and Transformation: Generative and Relational Principles in Biology*. Cambridge (UK): Cambridge University Press. ISBN:9780521207430.

Weinberg, G. M. 2011. *An Introduction to General Systems Thinking*, 2d ed. London: Dorset House. ISBN:9780932633491.

Weisberg, M. 2006. Forty years of 'The Strategy'. Levins on model building and idealization. *Biology and Philosophy* 21:623–45. DOI:10.1007/s10539-006-9051-9

Wells, J. and P. Nelson. 1997. Homology, a concept in crisis. *Origins & Design* 18. ISSN: 0748-9919. http://www.arn.org/docs/odesign/od182/hobi182.htm

Wenzel, J. W. 1992. Behavioral homology and phylogeny. *Annual Review of Ecology, Evolution, and Systematics* 23: 361–81. DOI:10.1146/annurev.es.23.110192.002045

Weston, P. 1988. Indirect and direct methods in systematics. In *Ontogeny and Systematics*, ed. J. C. Humphries, 27–56. New York (NY): Columbia University Press. ISBN:978812 1105545.

Whately, R. 1826. *Elements of Logic. Comprising the Substance of the Article in the Encyclopædia Metropolitana* [...]. London: J. Mawman. ISBN:9781432610418 (2004 reprint).

Wheeler, Q., and A. Hamilton. 2014. The New Systematics, the New Taxonomy, and the future of biodiversity studies. In *The Evolution of Phylogenetic Systematics*, ed. A. Hamilton, 287–301. Berkeley (CA): University of California Press. ISBN:9780520276581.

Wheeler, Q. D. 1986. Character weighting and cladistic analysis. *Systematic Zoology* 35:102–9. DOI:10.2307/2413294

Wheeler, Q. D. (ed.). 2008. *The New Taxonomy*. Boca Raton (FL): CRC Press. ISBN:97808 49390883.

Wheeler, W. C. 2001. Homology and DNA sequence data. In *The Character Concept in Evolutionary Biology*, ed. G. P. Wagner, 303–17. San Diego (CA): Academic Press. ISBN:9780127300559.

Wheeler W. M. 1929. Present tendencies in biological theory. *The Scientific Monthly* 28:97–109. ISSN:0096-3771.

Whitehead, A. N. 1925. *Science and the Modern World*. New York: Macmillan. ISBN:9780684836393 (1997 reprint).

Whittaker R. H. 1969. New concepts of kingdoms of organisms. *Science* 163:150–60. DOI:10.1126/science.163.3863.150

Whyte, L. L. 1969. Structural hierarchies: A challenging class of physical and biological problems. In: *Hierarchical Structures*, eds. L. L. Whyte, A. G. Wilson, and D. Wilson, 13–6. New York (NY): American Elsevier. ISBN:9780444000699.

Wilby, J. 1994. A critique of hierarchy theory. *Systems Practice* 7:653–70. DOI:10.1007/BF02173498

Wiley, E. O. 1979. An annotated Linnean hierarchy, with comments on natural taxa and competing systems. *Systematic Zoology* 28:308–37. DOI:10.1093/sysbio/28.3.308

Wiley, E. O. 1981. *Phylogenetics: The Theory and Practice of Phylogenetic Systematics*. New York (NY): John Wiley & Sons. ISBN:9780471059752.

Wiley, E. O. 2009. Patrocladistics, nothing new. *Taxon*, 58:2–6. DOI:10.1002/tax.581002

Wiley, E. O., and B. S. Lieberman. 2011. *Phylogenetics: Theory and Practice of Phylogenetic Systematics*, 2d ed. Hoboken (NJ): Wiley-Blackwell. ISBN:9780470905968.

Wiley, E. O., D. Siegel-Causey, D. R. Brooks, et al. 1991. *The Complete Cladist: a Primer of Phylogenetic Procedures*. Lawrence (KA): University of Kansas. ISBN: 9781332300662.

Wilkins, J. 1668 *An Essay Towards a Real Character, and a Philosophical Language [...]*. London: Royal Society.

Wilkins, J. S. 2007. The dimensions, modes and definitions of species and speciation. *Biology and Philosophy* 22:247–66. DOI:10.1007/s10539-006-9043-9

Wilkins, J. S. 2009a. *Species: A History of the Idea*. Berkeley (CA): University California Press. ISBN:9780520260856. (see also: Wilkins, J. S. 2018. *Species. The Evolution of the Idea*, 2nd ed.: Boca Raton (FL): CRC Press. ISBN:9780367657369).

Wilkins, J. S. 2009b. *Defining Species. A Sourcebook from Antiquity to Today*. New York (NY): Peter Lang. ISBN:9781433102165.

Wilkins, J. S. 2010. What is a species? Essences and generation. *Theory in Biosciences* 129:141–8. DOI:10.1007/s12064-010-0090-z

Wilkins, J. S. 2011. Philosophically speaking, how many species concepts are there? *Zootaxa* 2765:58–60. DOI:10.11646/zootaxa.2765.1.5

Wilkins, J. S. 2018. The reality of species: Real phenomena not theoretical objects. In *Routledge Handbook of Evolution and Philosophy*, ed. R. Joyce, 167–81. Abingdon: Routledge. ISBN:9780367573072.

Wilkins, J. S., and M. C. Ebach. 2014. *The Nature of Classification. Relationships and Kinds in The Natural Sciences*. Houndmills: Palgrave Macmillan. ISBN:9780230347922.

William, R. 2019. *Epiphenomenalism, In The Stanford Encyclopedia of Philosophy, eds.* E. N. Zalta, and U. Nodelman. https://plato.stanford.edu/archives/sum2023/entries/epiphenomenalism/

Williams, D. M., and M. C. Ebach. 2008. *Foundations of Systematics and Biogeography*. New York (NY): Springer Science. ISBN:9781441944450.

Williams, D. M., and Q. D. Wheeler. 2025. *The New Taxonomy. A Science Reimagined*. Boca Raton (FL): CRC Press. ISBN:9781003280743.

Williams, W. T. 1967. Numbers, taxonomy and judgment. *The Botanical Review* 33:379–86. DOI:10.1007/BF02858741

Williams, W. T., and M. B. Dale. 1965. Fundamental problems in numerical taxonomy. *Advances in Botanical Research* 2:35–68. DOI:10.1016/S0065-2296(08)60249-9

Williamson, T. 2000. *Knowledge and its Limits*, Oxford (UK): Oxford University Press. ISBN:9780198250432.

Willis, A., and R. Bell. 2018. Uncertainty in phylogenetic tree estimates. *Journal of Computational and Graphical Statistics* 27: 542–52. DOI:10.1080/10618600.2017.1391697

Willis J.C. 1922. *Age and Area. A Study in Geographical Distribution and Origin of Species*. Cambridge: Cambridge University Press.

Willner, W., K. Hubler, and M. A. Fischer. 2014. Return of the grades: To objectivity in evolutionary classification. *Preslia* 86:233–43. https://www.preslia.cz/P143Willner.pdf

Wills, M. A. 2007. Fossil ghost ranges are most common in some of the oldest and some of the youngest strata. *Proceedings of the Royal Society B* 274:2421–7. DOI:10.1098/rspb.2007.0357

Wilson, D. 1969. Forms of hierarchy: A selected bibliography. In: *Hierarchical Structures*, eds. L. L. Whyte, A. G. Wilson, and D. Wilson, 287–314. New York (NY): American Elsevier. ISBN:9780444000699.

Wilson, J. D., N. M. Koch, and M. J. Ramírez. 2022. Chronogram or phylogram for ancestral state estimation? Model-fit statistics indicate the branch lengths underlying a binary character's evolution. *Methods in Ecology and Evolution* 13:1679–89. DOI:10.1111/2041-210X.13872

Winsor, M. P. 2025. Linnaeus builds a search engine. *European Journal of Taxonomy* 981:300–5. DOI:10.5852/ejt.2025.981.2839

Winston, M. E., R. Chaffin, and D. Herrmann. 1987. A taxonomy of part-whole relations. *Cognitive Science* 11:417–44.

Winthrop, R. H. 1991. *Dictionary of Concepts in Cultural Anthropology*. New York (NY): Greenwood Press. ISBN:9780313242809.

Wippel, J. F. 2000. *The Metaphysical Thought of Thomas Aquinas*. Washington (DC): Catholic University of America Press. ISBN:9780813209838.

Wirkner, C. S., T. Göpel, J. Runge, et al. 2017. The first organ-based ontology for arthropods (Ontology of Arthropod Circulatory Systems—OArCS) and its integration into a novel formalization scheme for morphological descriptions. *Systematic Biology* 66:754–68. DOI:10.1093/sysbio/syw108

Wise, M. N. 2016. Goethe was right: 'The history of science is science itself'. In *Shifting Paradigms: Thomas S. Kuhn and the History of Science*, eds. A. Blum, K. Gavroglu, C. Joas, et al., 347–59. Berlin: Max Planck Open Access. ISBN:9783945561119.

Wolfe, C. T. 2024. Varieties of organicism: A critical analysis. In *Organization in Biology*, ed. M. Mossio, 41–58. Cham (Switzerland): Springer Nature Switzerland. ISBN:9783031389672.

Wolff, J. G. 2019. Information compression as a unifying principle in human learning, perception, and cognition. *Complexity* 2:1879746. DOI:10.1155/2019/1879746

Wood, S. W. 1994. A hierarchical theory of systematics. *Evolutionary Theory* 10:273–77. https://www.researchgate.net/publication/238703242_A_HIERARCHICAL_THEORY_OF_SYSTEMATICS

Woodger, J. H. 1937. *The Axiomatic Method in Biology*. Cambridge (UK): Cambridge University Press. ASIN B00085P2C8.

Woodger, J. H. 1942. On biological transformations. In *Essays on Growth and Form Presented to D'A. W. Thompson*, 2d. ed., eds. W. E. le Gros Clark, and P. B. Medawar, 95–120. Cambridge (UK): Cambridge University Press. ISBN:9780521066228.

Woodger, J. H. 1952. From biology to mathematics. *British Journal for the Philosophy of Science* 3:1–21. DOI:10.1093/bjps/III.9.1

Woodward, J. 2003. *Making Things Happen. Theory of Causal Explanation*. Oxford: Oxford University Press. ISBN:9780195189537.

Worm, A. 2014. Arbor autem humanum genus significat. Trees of genealogy and sacred history in the Twelfth century. In *The Tree: Symbol, Allegory, and Mnemonic Device in Medieval Art and Thought*, eds. P. Salonius, and W. Andrea, 35–67. Turnhout: Brepols Publ. ISBN:9782503548395.

Wray, G. A. 1999. Evolutionary dissociations between homologous genes and homologous structures. In *Homology*, eds. G. Bock, and G. Cardew, 189–203. Chichester: John Wiley & Sons. ISBN:0471984930.

Wray, G. A., and E. Abouheif. 1998. When is homology not homology? *Current Opinion in Genetics & Development* 8:675–80. DOI:10.1016/s0959-437x(98)80036-1

Wu, J. 1999. Hierarchy and scaling: Extrapolating information along a scaling ladder. *Canadian Journal of Remote Sensing* 25:367–80. DOI:10.1080/07038992.1999.10874736

Wu, J. 2013. Hierarchy theory: An overview. In *Linking Ecology and Ethics for a Changing World*, eds. R. Rozzi, S. T. A. Pickett, C. Palmer, et al., 281–301. Dordrecht: Springer Science+ Business Media. ISBN:9789400774698.

Wu, J. 2014. Hierarchy and scaling: Extrapolating information along a scaling ladder. *Canadian Journal of Remote Sensing* 25:367–80. DOI:10.1080/07038992.1999.10874736

Wu, J. and J. L. Vankat. 1995. Island biogeography: Theory and applications. In *Encyclopedia of Environmental Biology*, Vol. 2, ed. W. A. Nierenberg, 371–9. San Diego (CA): Academic Press. ISBN:9780080917788.

Yang, J.-L., L. Jun, L.-H. Dong, et al. 2011. Analysis on the reconstruction accuracy of the Fitch method for inferring ancestral states. *BMC Bioinformatics* 12:18. http://www.biomedcentral.com/1471-2105/12/18

Yardley, I. 2014. *The Universal Metaphor: The Universe as a Circle*. Oxford: Clarendon Press. ISBN:9780198242468.

Yochelson, E. L. 1966. Nomenclature in the machine age. *Systematic Zoology* 15:88–91. DOI:10.2307/sysbio/15.1.88b

Young, D. M. 1971. *Iterative Solution of Large Linear Systems*. New York (NY): Academic Press. ISBN:9781483248752.
Yudin, K. A. 1974. [On the notion of "character" and the levels of the development of animal taxonomy.] *In [Theoretical Issues in Animal Taxonomy and Phylogeny,]* 5–29. Leningrad: Zoological Institute of AN USSR. (in Russian).
Yule, G. U. 1924. A mathematical theory of evolution, based on the conclusions of Dr. J.C. Willis. *Philosophical Transactions of the Royal Society of London, Ser. B* 213: 21–87. DOI:10.1098/rstb.1925.0002
Zabrodin, V. Yu. 1989. [On the problem of naturalness of classifications: Classification and law.] *In [Classification in Modern Science]*, eds. A. I. Kochergin, and S. S. Mitrofanova, 59–73. Novosibirsk: Nauka. (in Russian).
Zachos, F. E. 2011. Linnean ranks, temporal banding, and time-clipping: Why not slaughter the sacred cow? *Biological Journal of the Linnean Society* 103:732–4. DOI:10.1111/j.1095-8312.2011.01711.x
Zachos, F. E. 2016. *Species Concepts in Biology. Historical Development, Theoretical Foundations and Practical Relevance*. Basel: Springer. ISBN:9783319831732.
Zadeh, L. A. 1965. Fuzzy sets. *Information and Control* 8: 338–53. ISSN:0019-9958.
Zadeh, L. A. 1971. Similarity relations and fuzzy orderings. *Information Sciences* 3: 177–200. ISSN:0020-0255.
Zadeh, L. A. 1992. Knowledge representation in fuzzy logic. In *An introduction to Fuzzy Logic. Applications in Intelligent Systems*, eds. R. R. Ygaer, and L. A. Zadeh, 1–25. New York (NY): Springer Science + Business Media. ISBN:9781461536406.
Zafeiris, A., and T. Vicsek. 2018. *Why We Live in Hierarchies? A Quantitative Treatise*. Cham (Switzerland): Springer International Publ. ISBN:9783319704814.
Zakharov, B. P. 2005. [*Transformational Typological Systematics*.] Moscow: KMK Scientific Press. ISBN:9785041154622. (in Russian).
Zalužanský, A. 1592 *Methodi Herbariae Libri Tres [...]*. Pragae: Georgij Daczennni.
Zander, R. H. 2010a. *Structuralism in phylogenetic systematics. Biological Theory* 5:383–394. DOI:10.1162/BIOT_a_00063
Zander, R. H. 2010b. Taxon mapping exemplifies punctuated equilibrium and atavistic saltation. *Plant Systematics and Evolution* 286:69–90. DOI:10.1007/s00606-010-0281-6
Zander, R. H. 2013. *A Framwork for Post-Phylogenetic Systematics*. St. Louis (MO): Zetetic Publ. ISBN:9781492220404.
Zander, R. H. 2024. Lineages of fractal genera comprise the 88-million-year steel evolutionary spine of the ecosphere. *Plants (Basel)* 13:1559. DOI:10.3390/plants13111559
Zander, R. H. 2025. On the origin of taxa. *Reprints.org*. DOI:10.20944/preprints202503.2285.v1
Zarenkov, N. A. 1976. [*Lectures on the Theory of Systematics*.] Moscow: Moscow State University Publ. (in Russian).
Zarenkov, N. A. 1983. [Biological systematics as a particular problem of the general science classification theory.] *In [Theory and Methodology of Biological Classifications,]* 29–46. Moscow: Moscow Society of Naturalists. (in Russian).
Zarenkov, N. A. 1988. [*Theoretical Biology*.] Moscow: Moscow State University Publ. (In Russian).
Zavadsky, K. M. 1968. [*Species and Speciation*.] Leningrad: Nauka. (in Russian).
Zavarzin, G. A. 1969. [Incompatibility of features and the theory of the biological system.] *Zhurnal Obshchei Biologii* 30:33–41. ISSN: 0044-4596. (in Russian).
Zavarzin, G. A. 1974. [*The Phenotypic Systematics of Bacteria. The Space of Logical Possibilities*.] Moscow: Nauka Publ. ISBN:9785951935274 (2023 reprint). (in Russian).
Zeki, S. 2022. The Paton prize lecture 2021: A colourful experience leading to a reassessment of colour vision and its theories. *Experimental Physiology* 107:1189–208. DOI:10.1113/EP089760

Zherikhin, V. V. 1986. [Biocenotic regulation of evolution.] *Paleontologichesky Journal* 1:3–12. ISSN:0031-031X. (in Russian).

Zherikhin, V. V., and A. S. Rautian. 1997. [Philocenogenesis and evolutionary crises.] *Zhurnal Obshchei Biologii* 58: 20–47. ISSN:0044-4596. (in Russian). *Zhurnal Obshchei Biologii* 65:187–92. ISSN:0044-4596. (in Russian).

Zimmermann, W. 1931. Arbeitsweise der botanischen Phylogenetik und anderer Gruppierungswissenschaften. In *Handbuch der biologischen Arbeitsmethoden, Abt. 3, T. 9*, ed. E. Abderhalden, 941–1053. Berlin: Urban & Schwarzenberg.

Zimmermann, W. 1934. Research on phylogeny of species and of single characters (Sippenphylogenetik und Merkmalsphylogenetik). *The American Naturalist* 68:381–4. DOI:10.1086/280558

Zipf, G. K. 1949. *Human Behavior and the Principle of Least-Effort: An Introduction to Human Ecology*. Cambridge (UK): Addison–Wesley. ISBN:9781614273127 (2012 reprint).

Zuev, V. V. 2002. [*The Problem of Reality in Biological Taxonomy*.] Novosibirsk: Novosibirsk State University Publ. ISBN:594356084X. (in Russian).

Zuev, V. V. 2015. [*An Introduction to the Theory of Biological Taxonomy*.] Moscow: Infra-M. ISBN:9785160106281. (in Russian).

Zylstra, U. 1992. Living things as hierarchically organized structures. *Synthese* 91:111–33. DOI:10.1007/BF00484972

Index

D

E

F

G

H

O

P

Q

R

T

U

V

W

X

Z

For Product Safety Concerns and Information please contact our EU
representative GPSR@taylorandfrancis.com
Taylor & Francis Verlag GmbH, Kaufingerstraße 24, 80331 München, Germany

www.ingramcontent.com/pod-product-compliance
Ingram Content Group UK Ltd.
Pitfield, Milton Keynes, MK11 3LW, UK
UKHW022218190926
475702UK00003B/433
9781032982120